RELATIVITY
AND
QUANTUM MECHANICS

RELATIVITY
AND
QUANTUM MECHANICS

By

Dr. Ajit Kumar Sharma
Deptt. of Physics
D.A.V. (P.G.) College
Bulandshahr
(U.P.)

DISCOVERY PUBLISHING HOUSE PVT. LTD.
NEW DELHI-110 002

Published by:
Tilak Wasan

DISCOVERY PUBLISHING HOUSE PVT. LTD.
4383/4B, Ansari Road, Darya Ganj
New Delhi-110 002 (India)
Phone : +91-11-23279245, 43596064-65
Fax : +91-11-23253475
E-mail : discoverypublishinghouse@gmail.com
sales@discoverypublishinggroup.com
web : www.discoverypublishinggroup.com

***First Edition:* 2011**

***Reprinted:* 2017**

ISBN: 978-81-8356-822-7

Relativity and Quantum Mechanics

Printed at:
Infinity Imaging Systems
Delhi

Preface

This book has been designed to over the syllabus of in Relativity and Quantum Mechanics required for the B.Sc/BSc Hons/M.Sc. students of the Indian Universities. I have compelled all the questions asked so for in different Universities as well as C.C.S. Universities Meerut.

I have arranged the subject matter in a continuous manner. Special emphasis has been laid on fundamental concept of the topics.

Care has been taken to make the treatment of the subject simple and accessible to the average students. The gradation of the solved and unsolved questions has been facilitated from simple to difficult.

It is believed that the book in the present form will be found to be useful by the student community and the teaching fralernity alike.

Suggestion for the improvement of the book will always be most welcome.

—*Author*

Contents

1

Elastic Stability

EULER'S BUCKLING LOAD

Consider a long slender columen subjected to an axial force P. If the column is perfectly straight and is ideal in every respect, then it will remain straight and will be in equilibrium. If now a small lateral force Q is applied in addition to the axial force P (Fig. 1), the member and will remain deflected as long as the lateral force Q is acting. When Q is removed, the member will return to its straight equilibrium position. However, there exists a critical axial load P_{cr}, such that under the action of P_{cr}, if the column is given a small lateral deflection by a force Q and the lateral force is removed, the column will continue to remain in the slightly buckled form and will be in equilibrium.

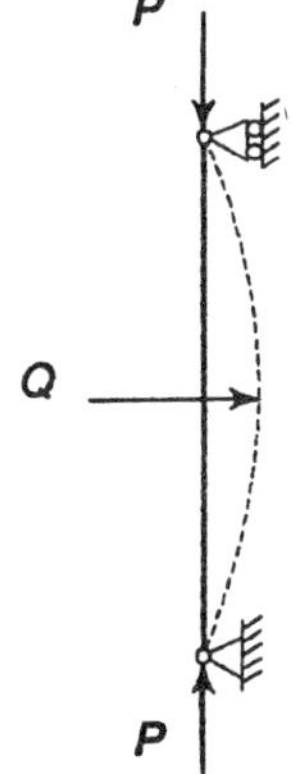

Fig. 1

The value of P_{cr}, known as the Euler's critical load or the buckling load can, therefore, be obtained by considering the equilibrium of a slightly buckled column. In elementary strength of materials, following this approach, Euler's critical loads for the columns shown in Fig. 2 have been obtained.

The critical loads for the first modes shown in Fig. 2 are as follows:

$$\frac{\pi^2 EI}{4L^2} \text{ for case (a)}, \qquad \frac{\pi^2 EI}{L^2} \text{ for case (b), and} \qquad \frac{4\pi^2 EI}{L^2} \text{ for case (c)}$$

The method followed in elementary strength of materials to derive the above formulas will be applied to the following problem, which is slightly more complicated than the above cases.

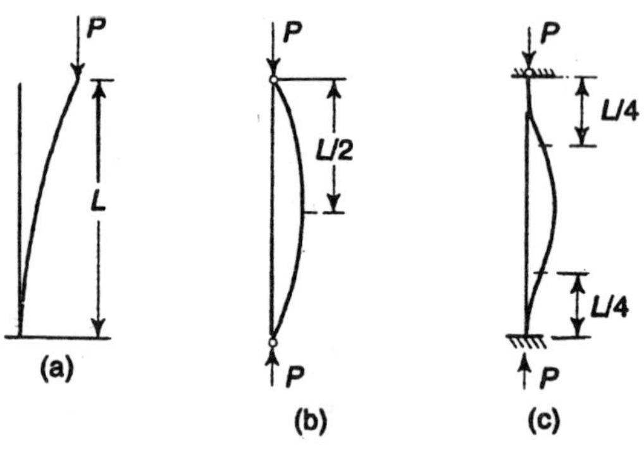

Fig. 2

Consider a centrally loaded column with the lower end built–in and the upper end hinged (Fig. 3). The critical value of the compressive load is that value of P_{cr} which can keep the strut in a slightly buckled shape. It may be observed that in order to keep point A in line with B, a lateral reaction R will be necessary.

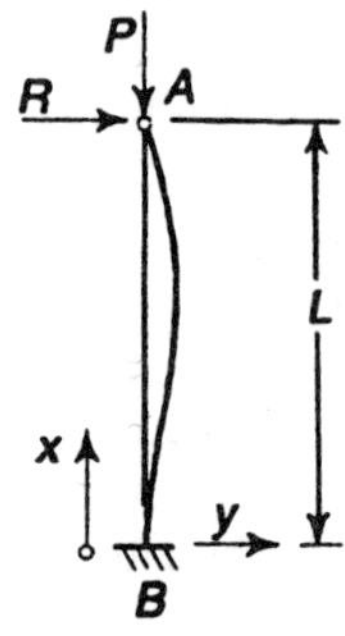

Fig. 3

The bending moment at any section x is

$$M = py - R(L - x)$$

Using the expression

$$M = -EL\frac{d^2y}{dx^2}$$

$$EI\frac{d^2y}{dx^2} = -py + R(L-x)$$

Let $$k^2 = \frac{P}{EI}$$

The differential equation then becomes

$$\frac{d^2y}{dx^2} = -k^2y + \frac{R}{EI}(L-x) \tag{1}$$

The general solution of this equation is

$$y = C_1 \cos kx + C_2 \cos kx + \frac{R}{P}(L-x)$$

The constants C_1 and C_2 and the reaction R will have to be determined from the boundary conditions. These are

$$y = 0 \text{ at } x = 0 \text{ and at } x = L, \quad \frac{dy}{dx} = 0 \text{ at } x = 0$$

Substituting these, we obtain the following equations:

$$C_1 + \frac{R}{P}L = 0$$

$$C_1 \cos kL + C_2 \sin kL = 0$$

$$kC_2 - \frac{R}{P} = 0$$

The trivial solution is $C_1 = C_2 = R = 0$, which means that the column remains straight. The non–trivial solution is

$$C_1 = -\frac{R}{P} = L, \qquad C_2 = \frac{1}{k}\frac{R}{P}$$

Substituting into the second equation, we obtain the transcendental equation

$$\tan kL = kL \tag{2}$$

A solution to this can be obtained from a graphical plot. The smallest value of kL satisfying this equation is $kL = 4.493$, which means

$$P_{cr} = k^2 EI = \frac{20.19\, EI}{L^2} \approx \frac{\pi^2 EI}{(0.699L)^2}$$

As another example, we shall analyse by the elementary method, a fairly general problem of column with a varying cross–section and with end as well as intermediate loading.

Example 1:

A column AB with hinged ends (Fig. 4) is compressed by two forces P_1 and P_2. The moment of inertia for the length L_1 of the column is I_1 and for the remaining length L_2, it is I_2. Determine the critical value of the force $P_1 + P_2$.

If the equilibrium position of the buckled column is as shown in Fig. 4, then to have zero moments at the hinged ends A and B, it is necessary to have a horizontal reaction R such that

$$RL = P_2\delta$$

or $$r = P_2\delta/L \tag{a}$$

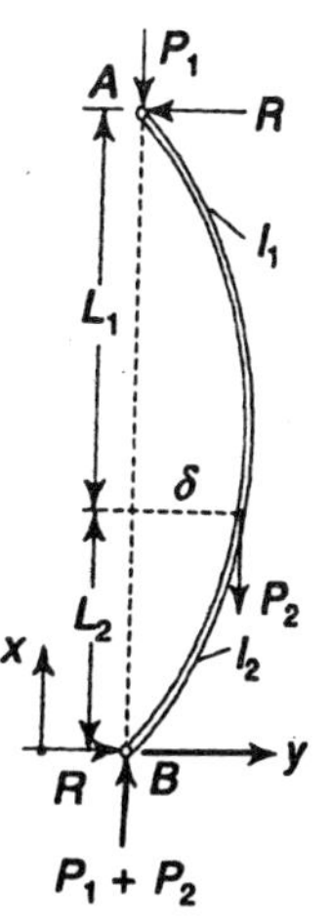

Fig. 4

Let y_1 be the deflection at any section of the L_1 portion and y_2 the deflection at any section of the L_2 portion.

For the L_1 portion, the moment is

$$M = P_1y_1 + R(L - x)$$

Using Eq. (a),

$$-EI\frac{d^2y_1}{dx^2} = P_1y_1 + \frac{\delta P_2}{L}(L-x)$$

and for the L_2 portion, the moment is

$$M = P_1y_2 + R(L - x) - P_2(\delta - y_2)$$

or $$-EI_2\frac{d^2y_2}{dx^2} = P_1y_2 + \frac{\delta P_2}{L}(L-x) - P_2(\delta - y_2)$$

Using the notations

$$\frac{P_1}{EI_1} = k_1^2,\quad \frac{P_2}{EI_2} = k_2^2,\quad \frac{P_1 + P_2}{EI_2} = k_3^2,\quad \frac{P_2}{EI_1} = k_4^2$$

the differential equations become

$$\frac{d^2y_1}{dx^2} = -k_1^2y_1 - \frac{\delta}{L}k_4^2(L-x)$$

$$\frac{d^2y_2}{dx^2} = -k_3^2y_2 - \frac{\delta}{L}k_2^2x$$

The solutions of the above equations are

$$y_1 = C_1 \sin k_1x + C_2 \cos k_2x - \frac{\delta}{L}\frac{k_4^2}{k_1^2}(L-x)$$

$$y_2 = C_3 \sin k_3x + C_4 \cos k_4x + \frac{\delta}{L}\frac{k_2^2}{k_3^2}x$$

The boundary conditions are

$$y_1 = 0 \text{ at } x = L,\quad y_1 = \delta \text{ at } x = L_2,\quad y_2 = \delta \text{ at } x = L_2,$$

$$y_2 = 0 \text{ at } x = 0,\quad \left(\frac{dy_1}{dx}\right) = \left(\frac{dy_2}{dx}\right) \quad \text{at } x = L_2$$

The first four conditions yield

$$C_1 = \frac{\delta(k_1^2 L - k_2^2 L_1)}{k_1^2 L(\sin k_1 L_2 - \tan k_1 L \cos k_1 L_2}$$

$$C_2 = -C_1 \tan k_1 L,\quad C_3 = \frac{\delta(k_3^2 L - k_2^2 L_2)}{k_3^2 L \sin k_3 L_2},\quad C_4 = 0$$

Substituting the values of these constants into the continuity condition, i.e.

$$\left(\frac{dy_1}{dx}\right) = \left(\frac{dy_2}{dx}\right) \qquad \text{at } x = L_2$$

the following transcendental equation is obtained :

$$\frac{k_4^2}{k_1^2} - \frac{k_1^2\, L + k_4^2\, L_1}{k_1 \tan k_1\, L_1} = \frac{k_2^2}{k_3^2} + \frac{k_3^2 L - k_2^2 L_2}{k_3 \tan k_3\, L_2}$$

For any particular case, the above equation can be solved to give the critical load. If, as an example, $L_1 = L_2$, $I_1 = I_2 = I$ and $P_1 = P_2$ are taken, we get

$$(P_1 + P_2)_{cr} = \frac{\pi^2 EI}{(0.87L)^2}$$

In this chapter, we shall discuss three specific topics:

(i) Beam columns;

(ii) Stability problem as an eigenvalue problem and

(iii) Energy methods to obtain approximate solutions to buckling Problems.

I. BEAM COLUMNS

BEAM COLUMN

In the theory of bending discussed in Chapter 6, it was found that stresses were directly proportional to the applied loads. Similarly, in determining the deflections, we could apply the principle of superposition because of the linear relationship between the load acting and the deflection produced. In these cases, it is assumed that the deformations produced by one load do not after the action of the other loads. Figure 5(a) shows a cantilever loaded by forces Q_1 and Q_2. If δ_1 is the deflection caused at point S due to Q_1 alone and δ_2 the deflection at the same point S due to Q_2 alone, then the deflection δ due to the combined action of Q_1 and Q_2 is $\delta_1 + \delta_2$.

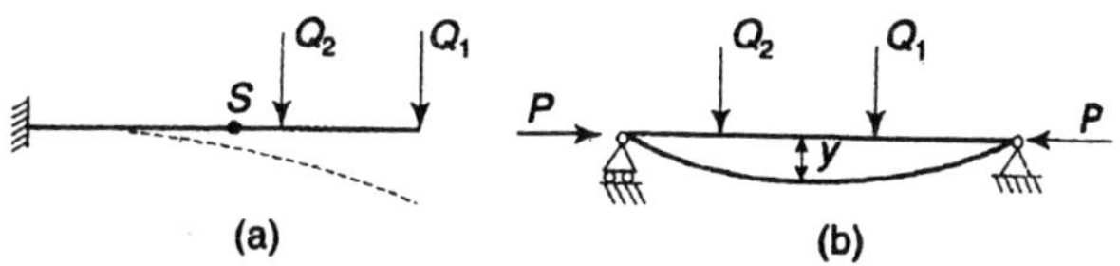

Fig. 5

In arriving at this result it is assumed that the deflection δ_1 caused by Q_1 does not affect the action of Q_2. However, in the case of a beam which is subjected to lateral forces Q_1 and Q_2 as well as to axial forces P as shown in Fig. 5(b), it can be seen that the bending moment caused by P depends on the deflection y produced by the lateral forces Q_1 and Q_2. In such cases, the principle of superposition cannot be applied without certain modifications. The beams that are subjected to axial loads in addition to lateral loads are known as beam columns. We shall restrict our analysis to beam columns having symmetrical cross–sections.

BEAM COLUMN EQUATIONS

Consider the beam shown in Fig. 6(a). The beam carries a distributed lateral load of intensity q, which is a function of x. In addition, the beam is subjected to an axial compressive force P. An elementary length Δx of the beam before deflection is considered. The lateral load q will be assumed to be positive when it is acting downward. The free body diagram of length Δx is shown in Fig. 6(b). The shearing force V and bending moment M acting

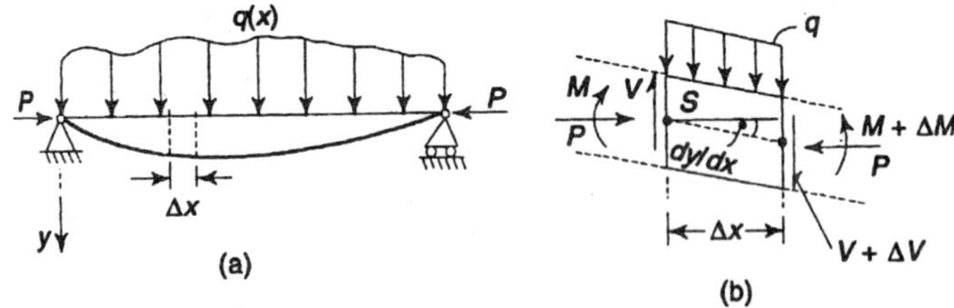

Fig. 6

on the sides of the element are assumed to be positive in the directions shown.

The relations among the load, shearing force V and bending moment M are obtained from the equilibrium considerations of the element. summing forces in Y directions

$$-V + q\,\Delta x + (V + \Delta V) = 0$$

or, in the limit as $\Delta x \to 0$

$$q = \frac{dV}{dx} \tag{3}$$

Taking moments about point S and assuming that the angle between the axis of the deformed beam and the horizontal is small, we get

$$M + q\,\Delta x\frac{\Delta x}{2} + (V + \Delta V)\Delta x - (M + \Delta M) + P\frac{dy}{dx}\Delta x = 0$$

or,
$$q\frac{\Delta x}{2} + V + \Delta V - \frac{\Delta M}{\Delta x} + P\frac{dy}{dx} = 0$$

In the limit as $\Delta x \to 0$

$$V = \frac{dM}{dx} - P\frac{dy}{dx} \tag{4}$$

As in the case of the bending of beams, we ignore the effects of shearing deformation and assume that the curvature of the beam axis is given by

$$EI\frac{d^2y}{dx^2} = -M \tag{5}$$

where E is the Young's modulus of the beam material and I is the moment of inertia about the neutral axis. Using Eq. (5), Eqs (4) and (3) can be written as

$$EI\frac{d^3y}{dx^3} + P\frac{dy}{dx} = -V \tag{6}$$

and

$$EI\frac{d^4y}{dx^4} + P\frac{d^2y}{dx^2} = q \tag{7}$$

Equations (3) – (7) are the basic differential equations for the bending of beam columns. These equations reduce to the familiar beam bending equations when P is equal to zero.

BEAM COLUMN WITH A CONCENTRATED LOAD

Consider a uniform beam of span L (Fig. 7) simply supported and carrying a load Q at distance a from the right hand support. The beam is subjected to an axial force P. The bending moment at any section x is due to Q as well as P. However, the bending moment due to P cannot be calculated until the deflection is determined. The beam column is therefore statically indeterminate.

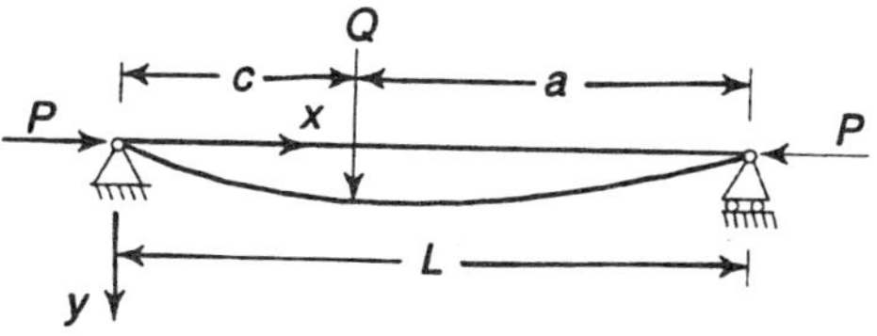

Fig. 7

The bending moment at any x is

$$M = \frac{Qa}{L}x + Py \quad \text{for } x \le (L-a)$$

$$M = \frac{Q(L-a)}{L}(L-x) + Py \quad \text{for } x \ge (L-a)$$

Hence, from Eq. (5)

$$EI\frac{d^2y}{dx^2} = -\frac{Qa}{L}x - Py \quad \text{for } x \le (L-a) \tag{8}$$

$$EI\frac{d^2y}{dx^2} = -\frac{Q(L-a)}{L}(L-a) - Py \quad \text{for } x \ge (L-a) \tag{9}$$

Putting $k^2 = \dfrac{P}{EI}$

the above equations become

$$\frac{d^2y}{dx^2} + k^2 y = -\frac{Qa}{EIL}x$$

The general solutions of these equations are

$$y = A\cos kx + B\sin kx - \frac{Qa}{PL}x \quad x \le (L-a)$$

$$y = C\cos kx + D\sin kx - \frac{Q(L-a)(L-x)}{PL} \quad x \ge (L-a)$$

The constants A, B, C and D are to be determined from the conditions of the beam. The conditions are

(i) $y = 0$ at $x = 0$ and at $x = L$

(ii) y at $x = (L-a)$ should be the same according to both solutions.

(iii) The tangent at $x = (L-a)$ would be the same according to both solutions.

From condition (i)

$$A = 0 \quad \text{and} \quad C = -D\tan kL$$

Conditions (ii) and (iii) give

$$B\sin k(L-a) - \frac{Qa}{PL}(L-a)$$

$$= D[\sin k(L-a) - \tan kL\cos kL(L-a) - \frac{Qa}{PL}(L-a)$$

$$Bk\cos k(L-a) - \frac{Qa}{PL}$$

$$= Dk[\cos k(L-a) + \tan kL\sin k(L-a) + \frac{Q(L-a)}{PL}$$

From the above two equations we get

$$B = \frac{Q\sin kx}{Pk\sin kL}, \qquad D = -\frac{Q\sin k(L-a)}{Pk\tan kL}$$

Substituting these constants, the solutions are

$$y = \frac{Q \sin ka}{Pk \sin kL} \sin kx - \frac{Qa}{PL} x \quad \text{for } x \le (L-a) \tag{10a}$$

$$y = \frac{Q \sin k(L-a)}{Pk \sin kL} \sin k(L-x) - \frac{Q(L-a)}{PL}(L-x) \quad \text{for } x \ge (L-a) \tag{10b}$$

By the differentiation of Eqs (10a) and (10b), we obtain the following formula, which are useful.

$$\frac{dy}{dx} = \frac{Q \sin ka}{P \sin kL} \cos kx - \frac{Qa}{PL} \quad 0 \le x \le L-a \tag{11}$$

$$\frac{dy}{dx} = \frac{Qk \sin k(L-a)}{P \sin kL} \cos k(L-a) + \frac{Q(L-a)}{PL} \quad x \ge L-a \le L$$

$$\frac{d^2 y}{dx^2} = \frac{Qk \sin ka}{P \sin kL} \sin kx \quad 0 \le x \le L-a \tag{12}$$

$$\frac{d^2 y}{dx^2} = \frac{Qk \sin k(L-a)}{P \sin kL} \sin k(L-a) \quad x \ge L-a \le L$$

As a particular case, if $a = L/2$, i.e. the load acts at midspan, then

$$\delta = y \text{ at } \frac{L}{2} = \frac{Q}{2Pk}\left[\tan \frac{kL}{2} - \frac{kL}{2}\right]$$

Putting $$u = \frac{kL}{2} = \frac{L}{2}\left(\frac{P}{EI}\right)^{1/2} \tag{13}$$

$$\delta = \frac{QL^3}{48EI} \frac{3(\tan u - u)}{u^3} \tag{14}$$

It is observed from the above equation that δ becomes infinite when $u = \pi/2$, i.w. when

$$\pi/2 = \frac{L}{2}\sqrt{\frac{P}{EI}}$$

or $$P = \frac{\pi^2 EI}{L^2} = P_{cr}$$

So, however small Q is, when P becomes equal to P_{cr}, the lateral deflections becomes very large. We should recall that P_{cr} given above is the Euler buckling load for a slender column with hinged ends.

BEAM COLUMN WITH SEVERAL CONCENTRATED LOADS

Equations (10a and b) show that deflection y is proportional to lateral load Q, whereas the relation between the deflection and axial force P is more complicated. Because of the linear relationship between deflection y and load Q, if Q is doubled (with P remaining unaltered), then the deflection also is doubled. Hence, the principle of superposition in a modified form can be used for the effect of the lateral load, provided the same axial force acts on the bar.

Consider the beam shown in Fig. 8, which is subjected to an axial force P and three lateral loads Q_1, Q_2 and Q_3 acting at distances a_1, a_2 and a_3 respectively from the right hand side support B.

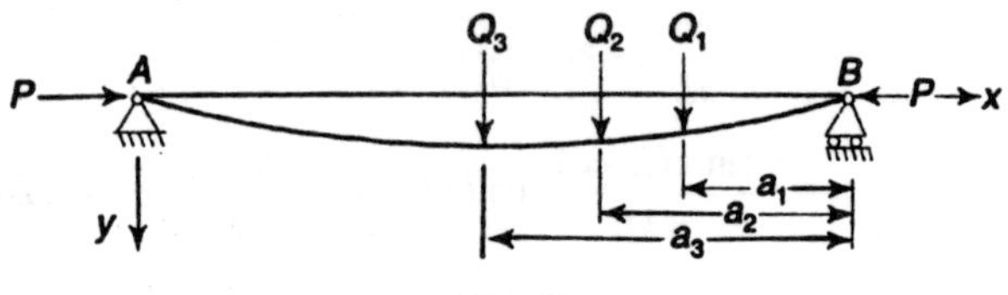

Fig. 8

At some section left of Q_3, let y_1 be the deflection due to Q_1 alone with P, y_2 the deflection at the same point due to Q_2 alone with P, and y_3 the deflection due to Q_3 alone with P. From Eq. (5) with each Q and P [(similar to Eq. (8)], we get the following:

$$EI\frac{d^2y_1}{dx^2} = -\frac{Q_1a_1}{L}x - Py_1$$

$$EI\frac{d^2y_2}{dx^2} = -\frac{Q_2a_2}{L}x - Py_2$$

$$EI\frac{d^2y_3}{dx^2} = -\frac{Q_3a_3}{L}x - Py_3$$

By adding these equations

$$EI\frac{d^2(y_1+y_2+y_3)}{dx^2} = -\frac{Q_1a_1}{L}x - \frac{Q_2a_2}{L}x - \frac{Q_3a_3}{L} - P(y_1+y_2+y_3) \qquad (15)$$

If Q_1, Q_2 and Q_3 are acting together with P, then the bending moment at section x is

$$M = \frac{Q_1a_1}{L}x + \frac{Q_2a_2}{L}x + \frac{Q_3a_3}{L}x + p(y_1+y_2+y_3)$$

From Eq. (5), therefore, we get

$$EI\frac{d^2(y_1+y_2+y_3)}{dx^2} = -\frac{Q_1a_1}{L}x - \frac{Q_2a_2}{L}x - \frac{Q_3a_3}{L}x - P(y_1+y_2+y_3)$$

This is identical to Eq. (15). Therefore, when there are several loads acting on a bar with an axial force P, the resultant deflection can be obtained by the superposition of the deflection produced by each lateral load acting in combination with axial force P.

Let a beam be acted upon by n lateral loads $Q_1, Q_2, \ldots, Q_m, Q_{m+1}, \ldots, Q_n$ at distances $a_1, a_2, \ldots, a_m, a_{m+1}, \ldots, a_n$ respectively from the right hand support. At some point x, which lies between Q_m and Q_{m+1}, the total deflection is obtained from Eq. (10) as

$$y = \frac{\sin kx}{Pk \sin kL} \sum_{i=1}^{m} Q_i \sin ka_i - \frac{x}{PL} \sum_{i=1}^{m} Q_i a_i$$

$$+ \frac{\sin k(L-x)}{pk \sin kL} \sum_{i=m+1}^{m} Q_i \sin k(L-a_i) - \frac{L-x}{PL} \sum_{i=m+1}^{m} Q_i (L-a_i) \quad (16)$$

In the above equation, we have made use of Eq. (10a) for loads $Q_1, Q_2, \ldots, Q_m$ lying to the right of x and Eq. (10b) for loads $Q_{m+1}, Q_{m+2}, \ldots, Q_n$ lying to the left of x.

CONTINUOUS LATERAL LOAD

The result obtained for a single load and the method of superposition can be used to solve the problem of a beam subjected to a continuously distributed load and an axial force, shown in Fig. 9. Let q be the intensity of loading. At distance a from the right hand support, the load on an elementary length Δa is $q\ \Delta a$.

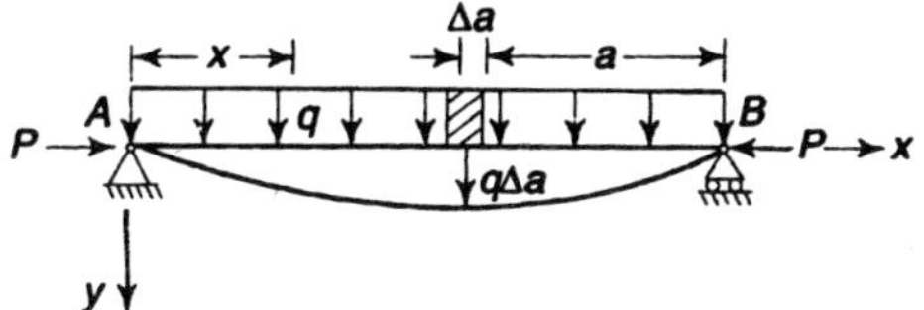

Fig. 9

At distance x from A, the deflection due to the load $q\ \Delta a$ from Eq. (10a) is

$$\Delta y = \frac{q\ \Delta a \sin ka}{Pk \sin kL} \sin kx - \frac{q\ a\ \Delta a}{PL} x$$

Assuming a to vary from 0 to $(L - x)$ from the right hand support, the deflection due to this part of the load, from Eq. (10a) and the principle of superposition, is

$$y_1 = \frac{\sin kx}{Pk \sin kL} \int_0^{L-x} q \sin ka\ da - \frac{x}{PL} \int_0^{L-x} qa\ da$$

Similarly, using Eq. (10b) and the principle of superposition, the deflection at x due to the loading to the left of x, i.e. for a varying from $(L - a)$ to L, is

$$y_2 = \frac{\sin k(L-x)}{Pk \sin kL} \int_{L-x}^{L} q \sin k(L-a)\, da - \frac{L-x}{PL} \int_{L-x}^{L} q\,(L-a)\, da$$

The total deflection y due to complete loading is $y = y_1 + y_2$. Summing the above two quantities and integrating with q as constant, we obtain the result, $u = \frac{1}{2}kL$ as

$$y = \frac{qL^4}{16EIu^4}\left[\frac{\cos(u - 2ux/L)}{\cos u} - 1\right] - \frac{qL^2}{8EI\,u^2}(L-x)x \tag{17}$$

Example 2:

Determine the deflection y, using the general differential equation for a beam–column given by Eq. (7), for a beam uniformly loaded laterally and subjected to an axial force P.

The general differential Eq. (7) is

$$EI\frac{d^4y}{dx^4} + p\frac{d^2y}{dx^2} = q$$

where q is a constant. The general solution of this equation is

$$y = A \sin kx + B \cos kx + Cx + D + \frac{qx^2}{2P}$$

where A, B, C and D are constants. The boundary conditions are that the deflection and bending moment are zero at $x = 0$ and $x = L$, i.e.

$$y = 0 \text{ and } \frac{d^2y}{dx^2} = 0 \quad \text{at } x = 0 \quad \text{and} \quad \text{at } x = L$$

These give

$$B = -D = \frac{q}{k^2P}; \quad A = \frac{q}{k^2P}\frac{1-\cos kL}{\sin kL}; \quad C = -\frac{qL}{2P}$$

Therefore,

$$y = \frac{q}{k^2P}\frac{1-\cos kL}{\sin kL}\sin kx + \frac{q}{k^2P}(\cos kx - 1) - \frac{qL}{2P}x + \frac{qx^2}{2P}$$

Putting

$$u = \frac{kL}{2} \quad \text{and} \quad P = \frac{4u^2EI}{L^2}$$

the above equation can be written as

$$y = \frac{qL^4}{16EIu^4}\left[\frac{1-\cos 2u}{\sin 2u}\sin\frac{2ux}{L} + \cos\frac{2ux}{L} - 1\right] - \frac{qL^2}{8EIu^2}(L-x)x$$

$$= \frac{qL^4}{16EIu}\left[\frac{\sin f - \cos 2u \sin f + \sin 2u \cos f}{\sin 2u} - 1\right] - \frac{qL^2}{8EIu^2}(L-x)x$$

where, we have put $f = 2ux/L$. Simplifying, we obtain

$$y = \frac{qL^4}{16EIu^4}\left[\frac{\cos(u-f)}{\cos u} - 1\right] - \frac{qL^2}{8EIu^2}(L-x)x$$

as in Eq. (17).

BEAM–COLUMN WITH END COUPLE

Consider the beam shown in Fig. 10, where a moment M_b is applied at support B.

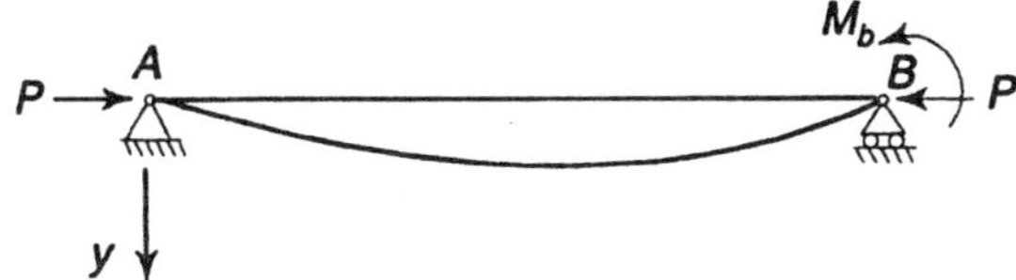

Fig. 10

The solution to this can be obtained from the equation for the deflection curve due to a single concentrated load. For this purpose, we assume that the distance a where that load Q is applied is made to approach zero, however, keeping the product $Qa = M_b$ constant. In this manner, we obtain moment M_b acting at support B. From Eq. (10a)

$$y = \frac{Q \sin ka}{Pk \sin kL} \sin kx - \frac{Qa}{PL} x$$

Now, the limit of $Q \sin ka$ as $a \to 0$ and $Q \to \infty$, so that $Qa = M_b$ remains constant is

$$Lt\, Q\left(ka \frac{k^3a^3}{3!} + \ldots\right) = k(Qa) = M_b k$$

Hence, $$y = \frac{M_b k}{Pk \sin kL} \sin kx - \frac{M_b}{PL} x$$

or $$y = \frac{M_b}{P}\left(\frac{\sin kx}{\sin kL} - \frac{x}{L}\right) \tag{18}$$

If two couples M_a and M_b are applied at the ends A and B of the bar, as shown in Fig. 11, the equation for the deflection curve can be obtained by applying the modified principle of superposition.

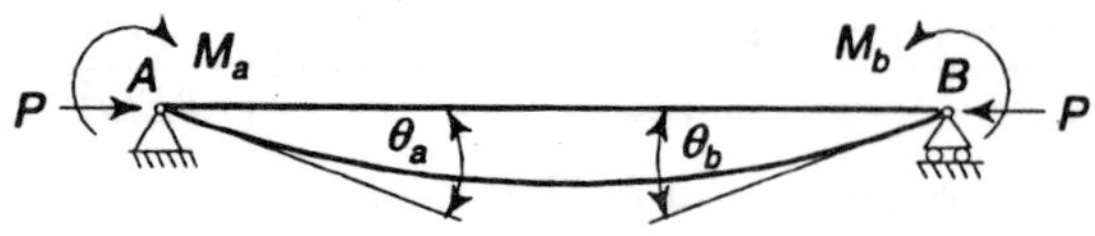

Fig. 11

Equation (18) gives the deflection produced by M_b. In this equation, if we substitute M_a for M_b and $(L - x)$ for x, we obtain the deflection produced by M_a. Adding these results, we get the deflection curve for M_a and M_b acting together. Thus,

$$y = \frac{M_b}{P}\left(\frac{\sin kx}{\sin kL} - \frac{x}{L}\right) + \frac{M_b}{P}\left[\frac{\sin k(L-x)}{\sin kL} - \frac{L-x}{L}\right] \tag{19}$$

The slopes θ_a and θ_b at A and B can be obtained by differentiating the above expression and putting $x = 0$ and $x = L$, i.e.

$$\theta_a = \left(\frac{dy}{dx}\right) at\ x = 0, \quad \text{and} \quad \theta_b = -\left(\frac{dy}{dx}\right) at\ x = L$$

The negative sign in θ_b expression is because of the sign convention adopted [Fig. (11)]. The slopes are

$$\theta_a = \frac{M_a L}{3EI}\psi(u) + \frac{M_b L}{6EI}\phi(u) \tag{20a}$$

$$\theta_b = \frac{M_b L}{3EI}\psi(u) + \frac{M_a L}{6EI}\phi(u) \tag{20b}$$

where $\psi(u) = \frac{3}{u}\left(\frac{1}{\sin 2u} - \frac{1}{2u}\right)$

$$\phi(u) = \frac{3}{2u}\left(\frac{1}{2u} - \frac{1}{\tan 2u}\right)$$

and $u = \frac{1}{2}kL = \frac{1}{2}L\left(\frac{P}{EI}\right)^{1/2}$

Example 3:

A beam–column carries a triangular load. Find the slopes at the ends of the column.

We make use of Eq. (16). The loading Q_i will now be equal to $\frac{q}{L}ada$ acting at distance a from B. Replacing the summation by integration, the equation becomes

$$y = \frac{\sin kx}{Pk \sin kL}\int_0^{L-x}\frac{qa}{L}\sin ka\,da - \frac{x}{PL}\int_0^{L-x}\frac{qa^2}{L}da$$

$$+\frac{\sin k(L-x)}{Pk\sin kL}\int_{L-x}^{L}\frac{qa}{L}\sin k(L-a)\,da-\frac{L-x}{PL}\int_{L-x}^{L}\frac{qa}{L}(L-a)\,da$$

Now, we make use of the formula

$$\frac{d}{d\alpha}\int_{U(\alpha)}^{V(\alpha)}F(\alpha,x)\,dx=\int_{U(\alpha)}^{V(\alpha)}\frac{dF(\alpha,x)}{d\alpha}dx+F(\alpha V)\frac{dV}{d\alpha}-F(\alpha,U)\frac{dU}{d\alpha}$$

Then,

$$\frac{dy}{dx}=\frac{k\cos kx}{Pk\sin kL}\int_{0}^{L-x}\frac{qa}{L}\sin ka\,da-\frac{\sin kx}{Pk\sin kL}\frac{q(L-x)}{L}\sin k(L-x)$$

$$-\frac{1}{PL}\int_{0}^{L-x}\frac{qa^2}{L}da+\frac{x}{PL}\frac{q(L-x)^2}{L}-\frac{k\cos k(L-x)}{Pk\sin kL}\times$$

$$\int_{L-x}^{L}\frac{qa}{L}\sin k(L-a)da+\frac{\sin k(L-x)}{Pk\sin kL}.\frac{q(L-x)}{L}\sin kx$$

$$+\frac{1}{PL}\int_{L-x}^{L}\frac{qa}{L}(L-a)da-\frac{L-x}{PL}\frac{q(L-x)}{L}x$$

$$\therefore\qquad\left\{\frac{dy}{dx}\right\}_{x=0}=\frac{q}{PL\sin kL}\int_{0}^{L}a\sin ka\,da-\frac{q}{PL^2}\int_{0}^{L}a^2\,da$$

$$=\frac{q}{3Pk^2L\tan kL}(3\tan kL-3kL-k^2L^2\tan kL)$$

Similarly,

$$\left\{\frac{dy}{dx}\right\}_{x=L}=\frac{q}{6pk^2L\sin kL}(6\sin kL-6kL-k^2L^2\sin kL)$$

II. GENERAL TREATMENT OF COLUMN STABILITY PROBLEMS

(AS AN EIGENVALUE PROBLEM)

GENERAL DIFFERENTIAL EQUATION AND SPECIFIC EXAMPLES

The general differential equation derived for a beam column, given by Eq. (7), can be used as a general equation to determine the critical loads of buckled bars. If the column is not subjected to lateral loads, then the general equation becomes, with $q = 0$,

$$EI=\frac{d^4y}{dx^4}+P\frac{d^2y}{dx^2}=0\qquad(21)$$

If EI varies along x, then the general equation can be derived on the same lines as in Sec. 3, giving

$$\frac{d^2}{dx^2}\left(EI\frac{d^2y}{dx^2}\right)+P\frac{d^2y}{dx^2}=0 \tag{22}$$

Equations (21) and (22) are the equilibrium equations of a slightly buckled beam subjected to axial load only. Hence, the axial load will represent the critical load. The boundary conditions, i.e. the end conditions, can be quite general. Hence, these equations represent the general differential equations for a column.

For the present, we shall assume that EI is constant along x and use Eq. (21). On using the notation $k^2=\frac{P}{EI}$, Eq. (21) becomes

$$\frac{d^4y}{dx^4}+k^2\frac{d^2y}{dx^2}=0 \tag{23}$$

The general solution of this equation is

$$y = A\sin kx + B\cos kx + Cx + D \tag{24}$$

The constants are determined from the end conditions of the bar.. We can consider the following particular cases:

(i) Column with Hinged Ends (Fundamental Case): In the case of a bar with hinged ends (Fig. 11), the deflection and bending moments are zero at the two ends, i.e.

$$y = 0, \quad \text{and} \quad \frac{d^2y}{dx^2}=0$$

at $\quad x = 0 \quad$ and $x = L$

These conditions give

$$B = 0, \qquad C = 0, \quad D = 0$$

and $\sin kL = 0$

Hence, $kL = n\pi$

The deflection curve is then obtained as

$$y = A\sin kx = A\sin\frac{n\pi x}{L} \tag{25}$$

where A is undetermined, i.e. in determining the load that keeps the column in a slightly buckled form, the amplitude of the deflection remains undetermined. For $n = 1$, $n = 2$ and $n = 3$, the shapes of the buckled bar are as shown in Fig. 12

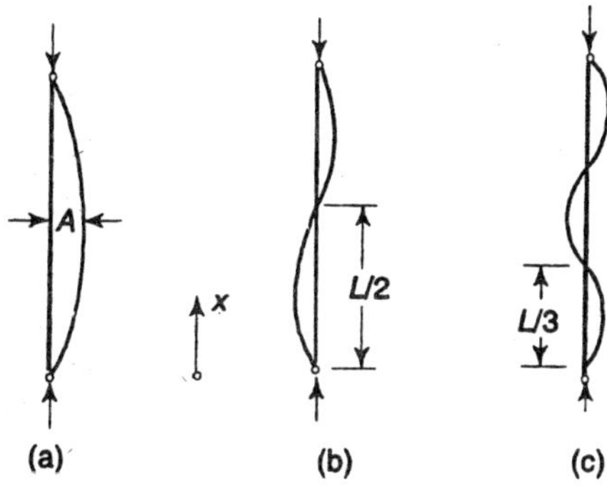

Fig. 12

The corresponding loads are obtained from the equation

$$k\frac{n\pi}{L}=\left(\frac{P}{EI}\right)^{1/2} \quad \text{or} \quad P_{cr}=\frac{n^2\pi^2 EI}{L^2}$$

(ii) Column with One End Fixed and the Other End Free: The end conditions are

at fixed end (i.e. at x = 0), y = 0 and dy/dx = 0

at free and (at $x = L$), moment and shear force are zero, i.e. d^2y/dx^2 = 0 and from Eq. (6)

$$EI\frac{d^3y}{dx^3}+P\frac{dy}{dx}=0$$

From these, the constants are determined as

$$B = -D \qquad C = -Ak,$$

$$A \sin kL + b \cos kL = 0, \qquad C = 0$$

Hence, $A = C = 0$ and $\cos kL = 0$

or $\quad kL = (2n-1)\dfrac{\pi}{2}$

The deflection curve is therefore

$$y = B(1 - \cos kx)$$

or $$y=B\left[1-\cos(2n-1)\frac{\pi x}{2L}\right] \tag{26}$$

With n = 1, we obtain

$$kL=\frac{\pi}{2} \quad \text{or,} \quad P_{cr}=\frac{\pi^2 EI}{4L^2}$$

This is the smallest load that can keep the column in a slightly buckled shape. When $n = 2, 3$, etc., we get the other critical loads as

$$P_{cr} = \frac{9\pi^2 EI}{4L^2}, \quad P_{cr} = \frac{25\pi^2 EI}{4L^2}, \ldots, \text{etc.}$$

The corresponding deflection curves are shown in Fig. 13.

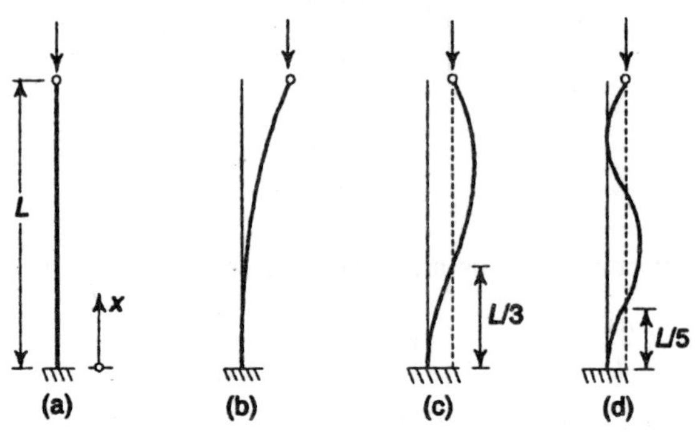

Fig. 13

(iii) Column with One End Fixed and Other End Pinned: This ease is shown in Fig. 14 and was discussed in Sec. 1. Since the top end is pinned, a lateral force R is necessary to keep the column in that position.

The end conditions are

$$y = 0, \quad \frac{dy}{dx} = 0 \quad \text{at } x = 0$$

$$y = 0, \quad \frac{d^2y}{dx^2} = 0 \quad \text{at } x = L$$

With these, the general solution yields the follc

$$B + D = 0$$

$$AK + C = 0$$

$$CL + D = 0$$

$$A \sin KL + B \cos KL = 0$$

P
R
L
x

Fig. 14

A trivial solution for the above set of equations in $A = B = C = D = 0$, which means that the deflection curve is a straight line (i.e. $y = 0$). For the existence of a non–trivial solution, the determinant of the coefficients should be equal to zero. The determinant is

$$\begin{vmatrix} 0 & 1 & 0 & 1 \\ k & 0 & 1 & 0 \\ 0 & 0 & L & 1 \\ \sin KL & \cos KL & 0 & 0 \end{vmatrix} = -\sin KL + LK \cos KL$$

For the existence of a non–trivial solution, the above quantity should be equal to zero, i.e.

$$-\sin KL + KL \cos KL = 0$$

or $$\tan KL = KL \tag{27}$$

The load which keeps the column in a slightly buckled form should, therefore, satisfy the above transcendental equation. The smallest root of this equation is $KL = 4.493$ and the corresponding critical load is

$$P_{cr} = \frac{20.19\, EI}{L^2} = \frac{\pi^2 EI}{(0.699L)^2}$$

(v) Column with Ends Fixed: For a column with both ends fixed, the boundary conditions are

$$y = 0, \quad \frac{dy}{dx} = 0 \quad \text{for } x = 0$$

$$y = 0, \quad \frac{dy}{dx} = 0 \quad \text{for } x = L$$

Substituting these in Eq. (24), we get

$$B + D = 0$$

$$AK + C = 0$$

$$A \sin KL + B \cos KL + CL + D = 0$$

$$AK \cos KL - BK \sin KL + C = 0$$

For the existence of a non–trivial solution, the following determinant should be equal to zero

$$\begin{vmatrix} 0 & 1 & 0 & 1 \\ k & 0 & 1 & 0 \\ \sin kL & \cos kL & L & 1 \\ k \cos kL & -\sin kL & 1 & 0 \end{vmatrix} = 0$$

i.e. $$2(\cos kL - 1) + kL \sin kL = 0$$

or $$\sin\frac{kL}{2}\left(\frac{kL}{2}\cos\frac{kL}{2} - \sin\frac{kL}{2}\right) = 0 \tag{28}$$

One solution is

$$\sin\frac{kL}{2} = 0$$

i.e. $kL = 2n\pi$ and hence $P_{cr} = \dfrac{4n^2\pi^2 EI}{L^2}$

Noting that $\sin kL = 0$ and $\cos kL = 1$, whenever $\sin kL/2 = 0$, the constants are found as

$$A = 0, \qquad C = 0, \qquad B = -D$$

and the deflection curve is therefore

$$y = B\left(\cos\frac{2n\pi x}{L} - 1\right) \tag{29}$$

A second solution to Eq. (28) is

$$\frac{kL}{2}\cos\frac{kL}{2} - \sin\frac{kL}{2} = 0$$

or $$\tan\frac{kL}{2} = \frac{kL}{2}$$

The lowest root of this transcendental equation is $kL/2 = 4.493$ and hence

$$P_{cr} = \frac{8.18\pi^2 EI}{L^2}$$

The deflection curves corresponding to these two critical loads are shown in Fig. 15

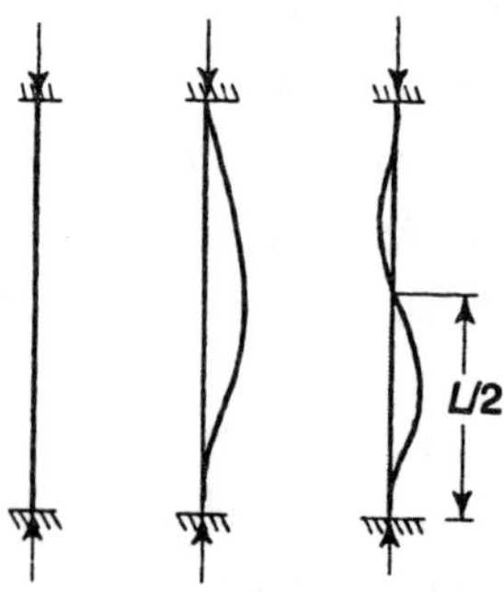

Fig. 15

(v) Column with Load Passing Through a Fixed Point: Consider a column with one end fixed and the other end loaded in such a manner that the load passes through a fixed point (Fig. 16). The load may be applied through the tension of a cable passing through the fixed point *O*.

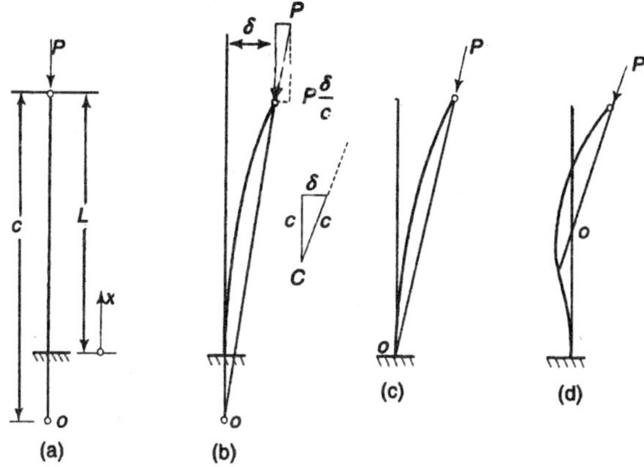

Fig. 16

During buckling, because the force P becomes inclined, a shearing force is developed at the top end. This shearing force is equal to the horizontal component of the inclined force P. Since the deflection is assumed to be very small, the vertical component will be almost equal to P and the horizontal component is

$$V = -P\frac{\delta}{c}$$

From Eq. (6)

$$EI\frac{d^3y}{dx^3} + P\frac{dy}{dx} = P\frac{\delta}{c}$$

or
$$\frac{d^3y}{dx^3} + k^2\frac{dy}{dx} = P\frac{k^2\delta}{c}$$

This is one of the boundary conditions, The other conditions are

$$y = 0 \quad \text{and} \quad \frac{dy}{dx} = 0 \quad \text{at } x = 0$$

$$\frac{d^2y}{dx^2} = 0 \quad \text{at } x = L$$

Substituting these in the general solution given by Eq. (24)

$$B + D = 0$$

$$Ak + C = 0$$

$$C = \frac{\delta}{c}$$

$A \sin kL + B \cos kL = 0$

Solving these equations, the constants are obtained as

$$A = -\frac{\delta}{kc}, \quad B = \frac{\delta}{kc}\tan kL = -D, \quad C = \frac{\delta}{c}$$

Substituting these, the deflection curve is obtained as

$$y = \frac{\delta}{kc}[\tan kL(\cos kL - 1) + kL - \sin kL]$$

or $$\tan kL = kL\left(1 - \frac{c}{L}\right) \tag{30}$$

The above equation gives the value of the critical load for any given ratio of c/L. For three specific values of c/L, the values of kL and $\frac{L^2 P_{cr}}{\pi^2 EI}$ are as follows:

$\frac{c}{L} = 0$	1	∞
$kL = 4.493$	π	$\frac{\pi}{2}$
$\frac{L^2 P_{cr}}{\pi^2 EI} = 2.05$	1	0.25

When $c = 0$, we get the case of a column pinned at the top and fixed at the bottom, which is the case discussed in (iii). When $c = L$, the critical load is obtained as

$$kL = \pi \quad \text{and} \quad P_{cr} = \frac{\pi^2 EI}{L^2}$$

which is the same as the one obtained for case (i), i.e. the fundamental case. This can be explained by the fact that when the line of action of P passes through the base point, the moment at the base is zero and the end behaves like a hinged end. The moment at the top end is also zero and consequently, the column acts as a hinged–end column. When C approaches infinity, the column behaves like it did in case (ii), where the load is always vertical.

BUCKLING PROBLEM AS A CHARACTERISTIC VALUE (EIGENVALUE) PROBLEM

In Sec. 8, the buckling problem was discussed, starting with the general differential equation of equilibrium of a slightly buckled column subjected to axial load only. The specific examples analysed, bring out some general characteristic features of the differential equation, which will be discussed now. These features give us a better insight into the problem and provide a basis for the application of energy methods to buckling problems.

The general differential equation of equilibrium of a slightly buckled column with general boundary conditions was given by Eq. (22) as

$$\frac{d^2}{dx^2}\left(EI_x\frac{d^2y}{dx^2}\right)+P\frac{d^2y}{dx^2}=0 \tag{31a}$$

It is assumed that the moment of inertia I_x can vary along the axis of the column. We can write

$$I_x = Ip\ (x)$$

where I is a constant moment of inertia and $p(x)$ is a dimensionless function of x. The differential equation then becomes

$$\frac{d^2}{dx^2}\left[EIp(x)\frac{d^2y}{dx^2}\right]+P\frac{d^2y}{dx^2}=0 \tag{31b}$$

Dividing by EI and using the notation $k^2=\frac{P}{EI}$, the above equation can be written as

$$\frac{d^2}{dx^2}\left[p(x)\frac{d^2y}{dx^2}\right]+k^2\frac{d^2y}{dx^2}=0 \tag{32}$$

This is a homogeneous differential equation with the following general solution:

$$y=C_1\phi_1(k,x)+C_2\phi_2(k,x)C_3\frac{x}{L}+C_4 \tag{33}$$

where C_1, C_2, C_3 and C_4 are constants and $\phi_2(k,\ x)$ are transcendental functions of x and k. When EI_x is constant, the general solution has the form given by Eq. (24). The constants have the dimensions of length and are determined from the boundary conditions. The most frequently encountered boundary conditions are

For freely supported end $\quad y=0$ and $\frac{d^2y}{dx^2}=0$

For a fixed end $\quad y=0$ and $\frac{dy}{dx}=0$ (34)

For a free end $\quad \frac{d^2y}{dx^2}=0$ and shear $=0$

The zero shear force condition is represented by an equation similar to Eq. (6) as

$$\frac{d}{dx}\left[p(x)\frac{d^2y}{dx^2}\right]+k^2\frac{dy}{dx}=0$$

These boundary conditions are linear and homogeneous equations and will, therefore, be referred to as homogeneous boundary conditions. Substituting these homogeneous boundary conditions for a specific cases in Eq. (33), we get a set of four linear homogeneous equations with the following general form

$$\begin{aligned}
\alpha_{11}C_1 + \alpha_{21}C_2 + \alpha_{31}C_3 + \alpha_{41}C_4 &= 0\\
\alpha_{12}C_1 + \alpha_{22}C_2 + \alpha_{32}C_3 + \alpha_{42}C_4 &= 0\\
\alpha_{13}C_1 + \alpha_{23}C_2 + \alpha_{33}C_3 + \alpha_{43}C_4 &= 0\\
\alpha_{14}C_1 + \alpha_{24}C_2 + \alpha_{34}C_3 + \alpha_{44}C_4 &= 0
\end{aligned} \tag{35}$$

Some of these coefficients αs are transcendental functions of the parameter k while others are constants. We have come across these kinds of equations. A trivial solution of Eq. (35) is that $C_1 = C_2 = C_3 = C_4 = 0$, representing a straight undeflected column ($y = 0$). However, a non–trivial solution, in which we are interested, exists if the determinant of the coefficients is zero, i.e.

$$\Delta = \begin{vmatrix} \alpha_{11} & \alpha_{21} & \alpha_{31} & \alpha_{41} \\ \alpha_{12} & \alpha_{22} & \alpha_{32} & \alpha_{42} \\ \alpha_{13} & \alpha_{23} & \alpha_{33} & \alpha_{43} \\ \alpha_{14} & \alpha_{24} & \alpha_{34} & \alpha_{44} \end{vmatrix} = 0 \tag{36}$$

The expansion of the determinant is Δ and equating it to zero yields an equation for the parameter k, which is the only unknown in this equation. In general, Eq. (36) is a transcendental equation providing an infinite number of roots k_i (i = 1, 2,...,). These are called the characteristic roots or values of the parameter k and for each k_i, there is a corresponding critical load P_i given by

$$k_i^2 = \frac{P_i}{EI}$$

Introducing one of the characteristic values k_i into the system of Eq. (35), we get four equations to determine the four constants C_{1i}, C_{2i}, C_{3i} and C_{4i}. However, as these equations are homogeneous, only the ratios of these constants can be determined. Let

$$\frac{C_{2i}}{C_{1i}} = \bar{C}_{2i}, \quad \frac{C_{3i}}{C_{1i}} = \bar{C}_{3i} \quad \frac{C_{4i}}{C_{1i}} = \bar{C}_{4i}$$

Substituting these in Eq. (33), we get the deflection curve corresponding to the load P_i as

$$y_i = C_{1i}\left[\phi_1(k_i,x)+\bar{C}_{2i}\,\phi_2(k_i,x)+\bar{C}_{3i}\,\phi_3(k_i,x)+\bar{C}_{4i}\,\phi_4(k_i,x)\right] \tag{37}$$

The constant C_{1i} remains undetermined. y_i is called characteristic function of the homogeneous differential Eq. (32) associated with the set of particular boundary conditions of the case under consideration.

The above analysis of the homogeneous differential Eq. (32) shows that there exists a set of values of the parameter k for which a deflected configuration of the column is possible. For each value of the parameter k a corresponding critical load to keep the column in that buckled shape is obtained from Eq. (37). The amplitude (i.e. the magnitude) of deflection however remains indeterminate. There is a close similarity between the analysis of a buckling problem and the analysis of a vibration problem connected is as follows:

Problem	**Equation Δ = 0**	**Characteristic values**	**Characteristic Function**
Buckling	Stability criterion	Buckling loads	Buckling mode
Vibrations	Frequency equation	Frequencies	Principal modes of vibration

THE ORTHOGONALITY RELATIONS

The characteristic functions y_i satisfying the homogeneous differential equation have an important property which play an important role in the energy methods. Consider the homogeneous differential equation

$$\frac{d^2}{dy^2}\left[P(x)\frac{d^2y}{dx^2}\right]+k^2\frac{d^2y}{dx^2}=0$$

This is satisfied by any characteristic function y_i and the associated characteristic value of the parameter k_i, i.e.

$$\frac{d^2}{dy^2}\left[P(x)\frac{d^2y_i}{dx^2}\right]+k_i^2\frac{d^2y_i}{dx^2}=0$$

Multiply this equation by any other characteristic functions y_k and integrate over the length L of the column, obtaining

$$\int_0^L\frac{d^2}{dy^2}\left[P(x)\frac{d^2y_i}{dx^2}\right]y_k\,dx+k_i^2\int_0^L\frac{d^2y_i}{dx^2}y_k\,dx=0 \tag{38}$$

Integrating the first term by parts, we obtain

$$\int_0^L \frac{d^2}{dy^2}\left[P(x)\frac{d^2y_i}{dx^2}\right]y_k\,dx$$

$$= \frac{d}{dy}\left[P(x)\frac{d^2y_i}{dx^2}\right]y_k\Bigg|_0^L - \int_0^L \frac{d}{dy}\left[P(x)\frac{d^2y_i}{dx^2}\right]\frac{dy_k}{dx}dx$$

Integrating once again by parts

$$= \frac{d}{dy}\left[P(x)\frac{d^2y_i}{dx^2}\right]y_k\Bigg|_0^L - \left[P(x)\frac{d^2y_i}{dx^2}\right]\frac{dy_k}{dx}\Bigg|_0^L + \int_0^L\left[P(x)\frac{d^2y_i}{dx^2}\right]\frac{d^2y_k}{dx^2}dx$$

Similarly, the second term in Eq. (38) yields

$$k_i^2\int_0^L \frac{d^2y_i}{dx^2}y_k\,dx = k_i^2\int_0^L\frac{dy_i}{dx}y_k\Bigg|_0^L - k_i^2\int\frac{dy_i}{dx}\frac{dy_k}{dx}dx$$

Substituting these in Eq. (38), we obtain

$$\left[\left\{\frac{d}{dy}P(x)\frac{d^2y_i}{dx^2} + k_i^2\frac{dy_i}{dx}\right\}y_k\right]_0^L - \left[P(x)\frac{d^2y_i}{dx^2}\frac{dy_k}{dx}\right]_0^L$$

$$\int_0^L\left[P(x)\frac{d^2y_i}{dx^2}\frac{d^2y_k}{dx^2}\right]dx - k_i^2\int_0^L\frac{dy_i}{dx}\frac{dy_k}{dx}dx = 0$$

The first term within the brackets vanish for any combination of the homogeneous boundary conditions given by Eq. (34). Consequently.,

$$\int_0^L P(x)\frac{d^2y_i}{dx^2}\frac{d^2y_k}{dx^2}dx - k_i^2\int_0^L\frac{dy_i}{dx}\frac{dy_k}{dx}dx = 0 \tag{39}$$

Since the above equation is valid for each combination of two characteristic functions, we can interchange y_i and y_k and obtain

$$\int_0^L P(x)\frac{d^2y_k}{dx^2}\frac{d^2y_i}{dx^2}dx - k_k^2\int_0^L\frac{dy_k}{dx}\frac{dy_i}{dx}dx = 0$$

Subtracting one from the other

$$\left(k_i^2 - k_k^2\right)\int_0^L\frac{dy_i}{dx}\frac{dy_k}{dx}dx = 0 \tag{40}$$

If *i* is different from *k* then in general $k_i^2 - k_k^2$ will be different from zero, and consequently,

$$\int_0^L\frac{dy_i}{dx}\frac{dy_k}{dx}dx = 0 \tag{41}$$

Using Eq. (41) in Eq. (39), we observe that

$$\int_0^L P(x)\frac{d^2 y_i}{dx^2}\frac{d^2 y_k}{dx^2}dx = 0 \tag{42}$$

If $i = k$, the integral

$$\int_0^L \left(\frac{dy_i}{dx}\right)^2 dx$$

cannot be equal to zero since the integrand is always a positive quantity and in Eq. (39), as k_i^2 is also different from zero, we observe that

$$\int_0^L P(x)\left(\frac{d^2 y_i}{dx^2}\right)^2 dx \neq 0 \quad \text{and} \quad \int_0^L \left(\frac{dy_i}{dx}\right)^2 dx \neq 0 \tag{43}$$

Equations (41) and (42) express the fundamental properties of the characteristic solutions of the differential Eq. (32) and these are known as the orthogonality relations of the characteristic functions y_i. A family of functions consisting of all of Eq. (32) with prescribed boundary conditions is said to constitute a complete system of orthogonal functions.

If I_x is independent of x, then in Eq. (32) $p(x) = 1$ and consequently, Eqs (41) and (42) reduce to

$$\int_0^L \frac{dy_i}{dx}\frac{dy_k}{dx}dx = \int_0^L \frac{d^2 y_i}{dx^2}\frac{d^2 y_k}{dx^2}dx = 0 \tag{44}$$

For example, the sequence of functions

$$y_i = \sin i\frac{\pi x}{L} \qquad (i = 1, 2, \ldots, \infty)$$

which are the terms of a Fourier expansion, form a complete system of orthogonal functions satisfying conditions given by Eq. (44). We may recall that for a hinged column, $\left(\text{i.e. } y = \dfrac{d^2 y}{dx^2} = 0 \text{ for } x = 0 \text{ and } x = L\right)$, the functions

$$y_n = \sin\frac{n\pi x}{L}$$

are the characteristic solutions which satisfy the orthogonality conditions. The advantage of representing a deflection curve by a series like

$$y = a_1 \sin\frac{\pi x}{L} + a_2 \sin\frac{2\pi x}{L} + a_3 \sin\frac{3\pi x}{L} + \ldots$$

will be demonstrated.

III. ENERGY METHODS FOR BUCKLING PROBLEMS

THEOREM OF STATIONARY POTENTIAL ENERGY

The energy method of analysing the problems of elastic stability is based on an extremum principle of mechanics. Consider an elastic body subjected to external surface and body forces. Let the body be in equilibrium. During the application of these forces, the body deforms and consequently, these forces do a certain amount of work W. The internal forces which are set up inside the elastic body also do work during the deformation process and this is stored as elastic strain energy. When external forces are applied gradually and no dissipation of energy takes place due to friction etc. the work done by the external forces should be equal to the internal elastic energy U, i.e.

$$W = U \tag{45}$$

Let portions of the body be given small virtual displacements. These are small displacements that are consistent with the constraints imposed on the body. For example, if a point of the body is fixed, then the virtual displacement there is zero. If a point of the body is constrained to lie on the surface of another body, then the virtual displacement there should be tangential to the surface of the contacting body. These virtual displacements being very small, the changes necessary in the external forces to bring about these virtual displacements will also be very small and will vanish in the limit. The work done by external surface and body forces P_i during these virtual displacements is

$$\delta W = \Sigma P_i \delta \Delta_i + \text{higher order terms} \tag{46}$$

where $\delta\Delta_i$ are the work absorbing components of the virtual displacements. It is convenient to define a potential V of the external forces in such a manner that the work done during virtual displacements is equal to $-\delta V$, i.e. a decrease in potential energy in the form of an equation

$$-\delta V = \Sigma P_i \delta\Delta_i = \delta W \tag{47}$$

In the above equation, we have neglected the higher order terms of Eq. (46). If a part of the body is subjected to distributed external forces, then over that part, the summation must be replaced by a surface integral.

From Eq. (47)

$$-\delta V - \delta W = 0$$

Using Eq. (45), the above equation can be written as

$$\delta (U + V) = 0 \tag{48}$$

The expression $U + V$ is known as the total potential of the system. Consequently, Eq. (48) can be stated as follows:

The first–order change in the total potential energy must vanish for every first–order terms are considered. Terms of higher order, as in Eq. (46), are ignored. Equation (48) is also stated as that in which the quantity $U + V$ assumes a stationary value, i.e.

$$U + V = \text{stationary} \tag{49}$$

It is shown in books on elasticity that for stable equilibrium, any virtual displacement will cause a positive change in the total potential energy of the system, which means that for a system in stable equilibrium, the total potential energy is minimum.

Example 4:

Fig. 17 shows a three--bar truss, the point D of which is subjected to P units of force. Applying the principal of minimum potential energy, determine the vertical and horizontal displacements of D due to the load. The members have equal cross–sectional areas.

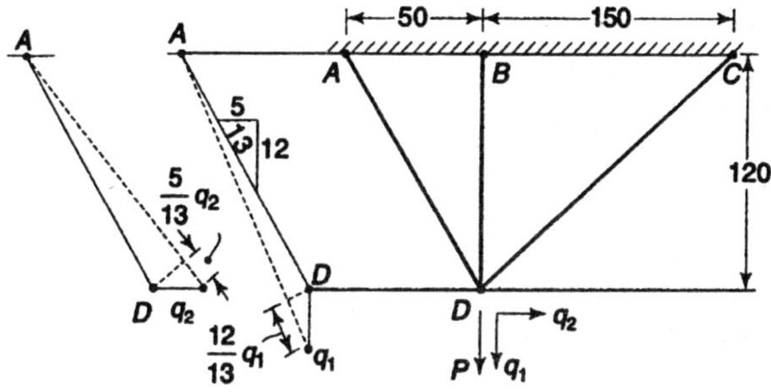

Fig. 17

Let the point D have a vertical displacement q_1 and a horizontal displacement q_2. The elongation caused in each member due to these displacements can be calculated geometrically. This is shown in the figure for member AD. The total extension of each member is

member $AD \dfrac{12}{13} q1 + \dfrac{5}{13} q_2$

member $BD \quad q_1$

member $CD \; \dfrac{4}{5} q_1 - \dfrac{3}{5} q_2$

If A is the cross–sectional area and E is the Young's modulus, then the elastic strain energy stored in a member of length L is

$$U = \frac{EA}{2L}\delta^2$$

where δ is the elongation. Hence, for the three members, the strain energies are

for AD $$U_1 = \frac{EA}{2(130)}\left(\frac{12}{13}q_1 + \frac{5}{13}q_2\right)^2$$

for BD $$U_2 = \frac{EA}{2(120)}q_1^2$$

for CD $$U_3 = \frac{EA}{2(150)}\left(\frac{4}{5}q_1 - \frac{3}{5}q_2\right)^2$$

The total elastic strain energy is the sum of the above three quantities.

Hence,

$$U = \left(958q_1^2 - 47q_1q_2 + 177q_2^2\right)\times 10^{-5}$$

Taking the undeformed position as the datum, the potential energy in the deformed configuration is

$$V = -Pq_1$$

Hence, the total potential energy is

$$U + V = EA\left(958q_1^2 - 47q_1q_2 + 177q_2^2\right)\times 10^{-5} - pq_1$$

For equilibrium position, the first–order variation of the above quantity should be equal to zero, i.e.

$$\delta\,(U + V) = EA\ [958(2q_1\ \delta q_1) - 47(q_1\delta q_2 + q_2\delta q_1) + 177(2q_2\ \delta q_2)\ ] \times 10^{-5} - P\delta q_1 = 0$$

or $$\left(1916q_1 - 47q_2 - \frac{P}{EA}\times 10^5\right)\delta q_1 + \left(-47q_1 + 354q_2\right)\delta q_2 = 0$$

Since δq_1 and δq_2 are arbitrary virtual displacements, the quantities inside the parentheses should vanish individually. Thus,

$$1916q_1 - 47q_2 - \frac{P}{EA}\times 10^5$$

$$-\ 47q_1 + 354q_2 = 0$$

Solving these two equations, we obtain

$$q_1 = 0.0523 \quad \text{and} \quad q_2 = 0.00695$$

It should be observed that we have not made use of any equation of statics in solving the problem.

COMPARISON WITH THE PRINCIPLE OF CONSERVATION OF ENERGY

It is important to realise that the principle of the minimum total potential is different from the law of conservation of energy. The latter principle states that in an equilibrium condition, the work done by all external forces during the loading process is equal to the internal elastic strain energy stored, i.e. Eq. (49) becomes

$$U - W = 0$$

If the loading is done gradually, the work done is equal to

$$W = \sum \frac{1}{2} P_i y_i$$

where y_i is the work absorbing component of the deflection at P_i. On the other hand, the virtual work done is

$$\delta W = \Sigma P_i \, \Delta y_i$$

There is no 1/2 factor in this case since the forces $p_i s$ are acting with full magnitude during the virtual displacements $\Delta y_i s$.

ENERGY AND STABILITY CONSIDERATIONS

In Example 4, we have demonstrated the use of the theorem of stationary potential energy in solving a statically indeterminate problem. Now we shall show, with reference to a specific problem, how energy considerations can be used to analyse stability problems. Consider a vertical bar hinged at one end and supported at the other end by a spring. It is assumed that the bar is infinitely rigid. It carries a centrally applied load P.

Let the bar be displaced through a small angle α. Because of this displacement, the load P is lowered by the amount

$$L - L\cos\alpha = L(1-\cos\alpha) \approx L\frac{\alpha^2}{2}$$

The decrease in potential energy is equal to the work done by P, i.e. $\frac{1}{2}PL\alpha^2$. At the same tine, the spring elongates by an amount αL and the energy stored due to this is $\frac{1}{2}S(\alpha L)^2$ where S is the spring constant. If the decrease in the potential energy is greater than the energy stored in the spring, i.e. if

$$\frac{1}{2}PL\alpha^2 > \frac{1}{2}S\alpha^2L^2$$

then the system is unstable. On the other hand, if

$$\frac{1}{2}PL\alpha^2 < \frac{1}{2}S\alpha^2L^2$$

then the system is stable. If

$$\frac{1}{2}PL\alpha^2 = \frac{1}{2}S\alpha^2L^2, \quad \text{i.e.} \quad P_{cr} = SL$$

then we get the value of the critical load which keeps the column in equilibrium in a slightly displaced configuration.

The same conclusion can be obtained by applying the principles of statics. For equilibrium in the displaced position, the moment about A should be zero. The end B of the column is subjected to a vertical load P and a horizontal force $S\alpha L$. For moment equilibrium about A,

$$P\alpha L = S\alpha L^2 \qquad \text{or} \qquad P_{cr} = SL$$

An analysis of stability problems in column buckling, using the above concept, will be taken up again.

APPLICATION TO BUCKLING PROBLEMS

We shall now discuss the application of the minimum total energy principle to column buckling problems. Consider the column shown in Fig. 18 carrying an axial load P. Let the moment of inertia I_x be variable. In calculating the strain energy, we shall consider only the bending energy. From the straight equilibrium configuration, let the column be moved to a neighbouring bent configuration.

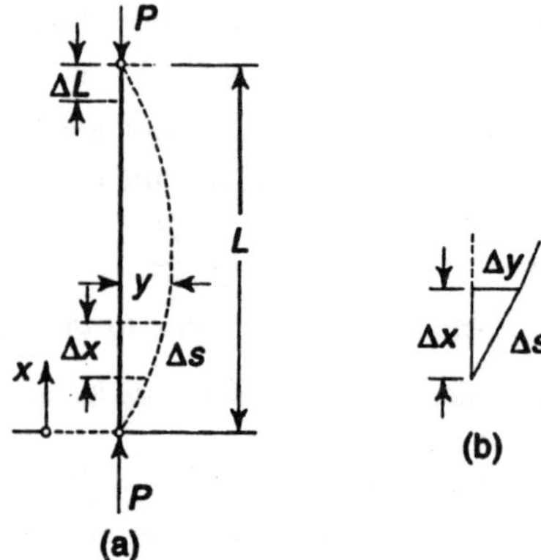

Fig. 18

Let the buckled form be expressed by $y = f(x)$. The elastic strain energy is

$$U = \frac{1}{2} E \int_0^L I_x \left(\frac{d^2 y}{dx^2} \right)^2 dx \tag{50}$$

Taking the undeflected position as datum, the potential energy in the buckled form is

$$V = - P\Delta L$$

To calculate ΔL, we observe from Fig. (20(b))

$$\Delta L = \int_0^L (\Delta s - \Delta x)$$

But $\quad \Delta S = (\Delta x^2 + \Delta y^2)^{1/2} = \Delta x + \frac{1}{2}\left(\frac{\Delta y}{\Delta x}\right)^2 \Delta x$

Hence, $\quad P\Delta L = \frac{1}{2} P \int_0^L \left(\frac{dy}{dx}\right)^2 dx \qquad (51)$

The total potential energy is, therefore, given by

$$U + V = \frac{1}{2} E \int_0^L I_x \left(\frac{d^2 y}{dx^2} \right)^2 dx - \frac{1}{2} P \left(\frac{dy}{dx}\right)^2 dx \tag{52}$$

For equilibrium, the variation of the above quantity should vanish, i.e.

$$\delta(U + V) = \delta \left[\frac{1}{2} E \int_0^L I_x \left(\frac{d^2 y}{dx^2} \right)^2 dx - \frac{1}{2} P \left(\frac{dy}{dx}\right)^2 dx \right] = 0 \tag{53}$$

The above equation permits the determination of the function $y = f(\mathrm{x})$ by applying the technique of the calculus of variations. While the mathematical procedure will not be discussed here, it may be mentioned that the final result agrees with the differential equation given by Eq. (31a). While this differential equation can be derived from static equilibrium considerations as was done in the derivation of Eq. (31a), the merit of the energy criterion for the solution of stability problems becomes evident in the Rayleigh–Ritz method discussed in the next section.

THE RAYLEIGH–RITZ METHOD

A direct solution to the extremum problem stated by Eq. (49) is obtained by the Rayleigh–Ritz method, revealing the importance of the energy criterion. We shall demonstrate the method with respect to the buckling problem discussed in the previous section. The deflection of the buckled column is expressed in the form of a finite series:

$$y = a_1\phi_1 + a_2\phi_2 + \ldots + a_n\phi_n \tag{54}$$

The ϕ terms are a set of arbitrarily chosen functions of x, such that each term satisfies the prescribed boundary conditions of the column. These are called coordinate functions. The coefficients a correspond to a set of parameters, as yet undetermined. With the value of y as given by Eq. (54), the elastic strain energy and the potential energy can be calculated, using Eqs (50) and (51). These lead to an expression involving the n parameters a, and having the form

$$U + V = F_1\ (a_1, a_2, \ldots, a_n) - PF_2\ (a_1, a_2, \ldots, a_n) \tag{55}$$

in which F_1 and F_2 are quadratic forms of the parameters a. If y, as given by Eq. (54), is to be a solution of the problem, then the parameters a must be chosen so as to make the total potential energy an extremum [Eq. (55)]. The problem has, therefore, been reduced to the familiar maximum–minimum problem involving the parameters $a_1, a_2, \ldots, a_n$. Hence, the conditions become

$$\frac{\partial(U+V)}{\partial a_i} - 0, \quad (i = 1, 2, \ldots, n) \tag{56}$$

i.e.

$$\begin{aligned} \frac{\partial F_1}{\partial a_1} - P\frac{\partial F_2}{\partial a_1} &= 0 \\ \frac{\partial F_1}{\partial a_2} - P\frac{\partial F_2}{\partial a_2} &= 0 \\ \frac{\partial F_1}{\partial a_n} - P\frac{\partial F_2}{\partial a_n} &= 0 \end{aligned} \tag{57}$$

The above set of equations involve only linear functions since these are derivatives of the quadratic expressions involved in $U + V$. Since Eq. (57) is a set of homogeneous equations, for the existence of a non–trivial solution, the determinant of the coefficients should be equal to zero, i.e,

$$\Delta = 0 \tag{58}$$

This is an equation of degree n in the unknown P and is the stability condition from which P can be determined. The smallest of the roots gives the critical load P_{cr}.

Introducing $P = P_{cr}$ in Eq. (57), a set of n linear homogeneous equations is obtained from which the ratios of the parameters a can be determined. Calling

$$\frac{a_2}{a_1} = \alpha_2, \quad \frac{a_3}{a_1} = \alpha_3, \quad \ldots, \quad \frac{a_n}{a_1} = \alpha_n$$

the buckling mode is obtained from Eq. (54) as

$$y = a_1(\phi_1\alpha_1 + \phi_2\alpha_2 + \ldots + \phi_n\alpha_n) \tag{59}$$

The importance of the Rayleigh–Ritz method lies in the fact that it offers a method of obtaining an approximate solution to the buckling problem. The method, in many cases, involves less labour than is involved in solving the differential equation and the associated eigenvalue (i.e. characteristic value problem). In the majority of cases, a few terms of the series in Eq. (54) give a sufficiently accurate result. Success or failure in applying the Rayleigh–Rize method to any problem depends largely on the proper choice of the coordinate functions. In the majority of cases, satisfactory results can be obtained only when the coordinate functions chosen form a system of orthogonal functions discussed in Sec. 10. This is the reason why Fourier Series play such an important role in the applications of the Rayleigh–Ritz method.

Example 5:

Consider a pin–ended column subjected to an axial compressive load P, as shown in Fig. 20. Assume that the buckled shape is given by

$$y = a \sin \frac{\pi x}{L}$$

where a is an unknown parameter. The coordinate function chosen satisfies the boundary conditions which are

$$y = 0 \quad \text{at } x = 0 \quad \text{and} \quad \text{at } x = L$$

$$\frac{d^2y}{dx^2} = 0 \quad \text{at } x = 0 \quad \text{and} \quad \text{at } x = L$$

Form Eq. (50), the strain energy is obtained as

$$U = \frac{1}{2} EI \int_0^L \left(\frac{d^2y}{dx^2}\right)^2 dx$$

$$= \frac{1}{2} EI \int_0^L a^2 \left(\frac{\pi}{L}\right)^4 \sin^2 \frac{\pi x}{L} dx$$

$$= \frac{1}{4} \pi^4 a^2 \left(\frac{EI}{L^3}\right)$$

Form Eq. (51), the potential energy is obtained as

$$V = -\frac{1}{2} P \int_0^L \left(\frac{dy}{dx}\right)^2 dx$$

$$= -\frac{1}{2} P \int_0^L a^2 \left(\frac{\pi}{L}\right)^2 \cos^2 \frac{\pi x}{L} dx$$

$$= -\frac{1}{4} P\pi^2 \left(\frac{a^2}{L}\right)$$

Thus, the total potential energy is

$$U + V = \frac{1}{4}\pi^4 a^2 \frac{EI}{L^3} - \frac{1}{4} P\pi^2 \frac{a^2}{L}$$

For this to be an extremum, we must have

$$= \frac{1}{2}\pi^4 a \frac{EI}{L^3} - \frac{1}{2} P\pi^2 \frac{a}{L} = 0$$

or $$\frac{1}{2}\pi^2 \frac{a}{L}\left(\pi^2 \frac{EI}{L^2} - P\right) = 0$$

The non–trivial solution is obtained when

$$P = P_{cr} = \frac{\pi^2 EI}{L^2}$$

We have been able to obtain an exact solution since the coordinate function we used happens to give the exact diflected shape for the column.

Example 6:

Consider a column fixed at one end and free at the other end

It is subjected to a compressive load P at the free end. Determine the approximate critical load assuming the deflection curve as

$$y = a_1\left(\frac{x}{L}\right)^2 + a_2\left(\frac{x}{L}\right)^3$$

The boundary conditions are

$$y = 0 \quad at\ x = 0, \qquad \frac{dy}{dx} = 0 \quad at\ x = 0$$

and $$\frac{d^2y}{dx^2} = 0 \quad at\ x = L$$

(i) Let us ignore the last condition for the time being. The first two conditions are satisfied by the coordinate functions. The strain energy is equal to

$$U = \frac{1}{2} EI \int_0^L \left(\frac{d^2y}{dx^2}\right)^2 dx$$

$$= \frac{1}{2} EI \int_0^L \left(\frac{2a_1}{L^2} + \frac{6a_2 x}{L^3}\right)^2 dx$$

$$= \frac{2EI}{L^3}\left(a_1^2 + 3a_1a_2 + 3a_2^2\right)$$

The potential energy is to equal to

$$V = -\frac{1}{2}P\int_0^L \left(\frac{dy}{dx}\right)^2 dx$$

$$= -\frac{1}{2}P\int_0^L \left(\frac{2a_1x}{L^2} + \frac{3a_2x^2}{L^3}\right)^2 dx$$

$$= \frac{P}{30L}\left(20a_1^2 + 45a_1a_2 + 27a_2^2\right)$$

Hence, the total potential energy is

$$U + V = a_1^2\left(\frac{2EI}{L^3} - \frac{2P}{3L}\right) + a_1a_2\left(\frac{6EI}{L^3} - \frac{3P}{2L}\right) + a_2^2\left(\frac{6EI}{L^3} - \frac{9P}{10L}\right)$$

For an extremum we should have

$$\frac{\partial(U+V)}{\partial a_1} = \left(\frac{4EI}{L^3} - \frac{4P}{3L}\right)a_1 + \left(\frac{6EI}{L^3} - \frac{3P}{2L}\right)a_2 = 0$$

For the existence of a non–trivial solution, the determinant D of the coefficients should be equal to zero. Hence,

$$\Delta = \left(\frac{4EI}{L^3} - \frac{4P}{3L}\right)\left(\frac{12EI}{L^3} - \frac{9P}{5L}\right) - \left(\frac{6EI}{L^3} - \frac{3P}{2L}\right) = 0$$

or $\quad 3P^2L^4 - 104PL^2EI + 240E^2I^2 = 0$

Therefore,

$$P = 2.49\frac{EI}{L^3} \quad \text{or} \quad 32.18\frac{EI}{L^2}$$

The smaller value is

$$P_{cr} = 2.49\frac{EI}{L^2}$$

Compared to the exact value $\dfrac{\pi^2 EI}{4L^2}$, the error is only +0.92%.

(ii) In the above analysis, we have ignored the third boundary condition, i.e. at $x = L$, $\dfrac{d^2y}{dx^2} = 0$. If we use this condition

$$\frac{d^2y}{dx^2} \text{ at } x = L = \frac{2a_1}{L^2} + \frac{6a_2L}{L_3} = 0$$

or $\quad a_1 = -3a_2$

Using this

$$y = a_1\left(\frac{x}{L}\right)^2 - \frac{1}{3}a_1\left(\frac{x}{L}\right)^3$$

Substituting $a_1 = -3a_2$ in the expressions for U and V

$$U = \frac{2EI}{L^2}\left(a_1^2 - a_1^2 + \frac{a_1^2}{3}\right) = \frac{2}{3}\frac{EI}{L^3}a_1^2$$

and $$V = \frac{P}{L}\left(\frac{2}{3}a_1^2 - \frac{2}{3}a_1^2 + \frac{1}{10}a_1^2\right) = \frac{4}{15}\frac{P}{L}a_1^2$$

Therefore,

$$U + V = \left(\frac{2}{3}\frac{EI}{L^3} - \frac{4}{15}\frac{P}{L}\right)a_1^2$$

For an extremum, the quantity inside the parentheses should be equal to zero, i.e.

$$\frac{4}{15}\frac{P}{L} = \frac{2}{3}\frac{EI}{L^3}$$

or $$P_{cr} = 2.5\frac{EI}{L^2}$$

which is almost identical with the previous solution but the solution has been obtained with comparative ease.

TIMOSHENKO'S CONCEPT OF SOLVING BUCKLING PROBLEMS

Consider a straight column subjected to an axial load P. If P is less than the critical load, then the column is in stable equilibrium, which means that if the column is slightly displaced from its straight equilibrium position by any transverse disturbing force, it will return to its vertical position as soon as the disturbing force is removed. In terms of energy, this means that when P is less than P_{cr}, in the slightly bent configuration, the elastic strain energy stored in the bent column is greater than the work done by the axial load in moving through a distance ΔL, i.e.

$$U - W > 0 \tag{60}$$

where U is the strain energy and $W = P\,\Delta L$. On the other hand, when P exceeds P_{cr}, if the column is slightly displaced, the work done by the external load P will exceed the strain energy in bending and the equilibrium becomes unstable. Consequently, the condition

$$U - W = 0 \tag{61}$$

characterises the state when the equilibrium configuration changes from stable to unstable.

Following the same procedure as in the Rayleigh–Ritz method, we can assume that the buckled column curve can be expressed by Eq. (54) as

$$y = a_1\phi_1 + a_2\phi_2 + \ldots + a_n\phi_n$$

The ϕ terms are functions of x so that each term satisfies the boundary conditions of the column. The constants a_1, a_2,..., define the amplitudes of the terms. The strain energy is given by

$$U = \frac{1}{2}EI\int_0^L\left(\frac{d^2y}{dx^2}\right)^2 dx = F_1(a_1, a_2, \ldots, a_n)$$

The work done by the external force during deformation is

$$W = \frac{1}{2}P\int_0^L\left(\frac{dy}{dx}\right)^2 dx = PF_2(a_1, a_2, \ldots, a_n)$$

Using Eq. (61)

$$P = EI\frac{\int_0^L (d^2y/dx^2)^2\, dx}{\int_0^L (dy/dx)^2\, dx} = \frac{F_1(a_1, a_2, \ldots, a_n)}{F_2(a_1, a_2, \ldots, a_n)} \tag{62}$$

Observing that for pin–ended column or a column with one end free

$$M = -EI\frac{d^2y}{dx^2}$$

and that $M = Py$

and $$EI\int\left(\frac{d^2y}{dx^2}\right)^2 dx = \frac{P^2}{EI}\int y^2\, dx$$

Eq. (62) can also be written as

$$P = \frac{EI\int_0^L (dy/dx)^2\, dx}{\int_0^L y^2\, dx} \tag{10.63}$$

Since we need the minimum value for the load P, the critical load is obtained when the expression in Eq. (62) or Eq. (63) is made a minimum. This requires that the derivatives of Eq. (62) or Eq. (63) with respect to each coefficient a_i must vanish. This yields

$$\frac{\partial F_1}{\partial a_i} - P\frac{\partial F_2}{\partial a_i} = 0, \quad (i = 1, 2, \ldots) \tag{64}$$

These are identical to Eq. (57). Since there are n homogeneous equations, a non–trivial solution exists when the determinant of the coefficients is equal to zero. This discussion is identical to that given in Sec. 15.

There is a fundamental difference between Eq. (62) and (63) though they appear to be equivalent. The elastic strain energy is obtained from the expression

$$U = \frac{1}{2EI}\int_0^L M^2\, dx$$

If we take the deflection curve as $y = y\ (x)$, then M could be expressed in two ways

$$M = -EI\frac{d^2y}{dx^2} \tag{65}$$

or $$M = Py$$

where P is the axial force acting on a pin–ended column or a column with one end free end the other end fixed. If we use the first expression in Eq. (65), we get the strain energy for any assumed form of the column. If we use the second expression, we take the external force also into account and consequently, the final result obtained for P_{cr} using Eq. (63) gives a slightly more accurate result. Equation (62) is generally referred to as the Rayleigh–Ritz formula and Eq. (63) as the Timoshenko formula.

COLUMNS WITH VARIABLE CROSS–SECTIONS

So for, in the examples considered, we have treated the moment of inertia I as independent of x. We shall consider a few problems where I varies with x. The energy method will be found to be very suitable to obtain fairly good solutions.

Example 7:

Consider a column with the moment of inertia of the cross–sectional area varying according to the equation

$$I = I_0\left(1+\sin\frac{\pi x}{L}\right)$$

The column is hinged at both ends. Assume that the deflection curve can be represented by the series

$$y = \sum a_n \sin\frac{n\pi x}{L}$$

Since the deflection curve must be symmetrical with respect to the middle point of the column (because the moment of inertia is symmetrical about the middle point), the even parameters in the above series vanish. The deflection equation then becomes

$$y = a_1 \sin\frac{\pi x}{L} + a_3 \sin\frac{3\pi x}{L} + a_5 \sin\frac{5\pi x}{L} + \ldots$$

We shall consider only two terms of the series. Thus

$$y = a_1 \sin\frac{\pi x}{L} + a_3 \sin\frac{3\pi x}{L}$$

$$U = \frac{1}{2}\int_0^L EI\left(\frac{d^2 y}{dx^2}\right)^2 dx$$

$$= \frac{1}{2}\int_0^L EI_0\left(1+\sin\frac{\pi x}{L}\right)\left[-a_1\left(\frac{\pi}{L}\right)^2 \sin\frac{\pi x}{L} - a_3\left(\frac{3\pi}{L}\right)^2 \sin\frac{3\pi x}{L}\right]^2 dx$$

$$= \frac{1}{2}EI_0\left(\frac{\pi}{L}\right)^3\left[\left(\frac{4}{3}+\frac{\pi}{2}\right)a_1^2 - \frac{24}{5}a_1 a_3 + \left(\frac{2916}{35}+\frac{81\pi}{2}\right)a_3^2\right]$$

and

$$\Delta L = EI\frac{1}{2}\int_0^L\left(\frac{dy}{dx}\right)^2 dx = \frac{1}{2}\left(\frac{L}{2}a_1^2 + \frac{L}{2}\times 9a_3^2\right)$$

Substituting in Eq. (62)

$$P = EI_0\frac{\pi}{L^2}\frac{(8/3+\pi)a_1^2 - (48/5)a_1 a_3 + (5832/35) + 81\pi)a_3^2}{a_1^2 + 9a_3^2} = \frac{F_1}{F_2}$$

For minimum P, we should have from Eq. (64)

$$\frac{\partial}{\partial a_1}(F_1 - PF_2) = 0 \quad \text{and} \quad \frac{\partial}{\partial a_2}(F_1 - PF_2) = 0$$

Thus, $$2\left(\frac{8}{3}+\pi-P^*\right)a_1 - \frac{48}{5}a_3 = 0$$

and $$-\frac{48}{5}a_1 + 2\left(\frac{5832}{35}+81\pi-9P^*\right)a_3 = 0$$

where $$P^* = \frac{PL^2}{\pi EI_0}$$

For the existence of a non–trivial solution, the determinant of the coefficients should be equal to zero. This gives

$$\Delta = 9P^{*2} - \left(\frac{5832}{35}+81\pi+24+9\pi\right)P^*$$

$$+\left(\frac{8}{3}+\pi\right)\left(\frac{5832}{35}+81\pi\right) - \left(\frac{24}{5}\right)^2 = 0$$

Solving,

$$P^* = 5.746 \text{ or } P = 18.05\frac{EI_0}{L^2}$$

USE OF TRIGONOMETRIC SERIES

In many instances, it will be useful to represent the deflection curve in the form of a trigonometric series. We have discussed in Sec. 11 that the functions satisfying Eqs (31a) and (31b) also satisfy orthogonality conditions. The trigonometric series which we shall consider now is made up of such functions. Let the deflection curve be represented by

$$y = a_1 \sin\frac{\pi x}{L} + a_2 \sin\frac{2\pi x}{L} + \ldots + a_n \sin\frac{n\pi x}{L} + \ldots \tag{66}$$

By properly determining the coefficients a_1, a_2,..., the above series can be made to represent any deflection curve. These coefficients may be calculated by a consideration of the strain energy of the beam or the column. The strain energy is given by

$$U = \frac{1}{2}EI\int_0^L \left(\frac{d^2y}{dx^2}\right)^2 dx$$

Now, $$\frac{d^2y}{dx^2} = -a_1\frac{\pi^2}{L^2}\sin\frac{\pi x}{L} - a_2\frac{2^2\pi^2}{L^2}\sin\frac{2\pi x}{L} - a_3\frac{3^2\pi^2}{L^2}\sin\frac{3\pi x}{L} - \ldots$$

Hence, the square of the above expression will involve terms of two kinds

$$a_n^2\frac{n^4\pi^4}{L^4}\sin^2\frac{n\pi x}{L}$$

and $$2a_m a_n\frac{n^4 m^2\pi^4}{L^4}\sin\frac{n\pi x}{L}\sin\frac{m\pi x}{L}$$

By direct integration it can be seen that

$$\int_0^L \sin^2\frac{n\pi x}{L}dx = \frac{L}{2}, \text{ and } \int_0^L \sin\frac{n\pi x}{L}\sin\frac{m\pi x}{L}dx = 0 \quad \text{for } n \neq m$$

These are the orthogonality relations expressed by Eqs (42) and (43). Hence, in the expression for strain energy, terms containing products like a_m a_n vanish and only terms like a_n^2 remain. Then

$$U = \frac{1}{2}EI\left[a_1^2\frac{\pi^4}{L^4}\frac{L}{2} + a_2^2\frac{2^4\pi^4}{L^4}\frac{L}{2} + a_3^2\frac{3^4\pi^4}{L^4}\frac{L}{2} + \ldots\right]$$

$$= \frac{EI\pi^4}{4L^3}\left(1a_1^2 + 2^4a_2^2 + 3^4a_3^2 + \ldots\right)$$

$$= \frac{EI\pi^4}{4L^3}\sum_{n=1}^{\infty} n^4 a_n^2 \tag{67}$$

Similarly, if we consider the expression

$$\Delta L = \frac{1}{2}\int_0^L \left(\frac{dy}{dx}\right)^2 dx$$

we find the integrand to consist of two kinds of terms

$$a_n^2 \frac{n^2\pi^2}{L^2}\cos^2\frac{n\pi x}{L}$$

and $$2a_m a_n \frac{mn\pi^2}{L^2}\cos\frac{m\pi x}{L}\cos\frac{n\pi x}{L}$$

By direct integration it can be shown that

$$\int_0^L \cos^2\frac{n\pi x}{L} = \frac{L}{2}$$

and $$\int_0^L \cos^2\frac{n\pi x}{L}\cos\frac{m\pi x}{L}dx = 0 \quad \text{for } n \neq m$$

Consequently,

$$\Delta L = \frac{\pi^2}{4L}\sum_{n=1}^{\infty} n^2 a_n^2 \tag{68}$$

With these expressions for U and ΔL, we can consider the following example.

Example 8:

A beam column is subjected to an axial force P and a lateral force Q at x = c (Fig. 19). Determine the deflection curve using the energy method.

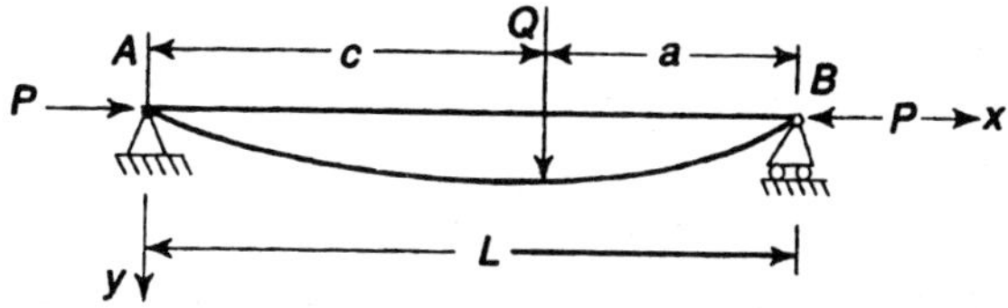

Fig. 19

Let the deflection curve be

$$y = a_1 \sin\frac{\pi x}{L} + a_2 \sin\frac{2\pi x}{L} + \ldots$$

Let a virtual displacement δy_n be given. This virtual displacement is obtained by changing one of the terms a_n sin $(n\pi x/L)$ to $(a_n + \delta a_n)$ sin $(n\pi x/L)$.

In other words, the deflection curve δa_n sin $(n\pi x/L)$ is wuperimposed on the original deflection curve. The work done by the external forces Q and P is

$$\delta W = Q\,\delta a_n \sin\frac{n\pi c}{L} + P\delta\,(\Delta L)$$

Using Eq. (68)

$$\delta W = Q\,\delta a_n \sin\frac{n\pi c}{L} + P\frac{\pi^2}{4L}2n^2 a_n \delta_n$$

The increase in strain energy is

$$\delta U = \frac{\partial U}{\partial a_n}\delta a_n$$

From Eq. (67)

$$\delta U = \frac{EI\pi^4}{L^3}n^4\, a_n\,\delta a_n$$

Since the increase in strain energy should be equal to the work done, we have

$$Q\sin\frac{n\pi c}{L}\delta a_n + P\frac{\pi^2}{2L}n^2 a_n \delta a_n = \frac{EI\pi^4}{2L^3}n^4 a_n\,\delta a_n$$

from which,

$$a_n = \frac{2QL^3}{EI\pi^4}\frac{1}{\left(n^2 - \frac{PL^2}{EI\pi^2}\right)}\sin\frac{n\pi c}{L}$$

If we use the notation

$$\beta = \frac{PL^2}{EI\pi^2}$$

then, $$a_n = \frac{2QL^3}{EI\pi^4}\frac{1}{n^2(n^2-\beta)}\sin\frac{n\pi c}{L}$$

The deflection curve is, therefore, given by

$$y = \frac{2QL^3}{EI\pi^4}\sum_{n=1}^{\infty}\frac{1}{n^2(n^2-\beta)}\sin\frac{n\pi c}{L}\sin\frac{n\pi x}{L}$$

Example 9:

Using an infinite series, determine the deflection curve for the beam column shown in Fig. 20.

In the solution of the previous example consider c to be very small. Then

$$\sin\frac{n\pi c}{L} \approx \frac{n\pi c}{L}$$

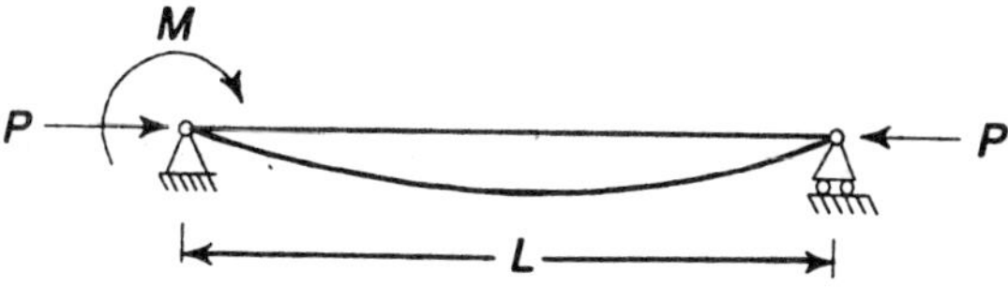

Fig. 20

and $$y = \frac{2L^3}{EI\pi^4}\frac{\pi}{L}Qc\sum_{n=1}^{\infty}\frac{n}{n^2(n^2-\beta)}\sin\frac{n\pi x}{L}$$

Let $c \to 0$ and $Q \to \infty$, such that $Qc = M =$ constant.

Then, $$y = \frac{2ML^3}{EI\pi^3}\sum_{n=1}^{\infty}\frac{1}{n^2(n^2-\beta)}\sin\frac{n\pi x}{L}$$

EXERCISES

1. A uniformly loaded beam is built–in at one end and simply supported at the other end. It is subjected to an axial force P. Determine the moment M_b at the built–in end.

$$\left[\begin{aligned} &\textit{Ans.}\quad M_b = +\frac{qL^2}{8}\frac{\beta(u)}{\phi(u)} \\ &\text{where, } \beta(u) = \frac{3}{u^3}(\tan u - u) \\ &\qquad\quad \phi(u) = \frac{3}{2u}\left(\frac{1}{2u} - \frac{1}{\tan 2u}\right) \end{aligned}\right]$$

2. A beam of uniform cross–section has an initial curvature given by the equation

$$y_0 = \delta \sin\frac{\pi x}{L}$$

It is subjected to end couples M_a and M_b and to an axial force P. Determine the deflection curve.

$$\left[Ans.\ y = -\frac{M_a}{P}\left[\frac{L-x}{L} - \frac{\sin k(L-x)}{\sin kL}\right] + \frac{M_b}{P}\left[\frac{\sin kx}{\sin kL} - \frac{x}{L}\right] + \frac{\delta\pi^2}{\pi^2 - k^2L^2}\sin\frac{\pi x}{L} \right]$$

3. The initial shape of a bar can be approximated by the series

$$y = \delta_1 \sin\frac{\pi x}{L} + \delta_2 \sin\frac{2\pi x}{L} + \ldots$$

If the bar is simply supported and subjected to axial force P only, show that the deflection curve due to P is given by

$$y_1 = \alpha\left(\frac{\delta_1}{1-\alpha}\sin\frac{\pi x}{L} + \frac{\delta_2}{2^2-\alpha}\sin\frac{2\pi x}{L} + \ldots\right), \text{ where } \alpha = \frac{k^2L^2}{\pi^2}$$

4. For a column with one end built–in and the other end free and carrying an axial load P, it is assumed that the deflection curve has the form

$$y = \frac{\delta x^2}{L^2}$$

where L is the length of the column and x is measured form the fixed end. Using the energy method, determine the critical load.

$$\left[Ans.\ P_{cr} = 2.5\frac{EI}{L^2} \right]$$

5. The deflection curve for a pin–ended column is represented by a polynomial as

$$y = ax^4 + bx^3 + cx^2 + dx + e$$

Determine the critical load by the energy method.

$$\left[Ans.\ P_{cr} = 9.88\frac{EI}{L^2} \right]$$

6. A prismatic bar with hinged ends (Fig. 26) is subjected to the action of a uniformly distributed axial load of intensity q and an axial compressive force P. Find the critical value of P by assuming, for the deflection curve, the equation

$$y = \delta \sin\frac{\pi x}{L}$$

7. Determine the critical load $(P_1 + P_2)$ by the energy method for the case shown in Fig. 27. The column has a moment of inertia I_1 for

half the length and moment of inertia I_2 for the other half. Assume the deflection curve in the form

$$y = \delta \sin \frac{\pi x}{L}$$

$$\left[\begin{aligned} &Ans.\ (P_1 + P_2)_{cr} = \\ &\frac{(\pi^2 EI_2 / L^2)(m+1)}{m + \frac{m}{6}\left(\frac{m-1}{m}\right)^2 - \frac{8}{\pi^2}(m-1) + n\left[\frac{1}{m} + \frac{m}{6}\left(\frac{m-1}{m}\right)^2 + \frac{8}{\pi^2}\frac{m-1}{m}\right]} \\ &\text{where } m = \frac{P_1 + P_2}{P_1} \text{ and } n = \frac{I_2}{I_1} \end{aligned} \right]$$

2

Hamiltonian Variational Principles and Methods

INTRODUCTION

Variational (minimal) principles has been the greatest fascination of old generation physicists and in this pursuit several and varied variational principles have been put forth in the various branches of physics, such as, in optics, the Fermat's principles of '*lest time*', in mechanics, the Gauss's principles of '*least constraint*', the Hertz's principle of '*lest curvature*' and most important of all the Hamilton's *variational principle'* which will be our main centre of attraction here in mechanics.

The great value of these variational principles; lies in their extreme economy of expression.

Let T, V be the kinetic and potential energies of a. Conservative holonomic dynamical system defined by n generalised co-ordinates $q_1, q_2, \ldots, q_n$ at time t, writing L = T – V, we know that Lagrange's equations for the motion of the system are:

$$(\partial L/\partial q_i) - (d/dt)(\partial L/\partial \dot{q}_i) \qquad (i = 1, 2, \ldots, n)$$

Now applying previous article we can say that these are the n Euler-Lagrange's equation arising from the variational problem.

$$\int_{t_1}^{t_1} L(q_1, q_2, \ldots, q_n; \dot{q}_1, \dot{q}_2, \ldots, \dot{q}_n; t)\ dt = \text{stationary } (\textit{i.e. } \text{stat})$$

where t_1, t_2 are fixed. Thus we have established that

During the motion of a 'conservative holonomic dynamical system over a fixed time interval, the time integral over that interval of the difference between the kinetic and potential energies is stationary. This is Hamilton's principle.

DERIVATION OF HAMILTON EQUATIONS FROM THE VARIATIONAL PRINCIPLE

Hamilton's principle implies

$$\delta I = \delta \int_{t_1}^{t_2} L dt = 0 \text{ or } \delta\, \delta \int_{t_1}^{t_2} [\sum_i p_i \dot{q}_i - H(q,p,t)]\, dt = 0$$

$$[\because L = \sum_i p_i \dot{q}_i - H(q,p,t)]$$

This implies $\delta \sum_i \int_{q_1}^{q_2} p_i d\, q_i - \delta \int_{t_1}^{t_2} H\, dt = 0.$ ...(1)

Equation (1) is called the modified Hamilton's principle.

Now $\delta I = \frac{\partial I}{\partial \alpha} d\alpha = d\alpha \frac{\partial}{\partial \alpha} \int_{t_1}^{t_2} (\sum_i p_i \dot{q}_i - H(q,p,t) dt = 0$ where $\delta = d\alpha(\partial / \partial_\alpha)$

$$\Rightarrow d\alpha \int_{t_1}^{t_2} \frac{\partial}{\partial \alpha} [\sum p_i, \dot{q}_i - H(q,p,t) dt = 0$$

[$\because$ the times t_1, t_2 are not varied not so they are not functions of α, thus the differentiation can be interchanged].

$$\Rightarrow d\alpha \int_{t_1}^{t_2} \sum_i \left[\frac{\partial p_i}{\partial \alpha} \dot{q}_i + \frac{\partial \dot{q}_i}{\partial \alpha_i} p_i - \frac{\partial H}{\partial q_i} \frac{\partial q_i}{\partial \alpha} - \frac{\partial H}{\partial p_i} \frac{\partial p_i}{\partial \alpha} \right] dt = 0 \qquad ...(2)$$

Further, we have $\int_{t_1}^{t_2} \frac{\partial \dot{q}_i}{\partial \alpha} p_i dt = \int_{t_1}^{t_2} p_i \frac{d}{dt}\left(\frac{\partial q_i}{\partial \alpha}\right) dt = p_i \left[\frac{\partial p_i}{\partial \alpha}\right]_{t_1}^{t_2} - \int_{t_1}^{t_2} \dot{p}_i \frac{\partial q_i}{\partial \alpha} dt.$

But all the varied paths have the same end points. Hence $\frac{\partial \dot{q}_i}{\partial \alpha}$ vanishes for t_1 and t_2. $\Rightarrow \int_{t_1}^{t_2} \frac{\partial q_i}{\partial \alpha} p_i dt = -\int_{t_1}^{t_2} \dot{p}_i \frac{\partial q_i}{\partial \alpha} dt = \int_{t_2}^{t_1} p_i \frac{\partial q_i}{\partial \alpha} dt.$

Making this substitution in equations (2), we have

$$d\alpha \int_{t_1}^{t_2} \sum \left[\frac{\partial q_i}{\partial \alpha} \dot{q}_i - \frac{\partial H}{\partial q_i} \frac{\partial q_i}{\partial \alpha} - \frac{\partial H}{\partial p_i} \frac{\partial p_i}{\partial \alpha} - \frac{\partial q_i}{\partial \alpha} \right] dt = 0 \qquad ...(2)$$

$$\Rightarrow \int_{t_1}^{t_2} \sum \dot{q}_i \frac{\partial \dot{q}_i}{\partial \alpha} q\alpha - \frac{\partial H}{\partial q_i} \frac{\partial q_i}{\partial \alpha} q\alpha - \frac{\partial H}{\partial p_i} \frac{\partial p_i}{\partial \alpha} d\alpha - p_i \frac{\partial q_i}{\partial \alpha} d\alpha\, dt = 0 \qquad ...(3)$$

But $\delta p_i = d\alpha.(\partial p_i / \partial \alpha)$ and $\delta q_i = (\partial q_i / \partial \alpha) d\alpha$

$$\therefore (3) \Rightarrow \int_{t_1}^{t_2} \Sigma \left[\delta p_i \left(\dot{q}_i - \frac{\partial H}{\partial p_i} \right) + \delta q_i \left(\frac{\partial H}{\partial q_i} - \dot{p}_i \right) \right] dt = 0.$$

The variation δq_i and δp_i are independent to each other, hence equation (4) holds good only when the coefficients of δp_i and δq_i vanish separately.

The (4) $\Rightarrow \dot{q}_i = (\partial H / \partial p_i), \dot{p}_i = -(\partial H / \partial q_i)$. These are Hamilton's equations.

Principal of Least Action

Principle of least action states that if T is kinetic energy, at time t, of a conservative, holonomic dynamical system specified by the generalised co-ordinates, then the integral $I = \int_{t_1}^{t_2} 2T dt$

has necessary an extreme value, minimum or maximum, on actual path as compared with varied path as the systems passes from one configuration at time t_0 *to another configuration at time* t_1.

We know that L = T – V, *i.e.* Lagrangian = K.E. – P.E. and T + V = E (const), since system is conservative.

But by Hamilton's principle, we know that

$$\int_{t_0}^{t_1} \delta L \, dt = 0 \Rightarrow \int_{t_0}^{t_1} \delta(T - V) dt = 0 \int_{t_0}^{t_1} \delta(2T - E) dt = 0$$

$$\Rightarrow \int_{t_0}^{t_2} \delta(2T) \, dt = 0 \Rightarrow \delta \int_{t_0}^{t_1} (2T) \, dt = 0. \qquad ...(1)$$

[$\because$ $\delta E = 0$ as E, the total energy is const.]

Result (1), is known as *Principle of least action.*

Equation (1) can also be written as $\delta A = 0$, where $A = \int_{t_0}^{t_1} 2T \, dt$ and is defined by action as follows:

This implies that principal of least action states that the action in the actual path is minimum compared with the varied path, as the system passes from one configuration to another.

DISTINCTION BETWEEN HAMILTON'S PRINCIPLE AND PRINCIPLE OF LEAST ACTION

In Hamilton's principle, the time of description $t_1 - t_0$ *is prescribed (fixed) as the body moves from one configuration to another configuration, while in* *the principle of least action there is no such restriction on the time $t_1 - t_0$ but the total energy between the end points A and B is prescribed.*

DEDUCTION OF LARGRANGE'S EQUATION FROM HAMILTON'S PRINCIPLE

By Hamilton's principle, we have $\int_{t_0}^{t_1} \delta L dt = 0$

$$\therefore (1) \Rightarrow \int_{t_0}^{t_1} \sum_{i=1}^{n} \left(\frac{L}{\partial q_i} \delta q_i + \frac{L}{\partial \dot{q}_i} \delta \dot{q}_i \right) dt = 0$$

$$\Rightarrow \int_{t_0}^{t_1} \sum_{i=1}^{n} \frac{\partial L}{\partial q_i} \delta q_i \, dt + \int_{t_0}^{t_1} \sum_{i=1}^{n} \frac{\partial L}{\partial \dot{q}_i} \frac{d}{dt} (\delta q_i) \, dt = 0$$

Now integrating by parts, we have

$$\int_{t_0}^{t_1} \sum_{i=1}^{n} \frac{\partial L}{\partial q_i} \delta q_i \, dt + \left\{ \sum_{i=1}^{n} \frac{\partial L}{\partial \dot{q}_i} \delta q_i \right\}_{t_0}^{t_1} - \int_{t_0}^{t_1} \left\{ \sum_{i=1}^{n} \frac{d}{dt} \left(\frac{\partial L}{\partial \dot{q}_i} \right) \delta q_i \right\} dt = 0$$

$$\Rightarrow \int_{t_0}^{t_1} \left[\sum_{i=1}^{n} \right] \left\{ \frac{d}{dt} \left(\frac{\partial L}{\partial \dot{q}_i} \right) - \frac{\partial L}{\partial q_i} \delta q_i \right\} dt = 0$$

[$\because$ middle term vanishes as all $\delta q_i = 0$ at t_0 and t_1]

$$\Rightarrow \sum_{i=1}^{n} \left\{ \frac{d}{dt} \left(\frac{\partial L}{\partial \dot{q}_i} \right) - \frac{\partial L}{\partial q_i} \right\} \delta q_i = 0.$$

Here δq_i's are arbitrary and independent to each other, so equating to zero their coefficients, we readily obtain

$$\frac{d}{dt} \left(\frac{\partial L}{\partial \dot{q}_i} \right) - \frac{\partial L}{\partial q_i} = 0$$

DEDUCTION OF HAMILTON'S PRINCIPLE

Assume that the conservative holonomic dynamical system moves from A to C, where A and C, are the initial and final configurations of the system at time t_1 and t_2 respectively. Let ABC be the actual path and AB'C. AB"C the two neighbouring paths out of infinite number of possibilities. *In order to deduce the principle, the following two conditions must be satisfied.*

1. δt must be equal to zero at A and B *i.e.* at t_1 the particle must be at A, at t_2 the particle must be at C.
2. δr must be equal to zero at A and C $\Rightarrow$ that the points A, and C are fixed in space.

Now assume that the system is acted upon by a number of forces represented by F.

Let the ith particle of system be acted upon by a force $F_i = m_i\ \ddot{r}_i$ is acceleration vector.

Again by D'A Alembert's principle, we have

$$\sum_i (F_i - m\ddot{r}_i)\cdot\delta r_i = 0,\ \textit{i.e.}\ \sum_i F_i\cdot\delta\dot{r}_i - \Sigma m_i\ \ddot{r}_i\cdot\delta r_i = 0 \qquad ...(1)$$

If there is a little variation along the actual and neighbouring paths, then

$$\delta r_i = r_i - r_i \quad \text{(say)}.$$

$$\Rightarrow (d/dt)(\delta r_i) = (d/dt)(r_i - r_i) = (dr_i'/dt) - (dr_i/dt)$$

$$= \delta\ (dr_i/dt) = \delta(d\dot{r}_i) \qquad ...(2)$$

where dashes have been used for neighbouring paths.

But $\ddot{r}_i\cdot\delta_i = (d/dt)\ddot{r}_i\cdot\delta r_i) - \dot{r}_i\cdot(d/dt)(\delta r_i)$. ...(3)

Using (2), (3) $\Rightarrow \ddot{r}_i\cdot\delta r_i = (d/dt)(\dot{r}_i\cdot\delta r_i) - \dot{r}_i\delta(\dot{r}_i)$ ...(4).

$$\therefore (1) \Rightarrow \sum_i F_i\cdot\delta r_i - \sum_i m_i\ [(d/dt)(r_i\cdot\delta r_i) - \dot{r}_i\cdot\delta\dot{r}_i] = 0$$

$$\text{or}\ \sum_i F_i\cdot\delta r_i - \sum_i m_i\ [(d/dt)(r_i\cdot\delta r_i) - \frac{1}{2}\delta(\ddot{r}_i^2 = 0$$

$$\text{or}\ \sum_i F_i\cdot\delta r_i + \sum_i \frac{1}{2} m_i\delta(\dot{r}_i^2 = \sum_i (d/dt)(m_i\ddot{r}_i\cdot\delta r_i)$$

$$\text{or}\ \sum_i F_i\cdot\delta r_i + \delta\left(\sum_i \frac{1}{2} m_i\dot{r}_i^2\right) = \sum_i (d/dt)(m_i r_i\cdot\delta r_i) \qquad ...(5)$$

But $\sum_i \frac{1}{2} m_i\dot{r}_i^2$ = kinetic energy of the system = T

and $\sum_i F_i\cdot\delta r_i$ = work done by the forces F_i during displacement δr_i δW (say).

$$\therefore \text{equation (5)} \Rightarrow \delta W + \delta T = \sum_i (d/dt)(m_i\dot{r}_i\cdot\delta r_i) \qquad ...(6)$$

Integrating (6) between the limits t_1 and t_2, we get

$$\int_{t_1}^{t_2}(\delta W + \delta T)\,dt = \int_{t_1}^{t_2}\sum_i \frac{d}{dt}(m_i\dot{r}_i\cdot\delta r_i)\,dt = \Sigma\int_{t_1}^{t_2} d\,(m_i r_i\cdot\delta r_i)$$

$$= \Sigma\left[m^2 r_i^2\cdot\delta r^2\right]_A^C = 0\ [\text{since}\ \delta r^2 = 0\ \text{at the end points A anc C.}$$

But we know that, for a conservative system, $\delta W = -\ \delta V$ where V is potential energy

$$\Rightarrow \int_{t_1}^{t_2}(-\delta V + \delta T)dt = 0, \text{i.e. } \delta\int_{t_1}^{t_2}(T - V)dt = 0$$

i.e. $\int_{t_1}^{t_2}(T - V)\,dt = 0$ or $\delta\int_{t_1}^{t_2}L\,dt = 0$

or $\delta\int_{t_1}^{t_2}L\,dt$ = stat = extremum. This is Hamilton's principle.

Extension of Hamilton's Principle to Non-conservative and Non-holonomic System

We can generalise Hamilton's principle to include non-conservative forces as well, so that an alternative form of the Lagrange's equation can be obtained. The extension of the Hamilton's principle is

$$\delta I = \delta\int_{1}^{2}(T + W)\,dt = 0 \qquad ...(1)$$

where $$W = \Sigma F_i \cdot r_i \qquad ...(2)$$

We know that the variations δq_i or δr_i are identical with virtual displacements of the co-ordinates as there is no variation of time. Thus we can consider the varied path in configuration space as built up by a succession of virtual displacements from the actual path of motion C. Again each virtual displacements takes place at some definite time and at that time the forces acting on the body have definite values. δW represents the amount of work done by the forces on the system during the period of virtual displacement from the actual to the varied path. Thus the Hamilton's principle given by (1) can be said as *the integral of the variation of kinetic energy together with the amount of virtual work involved in the variation must be zero.*

We can evaluate the variations δr_i in terms of δq_i by making use of the equations of transformation between r and q, where each q depends on the path chosen through a parameter say α:

$r_i = r\ \{q_k\ (\alpha,t)t\}$

We can abbreviated the process by using the equivalence of δr_i with a virtual displacement. We know that

$\sum_i F_i \cdot \delta r_i = \sum_k Q_k \delta q_k$.

thus (1) may be put as $\delta\int_{t}^{2}T\,dt + \int_{1}^{2}\sum_k Q_k \delta q_k dt = 0.$ ...(3)

Now it can be shown easily that (3) reduces to ordinary from of Hamilton's principle, in case Q_k's are derivable form the generalised potential. Therefore the integral of virtual work, under these conditions becomes,

$$\int_t^2 \sum_k Q_k \delta q_k \, \delta q_k \, dt = -\int_1^2 \sum_k \delta q_k \left(\frac{\partial V}{\partial q_k} - \frac{d}{dt}\frac{\partial V}{\partial \dot{q}_k} \right) dt$$

Now reversing by integration-paths procedure; the above integral can be put as

$$-\int_t^2 \sum_k \left(\frac{\partial V}{\partial q_k}\delta q_k + \frac{\partial V}{\partial \dot{q}_k}\delta q_k \right) dt = -\delta \int_1^2 V \, dt$$

Thus (3) reduces to

$$\delta \int_1^2 T \, dt = \int_1^2 V \, dt = \delta \int_1^2 (T - V) \, dt \; \delta \int_1^2 L dt = 0,$$

which is Hamilton's principle.

For more general problem, the variation of the first integral in (3) can be written at once, as T like L for conservative systems is a function of q_k and $\dot{q}_k$

$$\therefore \delta \int_1^2 T \, dt = \int_1^2 \sum_b \left(\frac{\partial T}{\partial q_k} - \frac{d}{dt}\frac{\partial T}{\partial \dot{q}_k} \right) \delta q_k \, dt \qquad ...(4)$$

$$\Rightarrow \int_1^2 \sum_k \left(\frac{\partial T}{\partial q_k} - \frac{d}{dt}\frac{\partial T}{\partial \dot{q}_k} + Q_k \right) \delta q_k \, dt = 0 \qquad ...(5)$$

Further it is assumed that the constraints are holonomic, so the integral (5) can vanish if and only if the separate coefficients vanish, *i.e.*

$$\frac{d}{dt}\frac{\partial T}{\partial \dot{q}_k} - \frac{\partial T}{\partial q_k} = Q_k$$

so that (1) represents the proper extension of Hamilton's principle, which yields that from of Lagrange's equation, in case the forces are not derived from a potential.

We can also extend the Hamilton's principle to cover certain categories of non-holonomic systems as well. While deriving Lagrange's equations from Hamilton's or D'Alembert's principle, application of holonomic constraints is made only in last step when the variation q_k are considered as independent of each other.

While in non-holonomic system the generalized co-ordinates are not independent to each other so they cannot be reduced further by means of equations of constraints of the form

$f\,[q_1, q_2, ... q_n; t] = 0$ Thus q_k's can not be as all independent.

We can further treat non-holonomic system, provided the equations of constraint can be putting the from

$$\sum_j q_{ij} da_j + a_{ij} dt = 0 \quad [j = 1,2, - ...m] \qquad ...(7)$$

which is a relation connecting the differentials of q's since in variation process used in Hamilton's principle, the time for each point on the path is taken constant, therefore the virtual displacements occurring in the variation must satisfy the equations of constraint of the form

$$\sum_j q_{ij} \delta q_j = 0, (j = 1,2,...m) \qquad ...(8)$$

The equations (8) can very well be used to reduce the number of virtual displacement to independent ones. The method used for eliminating these extra virtual displacements is that of Lagrange's *undetermined multipliers.*

Also from (8), we get

$$\lambda_l \sum_J a_{ij} \delta q_j = 0, (j = 1,2,...m) \qquad ...(9)$$

where λ_l are some undetermined constants, which in general, are functions of time.

Now first of all summing the equations (9) over l and then integrating the resulting equation from 1 to 2, we get

$$\int_1^2 \sum_{ji} \lambda_l a_{ij} \delta q_j dt = 0 \qquad ...(10)$$

Again the equations corresponding to (5) in the case of conservative system are given as

$$\int_1^2 dt \sum_j \left(\frac{\partial L}{\partial q_j} - \frac{d}{dt} \frac{\partial L}{\partial \dot{q}_j} \right) \delta q_j = 0 \qquad ...(11)$$

Combining the equations (10) with (11), we get

$$\int_1^2 dt \sum_j \left(\frac{\partial L}{\partial q_j} - \frac{d}{dt} \frac{\partial L}{\partial \dot{q}_j} + \sum_l \lambda_l a_{ij} \right) \delta q_j = 0 \qquad ...(12)$$

It is to be noted here that δq_j's are still not independent. They are connected by the m relations given by (8.) This is, while the first n – m of these equations may be chosen independently, the last m are then fixed by the equations (8). Since the values of λ_l's are at our disposal, we choose them to be such that

$$\frac{\partial L}{\partial q_j} - \frac{d}{dt} \frac{\partial L}{\partial \dot{q}_j} + \sum_l \lambda_l a_{ij} = 0, (j = n - m + 1, ...n) \qquad ...(13)$$

which are in the nature of equations of motion for the last m of the q_j variable.

Using the value of λ_l obtained from (13) in (12), we get

$$\int_1^2 dt \sum_{j=1}^{n-m} \left(\frac{\partial L}{\partial q_j} - \frac{d}{dt}\frac{\partial L}{\partial \dot{q}_j} + \sum_l \lambda_l a_{lj} \right) \delta q_j = 0 \qquad ...(14)$$

Since δq_j's involved here are independent ones, therefore we have

$$\frac{\partial L}{\partial q_j} - \frac{d}{dt}\frac{\partial L}{\partial \dot{q}_j} + \sum_l \lambda_l a_{lj} = 0, (j = 1, 2, ...n - m) \qquad ...(15)$$

Now combining (13) with (15), we get

$$\frac{d}{dt}\frac{\partial L}{\partial \dot{q}_j x} - \frac{\partial L}{\partial q_j} = \sum_l \lambda_l a_{lj}, (j = 1, 2, ...n) \qquad ...(16)$$

which are known as Lagrange's equations for non holonomic systems. We now observe that (16) gives us a total of only n equations, where as the unknowns involved are (n +m) in number, namely then n co-ordinates q_l and the m, λ_l's. The additional equationals required are exactly the equations of constraint *i.e.*

$$\sum_j a_{lj}\dot{q}_j + a_{lj} + 0 \qquad ...(17)$$

The equations (16) together with (17) constitute (n + m) equations for n + m unknowns.

It is to be noted here that equation (7) is not the most general type of non-holonomic constraint. For example, it does not include equations of constraints in the form of inequalities. More over it includes holonomic constraints. An equation of the form

$$f(q_1, q_2, q_3, ...q_n ; t) = 0 \qquad ...(18)$$

it is of the same form as the equation 97) with the coefficients

$$a_{lj} = \frac{\partial f}{\partial q_j} \text{ and } a_{ji} = \frac{\partial f}{\partial t} \qquad ...(19)$$

It means that method of Lagrange's undetermined multipliers can be used also for holonomic constraints, when it is not easy to reduce all the q's to independent co-ordinates.

VARIATIONAL METHODS

Calculus of Variations : The calculus of variations arose out of the quest for the mathematical requirements in the solution of problems like the study of:

(i) The path followed by a body falling freely under gravity, (brachistochrone) first studied by Newton,

(ii) The equilibrium shape of a freely hanging homogeneous flexible cord between two horizontal points (catenary), first studied by Bernoulli. Presently the technique of calculus of variations, serves as a mathematical preliminary (in the form of Euler-Lagrange's equation) in the study of a wide range of physical problems such a geodesics and minimal surface in Riemannian and differential geometrics and the various *variational (minimal)* principles in the different branches of physics, The technique is co-ordinates invariant and there in lies its great power. First we develop here this technique in a purely mathematical form.

Suppose A and B are fixed points (x_1, y_1), (x_2, y_2) in a cartesian plane.

Also suppose that (x, y, y′) is known functional form of the variables x, y, y′ {= dy/dx)}. Then if C is a curve joining A and B and having equation y = y (x), Then the integral

$$I = \int_{x_1}^{x_2} f[x, y, y']\, dx \qquad ...(1)$$

has a definite value whenever the function y (x) is prescribed. The value of I will change as we vary the from of the curve C through A, B. Consequently we may consider that in general there will be some curve

C through these fixed points such that the value of I *taken along it is stationary (in general a* maximum or minimum) compared with the value along neighbouring paths C. Calculus of variations, the branch of mathematics of concerned amongst other things with finding the form of y (x) from which this stationary property holds. Let the path $\bar{C}$ have the equation y = $\bar{y}$ (x) and let the equation of a neighbouring curve C by y = $\bar{y}$ (x) + εη (x), and $\bar{P}X = \bar{y}(x) \Rightarrow P\bar{P} = \varepsilon\eta(x)$.

Hence the values of I taken along is thus a function of ε of the form

$$I(\varepsilon) = \int_{x_1}^{x_2} f[x, \bar{y} + \varepsilon\eta, \bar{y}' + \varepsilon\eta']\, dx = \int_{x_1}^{x_2} f[x, y, y']\, dx \qquad ...(2)$$

where $y = \bar{y} + \varepsilon\eta$ and $y' = \bar{y}' + \varepsilon\eta'$

But the curve $\bar{C}$, for which ε = 0, makes I stationary, so we must have I′ stationary, so we must have I′ (0) = 0. Now differentiating (2) w.r.t. ε, we get

$$I'(\varepsilon) = \int_{x_1}^{x_2} \left[\frac{\partial f}{\partial x}\frac{\partial x}{\partial \varepsilon} + \frac{\partial f}{\partial y}\frac{\partial y}{\partial \varepsilon} + \frac{\partial f}{\partial y'}\frac{\partial y'}{\partial \varepsilon}\right] dx \qquad [\because f = f[x, y, y']] \qquad ...(3)$$

$$\Rightarrow I'(\varepsilon) = \int_{x_1}^{x_2} \left[\frac{\partial f}{\partial y} \eta(x) + \frac{\partial f}{\partial y'} \eta'(x) \right] dx \qquad [\because y = \bar{y} + \varepsilon\eta(x)]$$

$$= \int_{x_1}^{x_2} [\eta f_y . \eta(x) + f_{y'} . \eta'(x)]\, dx$$

$$I'(\varepsilon) = \int_{x_1}^{x_2} [\eta f_y (x, \bar{y} + \varepsilon\eta, \bar{y}' + \varepsilon\eta')] \; + \eta' f_{y'} (x, \bar{y} + \varepsilon\eta, \bar{y} + \varepsilon\eta')] dx \qquad \text{...(4)}$$

where $f_{y'} (x, \bar{y} + \varepsilon\eta, \bar{y} + \varepsilon\eta')] dx$

$= (\partial / \partial y)\, f\, (x, y, y')$ when $y = \bar{y} + \varepsilon\eta,\ y' = \bar{y}' + \varepsilon\eta,$

with a similar meaning for $f_{y'}\ (x, \bar{y} + \varepsilon\eta, \bar{y}' + \varepsilon\eta')$. Hence the stationary condition give $I'(0) = 0$.

$$\Rightarrow \int_{x_1}^{x_2} [\eta f_{y'} (x, \bar{y}, \bar{y}') + \eta' f_{y'}\ (x, \bar{y}, \bar{y}')\, dx = 0 \qquad \text{...(5)}$$

Now $\int_{x_1}^{x_2} \eta' f_{y'} (x, \bar{y}, \bar{y}')\, dx$

$$= \left[\eta f_{y'} (x, \bar{y}, \bar{y}') \right]_{x_1}^{x_2} - \int_{x_1}^{x_2} n \frac{d}{dx} f_{y'}\ (x, \bar{y}, \bar{y}')\, dx$$

$$= - \int_{x_1}^{x_2} n \frac{d}{dx} f_{y'}\ (x, \bar{y}, \bar{y}')\, dx\ (\because \eta(x_1) = \eta(x_2) = 0) \qquad \text{...(6)}$$

Using (6), equation (5) implies

$$\int_{x_1}^{x_2} \eta(x) \left[f_y (x, \bar{y}, \bar{y}' - \frac{d}{dx} f_{y'} (x, \bar{y}, \bar{y}') \right] dx = 0 \qquad \text{...(7)}$$

But η (x) is arbitrary, subject to its being differentiable and vanishing at A, B; (7) implies that

$$f_y (x, \bar{y}, \bar{y}') - (d / dx)\, f_{y'} (x, \bar{y}, \bar{y}') = 0 \qquad \text{...(8)}$$

or $\qquad (\partial f / \partial y) - (d / dx)\, (\partial f / \partial y') = 0. \qquad$ ...(9)

This is called *Euler's Lagrange's equation.* Equation (9) can be proved from (7) by using the following lemma.

Lemma: If x_1 and x_2 ($>x_1$) *are fixed constants and* ϕ *(x) is a particular continuous function for* $x_1 \leq \phi \leq x_2$ and if

$$\int_{x_1}^{x_2} \eta(x)\phi(x)\,dx = 0 \qquad ...(10)$$

for every choice of continuously differentiable function h(x) for which $\eta(x_1) = \eta(x_2) = 0$ *then* $f(x) = 0$ *indentically in* $x_1 \le x \le x_2$.

Proof:

Let the lemma be not true for all x in $x_1 \le x \le x_2$. Let x' be a point where $\phi(x') \neq 0$ and > 0. But $\phi(x)$ is continuous in $x_1 \le x \le x_2$ and in particular is continuous at $x = x'$. Hence there exist an interval $x'_1 \le x \le x'_0$ around x' where $\phi(x) > 0$. For the other x's $\phi(x)$ any both vanish; then the equation $\int_{x_1}^{x_2} \eta(x)\phi(x)dx = 0$ could be written as $\int_{x_1}^{x_{2'}} \eta(x)\phi(x)dx = 0$.

But $\eta(x)$ is at our disposal; we could choose it as greater than zero for $x'_1 \le x \le x'_2$ and $\eta(x) = 0$ for other values of x. Thus $\eta(x)\ \phi(x) > 0$ for $x'_1 \le x \le x'_2$; consequently integrand cannot be zero. Thus we arrive at the contradiction. Hence $\phi(x) = 0$ in $x_1 \le x \le x_2$.

Certain Remarks About Euler-Lagrange's Equation

(A) We have $(\partial f/\partial y) - (d/dx)(\partial f/\partial y') = 0$

where $f = f(x, y, y')$

which can be re-written as

$$(\partial f/\partial y) - (dp/dx) = 0 \text{ where } p = (\partial f/\partial y') \qquad ...(11)$$

But $\left(\dfrac{dp}{dx}\right) = \left(\dfrac{\partial p}{\partial x}\right) + \left(\dfrac{\partial p}{\partial y}\right)\left(\dfrac{dy}{dx}\right) + \left(\dfrac{\partial p}{\partial y'}\right)\left(\dfrac{dy'}{dx}\right) = \left(\dfrac{\partial^2 f}{\partial x \partial y'}\right) +$

$(\partial^2 f/\partial y \partial y')(dy/dx) + (\partial^2 f/\partial y'^2)(d^2y/dx^2)$. Hence (9) give

$$(\partial f/\partial y) - [(\partial/\partial x)(\partial f/\partial y') + (\partial/\partial y)((\partial f/\partial y')(dy/dx)$$

$$+ (\partial^2 f/\partial y'^2)(d^2y/dx^2)] = 0$$

or $$(\partial^2 f/\partial y'^2)(d^2y/dx^2) + (\partial^2 f/\partial y\,\partial y')(dy/dx) - (\partial f/\partial y) + (\partial^2 f/\partial x\,\partial y') = 0 \qquad ...(12)$$

This is a second order differential equation for determining y as a function of x. The solution will contain two arbitrary constants which could be determined from the conditions:

$$y = y_1 \text{ at } x = x_1 \text{ and } y = y_2 \text{ at } x = x_2 \qquad ...(13)$$

(B) Consider $\frac{d}{dx}\left[f - y'\frac{\partial f}{\partial y'}\right] = \frac{\partial f}{\partial y'}$

i.e. $\frac{dQ}{dx} = \frac{\partial f}{\partial x}$ which $Q = f - y'\ \frac{\partial f}{\partial y'}$ and f is a function of x, y, y¢

$$\Rightarrow \frac{\partial f}{\partial x}\left[f - y'\frac{\partial f}{\partial y'}\right] + \frac{\partial}{\partial y}\left[f - y'\frac{\partial f}{dy'}\right]\frac{dy}{dx} + \frac{\partial}{\partial y'}\left[f - y'\frac{\partial f}{\partial y'}\right]\frac{d^2y}{dx^2} = \frac{\partial f}{\partial x}$$

$$\Rightarrow - y'\frac{\partial^2 f}{\partial x \partial y'} + \frac{\partial f}{\partial y}\frac{dy}{dx} - y'\frac{\partial^2 f}{\partial y \partial y'}\frac{dy}{dx} + \frac{\partial f}{\partial y'}\frac{d^2y}{dx^2}\frac{\partial f}{\partial y'}\frac{\partial^2 y}{\partial x^2} - y'\frac{\partial^2 f}{\partial y'^2}\frac{d^2y}{dx^2} = 0$$

$$\Rightarrow y'\left[\frac{\partial f}{\partial y} - \frac{\partial^2 f}{\partial x \partial y'} - y'\frac{\partial^2 y}{\partial y \partial y'} - y''\frac{\partial^2 f}{\partial y'^2}\right] = 0$$

$$\Rightarrow f_y - f_{xy'} - y'\, fyy' - y'' f_{yy'} = 0 \text{ (if } y' \neq 0) \qquad ...(15)$$

which is same as (12) and shows that if Euler's equation is satisfied equation (14) is also satisfied but if (14) is satisfied, Euler's equation may not be true.

(C) If f does not contain x explicity, i.e. $l = \int_{x_1}^{x_2} f(y, y')dx$,

then from (14), we have

$$(d/dx)\{f - y'(\partial f/\partial y')\} = 0 \qquad \{\because (df/x) = 0\}$$

$$\Rightarrow f - y'(\partial f/\partial y') = \text{constant} = C \text{ (say.)} \qquad ...(16)$$

Obviously (14) is a first order differential equation.

(D) If f does not contain y explicity, then we have $l = \int_{x_1}^{x_2} f(x, y')dx$.

$\therefore$ from the equation

$(\partial f/\partial y) - (d/dx)(\partial f/\partial y') = 0$ we obtain $(d/dx)(\partial f/\partial y') = 0$

$$\Rightarrow (\partial f/\partial y') = C \text{ (say).} \qquad ...(17)$$

(E) If we assume that the end points are not fixed.

i.e. $\quad \eta(x_1) \neq 0$ and $\eta(x_2) \neq 0$.

Euler's equation still holds in this case, if $(\partial f/\partial y') = 0$ when $x = x_1, x_2$ This becomes the boundary condition of the problem.

(F) Euler's equation gives only a necessary condition; even if it is satisfied, there may be no extremum.

(G) The conditions to be imposed on f are such that its partial derivative are continuous and differentiable.

(H) The conditions to be imposed on the curves are that they should be continuous curves.

Brachistochrone Problem

Given two points A and B in a vertical plane, to find for the moveable particle M, the path A MB, descending along which by its own gravity and beginning to be urged from the point A, it may in the shortest time reach the point B. It is a tacit in the statement that the particle descends without friction.

Let A be the origin, then velocity of the particle of mass M at any time t is given by

$$v = (ds/dt) = \sqrt{(2gy)}$$

$$\Rightarrow dt = (ds/\sqrt{2gy})$$

$$\Rightarrow t = \int_A^B \frac{ds}{\sqrt{(2gy)}} = \int_A^B \frac{\sqrt{(1+y'^2}}{\sqrt{(2gy)}} dx$$

$$\frac{1}{\sqrt{(2g)}} \int_A^B \frac{\sqrt{1+y'^2}}{\sqrt{2}} dx = \frac{1}{\sqrt{(2g)}} \int_A^B f(y, y') dx,$$

where $f(y, y') = \dfrac{\sqrt{(1+y'^2)}}{\sqrt{y}}$

As x is absent in f, so we have $f - y'(\partial f/\partial y') A$

$$\Rightarrow \frac{\sqrt{(1+y'^2)}}{\sqrt{y}} - y' \frac{y'}{\sqrt{\{y(1+y'^2\}}} = A$$

$$\Rightarrow \frac{1}{\sqrt{\{y(1+y'^2)\}}} = \text{constant} \Rightarrow y(1+y'^2) = \text{constant} = 2c, \text{(say)}$$

But $(dy/dx) = \tan\psi \Rightarrow y' = \tan\psi$

$\therefore y\sec^2\psi = 2c$ or $y = 2c\cos^2\psi = c(1+\cos^2\psi)$

Also $dx = \cot\psi\, dy = -2c\sin 2\psi \cot\psi\, d\psi$

$= -4c\cos^2\psi\, d\psi = -2c\,(1+\cos^2\psi)\, d\psi$

$$\Rightarrow x = a - 2c\left(\psi + \frac{\sin^2\psi}{2}\right) = a - 2c(\psi + \sin\psi\cos\psi)$$

Hence the required path is

$x = a - c\,(2\psi + \sin 2\psi),\ y = c\,(1 + \cos^2\psi)$ (cycloid)

Extension of the Variational Method

Suppose the n-co-ordinates $q_1, q_2, \ldots. q_n$ are each functions of an independent variable t and that we required the solution of the variational problem

$$I = \int_{t_1}^{t_2} f(q_1, q_2, \ldots, q_n; \dot{q}_1 \dot{q}_2, \ldots, \dot{q}_n; t)dt$$

$$= \int_{t_1}^{t_2} f(q, \dot{q}, t)\ dt = \text{stationary value}$$

where f is of known functional from, t_1 and t_2 are fixed and each q_i is to be determined.

Let $q_i = q_i(t)$ give the stationary value to I and let $q_i = q_i(t) + \varepsilon_i \eta_i(t)$ be the neighbouring path.

$$\Rightarrow I(\varepsilon_1, \varepsilon_2, \ldots, \varepsilon_n) = \int_{t_1}^{t_2} f\{q_i(t) + \varepsilon_i\eta_i(t); \dot{q}_i(t) + \varepsilon_i\eta_i(t), t\}\ dt$$

$$= \int_{t_1}^{t_2} f(q_i\dot{q}_i; t)dt + \Sigma\varepsilon_i \int_{t_1}^{t_2} \left\{ \dot{\eta}_i(t)\frac{\partial f}{\partial \dot{q}_i} + \dot{\eta}_i(t)\frac{\partial f}{\partial \dot{q}_i} \right\} dt + o(\varepsilon_i^2)$$

$$\text{First variation}\quad \delta I = \Sigma\varepsilon_i \int_{t_1}^{t_2} \left\{ \dot{\eta}_i(t)\frac{\partial f}{\partial \dot{q}_i} + \dot{\eta}_i(t)\frac{\partial f}{\partial \dot{q}_i} \right\} dt$$

$$= \Sigma\varepsilon_i\ \dot{\eta}_i(t)\frac{\partial f}{\partial \dot{q}_i}\ dt + \Sigma\varepsilon_i \int_{t_1}^{t_2} \dot{\eta}_i(t)\frac{\partial f}{\partial \dot{q}_i} dt$$

$$= \Sigma\varepsilon_i \int_{t_1}^{t_2} \dot{\eta}_i(t)\frac{\partial f}{\partial q_i} dt + \Sigma e_i \left[\left\{ n_i(t)\frac{\partial f}{\partial \dot{q}_i} \right\}_{t_1}^{t_2} - \int_{t_1}^{t_2} \dot{\eta}_i(t)\frac{d}{dt}\left(\frac{\partial f}{\partial \dot{q}_i}\right) dt \right]$$

$$= \Sigma\varepsilon_i \int_{t_1}^{t_2} \dot{\eta}_i(t)_i \left[\frac{\partial f}{\partial q_i} - \frac{d}{dt}\left(\frac{\partial f}{\partial \dot{q}_i}\right) \right] \qquad \{\because 0 = \eta_i(t_1) = \eta_i(t_2)\}.$$

As η_1's are arbitrary and I is stationary, so we have

$$(\partial f / \partial q_i) - (d/dt)(\partial f / \partial \dot{q}_i) \qquad (i = 1, 2, \ldots, n)$$

HAMILTON'S FORM OF THE EQUATIONS OF MOTION

Here we shall obtain the differential equations of motion of a conservative holonomic dynamical system in a form which constitutes the basis of most of the advanced theory of dynamics. Let $(q_1, q_2, \ldots\ldots\ldots, q_n)$ be the generalised co-ordinates and let $L(q_1, q_2, ... q_n; \dot{q}_1, \dot{q}_2, ... \dot{q}_n; t)$, the kinetic potential of the system, so that the equations of motion in the Lagrangian form are

$$\frac{d}{dt}(\partial L/\partial \dot{q}_i) - (\partial L/\partial q_i) = 0; (i = 1, 2, ..., n) \quad ...(1)$$

writing $p_i = (\partial L/\partial \dot{q}_i)$ we get $\dot{p}_i = (\partial L/\partial q_i)\,(i = 1, 2, ..., n)$ ...(2)

hence from the former of these sets of equations we can regard either of the sets of quantities $(q_1, q_2 ... a_n)$ or $(p_1, p_2, ..., p_n)$ as functions of the other set.

Now, let δ denote the increment in any function of the variables $(q_1, q_2 ..., q_n; p_1, p_2, .., p_n)$ or $(\dot{q}_1, q_2 ..., q_n; \dot{q}_1, \dot{q}_2, .., \dot{q}_n)$; then we get

$$dL = \sum_{i=1}^{n}\left(\frac{\partial L}{\partial q_i}dq_i + \frac{\partial L}{\partial \dot{q}_i}d\dot{q}_i\right) + \frac{\partial L}{\partial t} \text{(when L contains t explicitly)}$$

$$dL = \sum_{i=1}^{n}\left((\dot{p}_1 dq_i + d\dot{q}_i)\frac{\partial L}{\partial t}\right)dt$$

$$= d\left(\sum_{i=1}^{n} p_i \dot{q}_i\right) + \sum_{i=1}^{n}(\dot{q}_1 dq_i - p_i dp_i) + \frac{\partial L}{\partial t}dt$$

$$\Rightarrow d\left[\sum_{i=1}^{n}(p_1\dot{q}_1) - L\right] = \sum_{i=1}^{n}(\dot{q}_i dp_i - p_i dq_i) - (\partial L/\partial t)dt.$$

Thus if the quantity $\sum_{i=1}^{n}(p_i \dot{q}_i = L)$ when expressed in terms of $(q_1, q_2 ... q_n; p_1, p_2, ..., p_n; t)$ be denoted by H, we have

$$dH = \sum_{i=1}^{n}(q_1 dp_i - \dot{p}_i\, dq_i) - \frac{\partial L}{\partial t}dt$$

$$\Rightarrow \sum_{i=1}^{n}\frac{\partial H}{\partial q_i}dq_i + \sum_{i=1}^{n}\frac{\partial H}{\partial p_i}dp_i + \frac{\partial H}{\partial t}dt = \sum_{i=1}^{n}(\dot{q}_1\, d\,p_i - \dot{p}_i\, d\,q_i) - \frac{\partial L}{\partial t}dt$$

$$\Rightarrow \dot{p}t = -\frac{\partial H}{\partial pi}, \dot{p}i = -\frac{\partial H}{\partial qi}, \frac{\partial H}{\partial t} = -\frac{\partial L}{\partial t}$$

If H does not contain t explicity (*i.e.* does not contain t explicity) we have $\dot{p}i = -(\partial H/\partial qi)$ and $\dot{q}i = (\partial H/\partial pi)$. ...(4)

These equations are called as Hamilton's equations, or Hamilton's canonical equations and the function H is called Hamiltonian.

The total order of Hamilton's equations is the same as the total order of Lagrange's equations, namely 2n. But whereas Lagrange's equations present us with n equations each of the second order. Hamilton's equation are 2n equations, each of the first order. Hamilton's equations can also be written as

$$\frac{dpi}{-(\partial H/\partial qi)} = \frac{dqi}{(\partial H/\partial pi)} = dt.$$

PHYSICAL SIGNIFICANCE OF THE HAMILTONIAN

If the Hamiltonian H is independent of t explicity, prove that it is (a) constant and (b) equal to the total energy of the system.

Proof:

(a) we have $\frac{dH}{dt} = \sum_{i=1}^{n} \frac{\partial H}{\partial qi}\frac{dqi}{dt} + \sum_{i=1}^{n} \frac{\partial H}{\partial pi}\frac{dpi}{dt}$

$= \sum_{i=1}^{n} -(\dot{p}_i)\dot{q}_i + \sum_{i=1}^{n} \dot{q}_i\dot{p}_i \left(\because \dot{p}_i = -\frac{\partial H}{\partial qi} \text{ and } \dot{q}_i = \frac{\partial H}{\partial pi}\right) = 0$

$\Rightarrow$ H = constant, say E.

(b) By Euler's theorem on homogeneous function, we have

$\Sigma\dot{q}_i \frac{\partial T}{\partial \dot{q}_i} = 2T$, where T is the K.E. of the system.

But $L = T - V$, $\therefore \frac{\partial L}{\partial \dot{q}_i} = \frac{\partial (T-V)}{\dot{q}_i} = \frac{\partial T}{\partial \dot{q}_i}$ (V does not depend on $\dot{q}_i$)

or $\Sigma\dot{q}_i \frac{\partial L}{\partial \dot{q}_i} = 2T \Rightarrow \Sigma\dot{q}_i\, \dot{p}_i = 2T\left(\because p_i = \frac{\partial L}{\partial \dot{q}_i}\right)$

$\therefore H = \Sigma p_i\dot{q}_i - L = 2T - (T - V) = T + V = E.$

PASSAGE FROM THE HAMILTONIAN TO THE LAGRANGIAN

Suppose that we are given a function *H (q, p, t) and are told that the motion of the system satisfies the canonical equations

$\dot{p}_i = -(\partial H/\partial qw_i)$ and $\dot{q}_i = (\partial H/\partial qw_i)$ ...(1)

Then we want to find a function

$L\,(p_1, p_2, ..., p_n; q_1, q_2, ..., q_n\,; t)$, *i.e.* L (p, q, t)

Such that the motion also satisfies the equations

$(d/dt)\ (\partial L/\partial \dot{q}_i) - (\partial L/q_i) = 0.$ ...(2)

Solve the first set of equations in (1) for the p's in terms of the q's, the $\dot{q}$'s and t.

Then write $L = \sum_{i=1}^{n} \dot{q}_i p_i - H$ and express L as a function of the q's, the $\dot{q}$'s and t. This is the required Lagrangian.

$$\therefore (\partial L/\partial q_i) = p_i\left(\text{using } L = \sum_{i=1}^{n} p_i\dot{q}_i - H\right)$$

$$\Rightarrow (d/dt)(\partial L/\partial \dot{q}_i) = \dot{p}_i \text{ and } (\partial L/\partial q_i) = -(\partial H/\partial q_i)$$

$$\Rightarrow \frac{d}{dt}(dL/\partial \dot{q}_i) - (\partial L/\partial q_i) = \dot{p}_i + (\partial H/\partial q_i) = 0$$

i.e. L satisfies (2) assuming (1).

MISCELLANEOUS EXAMPLES

Example 1:

Obtain Euler's equations from Hamilton's equations.

Solution:

We know that $2T = (A\omega_1^2 + B\omega_2^2 + C\omega_3^2)$,

$$\Rightarrow L = T - V = \frac{1}{2}(A\omega_1^2 + B_2^2 + C\omega_3^2) - V$$

Also Euler's geometrical relations give

$$\omega_1 = \dot{\theta}\sin\psi\sin\theta\cos\psi;$$

$$\omega_2 = \dot{\theta}\cos\psi + \dot{\phi}\sin\theta\sin\psi; \text{ and}$$

$$\omega_3 = \psi + \phi\cos\theta.$$

$$\text{Now } H = T + V = \frac{1}{2}(A\omega_1^2 + B\omega_2^2 + C_3^2) + V$$

$$P_\phi = \frac{\partial L}{\partial \dot{\theta}} = \frac{\partial L}{\partial \omega_1}\frac{\partial \omega_1}{\partial \dot{\theta}} + \frac{\partial L}{\partial \omega_2}\frac{\partial \omega_2}{\partial \dot{\theta}} + \frac{\partial L}{\partial \omega_3}\frac{\partial \omega_3}{\partial \dot{\theta}}.$$

$$A\omega_1\sin\psi + B\omega_2\cos\psi + C\omega_3 . 0$$

$$p_\theta = \frac{\partial L}{\partial \dot{\phi}} = -A\omega_1\sin\theta\cos\psi + B\omega_2\sin\theta\sin\psi + C\omega_3\cos\theta \qquad ...(1)$$

and $p_\psi = (\partial L/\partial \dot{\psi}) = C\omega_3$

Solving the three equations for ω_2, ω_2, ω_3 we have

$$\omega_1 = \frac{1}{A}\left[p_\theta\sin\psi + (p_\psi\cos\theta - p_\phi)\frac{\cos\psi}{\sin\theta}\right]$$

$$\omega_2 = \frac{1}{A}\left[p_\theta\cos\psi - (p_\psi\cos\theta - p_\phi)\frac{\sin\psi}{\sin\theta}\right]; \text{and } \omega_3 = \frac{1}{C}.p_\psi$$

Also, Hamilton's equations are $\dot{p}_\psi = -\dfrac{\partial H}{\partial \psi}$ and $\dot{\psi} = \dfrac{\partial H}{\partial p_\psi}$

Now $\dot{p}_\psi = -\frac{\partial H}{\partial \psi}$

$$\Rightarrow C\dot{\omega}_3 = -\left[\frac{\partial H}{\partial \omega_1}\frac{\partial \omega_1}{\partial \psi} + \frac{\partial H}{\partial \omega_2}\frac{\partial \omega_2}{\partial \psi} + \frac{\partial H}{\partial \omega_3}\frac{\partial \omega_3}{\partial \psi}\right] - \frac{\partial V}{\partial \psi}$$

$$= -\left[A\omega_1.\frac{1}{A}B\omega_2 + B\omega_2\left(\frac{A}{B}\omega_1\right) + C\omega_3.0\right] - \frac{\partial V}{\partial \psi}$$

$$= (A - B)\omega_1\omega_2 - \frac{\partial V}{\partial \psi}$$

$$\Rightarrow C\frac{d\omega_1}{dt} - (A - B)\omega_1\omega_2 = N\left(\because -\frac{\partial V}{\partial \phi} = N\right)$$

This is Euler's third familiar dynamical equation

Also, $\dot{\psi} = (\partial H / \partial p_\psi) = \frac{\partial H}{\partial \omega_1}\frac{\partial \omega_1}{\partial p_\psi} + \frac{\partial H}{\partial \omega_2}\frac{\partial \omega_2}{\partial p_\psi} + \frac{\partial H}{\partial \omega_3}\frac{\partial \omega_3}{\partial p_\psi}$

$= (\omega_1 \cos\psi - \omega_2 \sin\psi)\cot\theta + \omega_3 = -\phi \sin\theta \cot\theta + \omega_3$

i.e. $\dot{\psi} = -\dot{\phi}\cos\theta + \omega_3 \Rightarrow \omega_3 = \dot{\psi} + \dot{\phi}\cos\theta$

This is Euler's third geometrical equation.

One the same lines, we can deduce Euler's other equations (dynamical and geometrical).

Example 2(a):

Prove that

$\left(\frac{\partial H}{dt}\right) = \left(\frac{\partial H}{dt}\right)$ *where H is the Hamilton's function.*

Solution:

Let $q_1, q_2 ..., q_n$ be the generalised co-ordinates then Hamilton's equation are given by

$$\dot{p}_i = -\frac{\partial H}{dq_i} \text{ and } \dot{q}_i = \frac{\partial H}{dp_i}(i = 1,2,...,n) \qquad ...(1)$$

But Hamiltonian H is a function of q's and ps'

$$\therefore \frac{\partial H}{dt} = \frac{\partial H}{\partial t} + \sum_{i=1}^{n}\frac{\partial H}{\partial q_i}\dot{q}_i + \sum_{i=1}^{n}\frac{\partial H}{\partial p_i}\dot{p}_i$$

$$= \frac{\partial H}{\partial t} + \sum_{i=1}^{n}\frac{\partial H}{\partial q_i}(-\dot{p}_i)\dot{q}_i + \sum_{i=1}^{n}\dot{q}_i\dot{p}_i = \frac{\partial H}{\partial t} \qquad \text{[using (1)]}$$

Example 2(b):

Using cylindrical coordinates, write the Hamiltonian and Hamilton's equations for a particle of mass moving on the inside of a frictionless cone $x^2 + y^2 = z^2 \tan^2 \alpha$

Solution:

Like previous example, we have

$$T = \frac{1}{2}m(\dot{x}^2 + \dot{y}^2 + \dot{z}^2) = \frac{1}{2}m(\dot{\rho}^2 + \rho^2\dot{\phi}^2 + \rho^2 \cot^2 \alpha)$$

$$[\because x = \rho\cos\phi, y = \rho\sin\phi, z = \rho\cot\alpha]$$

$$= \frac{1}{2}m(\dot{\rho}^2 \operatorname{cosec}^2\alpha + \dot{\rho}^2\dot{\phi}^2) \qquad ...(1)$$

and $V = -W = -mgz = mg\rho\cot\alpha,$

$$[\because \text{the particle is above the vertex (origin)}$$

$$\Rightarrow L = T - V \frac{1}{2}m(\rho^2 \operatorname{cosec}^2\alpha + \rho^2\dot{\phi}^2) - mg\rho\cot\alpha \qquad ...(2)$$

This gives, $p_\rho = \dfrac{\partial L}{\partial \dot{\rho}} = m\rho \operatorname{cosec}^2\alpha, p_\phi = \dfrac{\partial L}{\partial \dot{\phi}} = m\rho^2\dot{\phi}$ ...(3)

Again, L does not involve t explicitly, therefore Hemiltonian H is given by,

$$H = T + V \frac{1}{2} m(\dot{\rho}^2 \operatorname{cosec}^2\alpha + \rho^2\dot{\phi}^2) + mg\,\rho\cot\alpha$$

$$= \frac{1}{2}m\left[\frac{p_\rho^2}{m^2 \operatorname{cosec}^2\alpha} + \frac{p_\phi^2}{m^2\rho^2}\right] + mg\,\rho\cot\alpha + \frac{1}{2m}\left[\frac{p_\rho^2}{\operatorname{cosec}^2\alpha} + \frac{p_\phi^2}{\rho^2}\right] + mg\,\rho\cot\alpha$$

Thus Hamilton's equations are given by:

$$\dot{p}_\rho = -\frac{\partial H}{\partial \phi} = \frac{p_\phi^2}{m\rho^2} - mg\cot\alpha; \dot{\rho} = \frac{\partial H}{\partial p_\rho} = \frac{p_\rho}{m \operatorname{cosec}^2\alpha}.$$

$$\dot{p}_\phi = -\frac{\partial H}{\partial \phi} = 0 \qquad \dot{\phi} = \frac{\partial H}{\partial p_\phi} = \frac{p_\phi}{m\rho^2}.$$

Example 3:

Use the variational method to show that the shortest curve joining two fixed is a straight line

Solution:

We have

$$\frac{ds}{dx} = \sqrt{\left\{1+\left(\frac{dy}{dx}\right)^2\right\}}$$ for any curve, joining two fixed points say

$A \equiv (x_1, y_1)$ and $B \equiv (x_2, y_2)$.

$$\Rightarrow s = \int_{x_1}^{x_2} \sqrt{(1+y'^2)}\ dx = \text{length of the curve joining A to B.}$$

$$= \int_{x_1}^{x_2} f(y')\, dx \text{ where } f(y') = \sqrt{(1+y'^2)} \qquad ...(1)$$

In equation (1), y is absent from f, therefore for s to be stationary (here minimum), we have from Euler-Lagrange's equation

$$\frac{\partial f}{\partial y'} = (\text{constant}) \Rightarrow \frac{\partial}{\partial y'}[(\sqrt{1+y'^2})] = \text{cpmst.} \because f = (1+y'^2)^{1/2}$$

$$\Rightarrow \frac{y'}{\sqrt{(1+y'^2)}} = (\text{constant}) \textit{ i.e. } y' = b \text{ where b is constant.}$$

$$\Rightarrow \frac{dy}{dx} = b \text{ or } y = a + b\,x \text{ where a is a constant.} \qquad ...(2)$$

$\Rightarrow$ the shortest path joining A to B is the curve whose equation is $y = a + bx$ and this is evidently a line.

Example 4(a):

Show that the area of the surface of revolution of a curve $y = y(x)$ is

$$2\pi \int_{x_1}^{x_2} y\sqrt{(1+y'^2)}dx.$$

Hence show that for this to be a minimum,, the curve must be a catenary.

Solution:

Consider any curve, and let $A \equiv (x_1, y_1)$ and $B \equiv (x_2, y_2)$ be its extremities, then it this curve revolves about x-axis the surface of revolution is given by

$$S = \int 2\pi y\, ds = \pi \int y \frac{ds}{dx} dx = 2\pi \int_{x_1}^{x_2} y\sqrt{(1+y'^2)}dx$$

$$\Rightarrow S = 2\pi \int_{x_1}^{x_2} f(y, y')\, dx, \text{ where } f = f(y, y') = y\sqrt{(1+y'^2)}$$

Evidently, x is absent from f, so for S to be stationary (here a minimum) we have:

$$f - y'\frac{\partial f}{\partial y'} = \text{(constant) } i.e.\ y\sqrt{(1+y'^2)} - y'\frac{yy'}{\sqrt{(1+y')}} = \text{const.}$$

$$\Rightarrow \frac{y}{\sqrt{(1+y'^2)}} = c \text{ where c is constant } i.e.\ y = c\sqrt{(1+\tan^2\psi}$$

$$[\because y' = \tan\psi]$$

$\therefore$ y = c sec y and this represents catenary.

Example 4(b):

Use Hamilton's equations to find the equations of motion of a projectile in space.

Solution:

Let (x,y,z) be the co-ordinates of the projectile in space at time t, then we have

$$T = \frac{1}{2}m(\dot{x}^2 + \dot{y}^2 + \dot{z}^2), V = mgz$$

$$\therefore L = T - V = \frac{1}{2}m\,(\dot{x}^2 + \dot{y}^2 + \dot{z}^2) - mgz$$

$$\Rightarrow p_x = \frac{\partial L}{\partial \dot{x}} = m\dot{x},\ p_y = \frac{\partial L}{\partial \dot{y}} = m\dot{y}, p_z = \frac{\partial L}{\partial \dot{z}} = m\dot{z}$$

But L does not involve t explicity therefore Hamiltonian H is given by

$$H = T + V = \frac{1}{2}(\dot{x}^2 + \dot{y}^2 + \dot{z}^2) + mgz$$

$$= \frac{1}{2}m\left(\frac{p_x^2}{m^2} + \frac{p_y^2}{m^2} + \frac{p_z^2}{m^2}\right) + mgz = \frac{1}{2m}(p_x^2 + p_y^2 + p_z^2) + mgz$$

Now Hamilton's equations are given by

$$\dot{p}_x = -\frac{\partial H}{\partial x} = 0 \qquad ...(1)$$

$$\dot{x} = \frac{\partial H}{\partial p_x} = \frac{p_x}{m} \qquad ...(2)$$

$$\dot{p}_y = -\frac{\partial H}{\partial_y} = 0 \qquad ...(3)$$

$$\dot{y} = \frac{\partial H}{\partial p_y} = \frac{p_y}{m} \qquad ...(4)$$

$$\dot{p}_z = -\frac{\partial H}{\partial_z} = -\,mg \qquad ...(5)$$

$$\dot{z} = \frac{\partial H}{\partial p_z} = \frac{p_z}{m} \qquad ...(6)$$

Using (1) and (2), we have $\ddot{x} = 0$ ...(7)

Using (3) and (4), we have $\ddot{y} = 0$ ...(8)

Again making use of (6) and (5), we have

$$m\ddot{z} = \dot{p}_z = 0\,mg \text{ or } \ddot{z} = -\,g. \qquad ...(9)$$

These (7, 8, 9) are the equations of motion of the projectile in space.

Example 4(c):

Using cylindrical coordinates (ρ, ϕ, z) *write the Hamiltonian and Hamilton's equations for a particle of mass m moving in a force field of potential* $V(\rho, \phi, z)$.

Solution:

In cylindrical coordinates, co-ordinates of any points are

$$x = \rho\cos\phi, y = \rho\sin\phi, z = z \qquad ...(1)$$

$$\therefore T = \frac{1}{2}m(\dot{x}^2 + \dot{y}^2 + \dot{z}^2) = \frac{1}{2}m(\dot{\rho}^2 + \rho^2\dot{\phi}^2 + \dot{z}^2) \qquad ...(2)$$

$$\Rightarrow L = T - V = \frac{1}{2}m(\dot{\rho}^2 + \rho^2\dot{\phi}^2 + \dot{z}^r) - V(\rho, \phi, z) \qquad ...(3)$$

$$\Rightarrow p_p = \frac{\partial L}{\partial \dot{\rho}} = m\dot{\rho}, p_\phi = \frac{\partial L}{\partial \phi} = mp^2\dot{\phi} \text{ and } p_z = \frac{\partial L}{\partial \dot{z}} = m\dot{z}$$

Evidently, L does not involve t explicity, therefore Hamiltonian H is given by

$$h = T + V\frac{1}{2}m(\dot{\rho}^2 + \rho^2\dot{\phi}^2 + z^2) + V$$

$$= \frac{1}{2}\left[\frac{p_\rho^2}{m^2} + \frac{p_\phi^2}{m^2\rho^2} + \frac{p_z^2}{m^2}\right] + V = \frac{1}{2m}\left[p_\rho^2 + \frac{p_\phi^2}{\rho^2} + p_z^2\right] + V$$

Hence, Hamilton's are given by:

$$\dot{p}_\rho = -\frac{\partial H}{\partial \rho} = \frac{p_\phi^2}{m\rho^2} - \frac{\partial V}{\partial \rho}; \dot{\rho} = \frac{\partial H}{\partial p_\rho} = \frac{p_\rho}{m_\phi}$$

$$\dot{p}_\phi = -\frac{\partial H}{\partial \phi} = -\,\frac{\partial V}{\partial \phi}; \dot{\phi} = \frac{\partial H}{\partial p_\phi} = \frac{p_\phi}{m\rho^2}$$

$$\dot{p}_z = -\frac{\partial H}{\partial z} = -\frac{\partial V}{\partial z}; \dot{z} = \frac{\partial H}{\partial p_z} = \frac{p_z}{m}$$

Example 5:

A particle of unit mass is projected so that its total energy is h in a field of force of which the potential energy is ϕ (r) at a distance r from the origin.

Deduce from the principle of energy and least action that the differential equation of the path is

$$c^2\left[r^2 + \left(\frac{dr}{d\theta}\right)^2\right] = r^4[h - \phi(r)].$$

Solution:

Let T and V be the kinetic are potential energies, then we have T + V = h
$\Rightarrow T = h - V$

$= h - \phi\,(r)$, $\because\ V = \phi\,(r)$ given

$\Rightarrow \frac{1}{2}v^2 = h - \phi\,(r)$ where v is velocity $\left(\because T = \frac{1}{2}mv^2 = \frac{1}{2}v^2\right)$

$\Rightarrow v = \sqrt{2}\{h - \phi(r)\}^{1/2}$

Whence, the action $A = \int_{t_0}^{t_1} 2T\ dt$, by definition given earlier

$= \int_{t_0}^{t_1} 2.\frac{1}{2}v^2 dt = \int_{t_0}^{t_1} v\,d\,s$ since v = (ds/dt) = $\sqrt{2}\int_{t_0}^{t_1}\{h - \phi(r)\}^{1/2} ds$

$= \sqrt{2}\int_{t_0}^{t_1}\{h - \phi(r)\}^{1/2}\left\{1 + r^2\left(\frac{d\theta}{dr}\right)^2\right\}^{1/2} dr$ $\quad \frac{ds}{dr} = \left\{1 + r^2\left(\frac{d\theta}{dr}\right)^2\right\}^{1/2}$

$= \sqrt{2}\int_{t_0}^{t_1}[h - \phi(r)\}^{1/2}\{1 + r^2\theta'^2\}^{1/2}]\ dr$ where $\theta' = \frac{d\theta}{dr}$

$= \sqrt{2}\int_{t_0}^{t_1} f\,(\theta', r)dr$, where $f(\theta', r) = \{h - \phi(r)\}^{1/2}\{1 + r^2\theta'^2\}^{1/2}$

Evidently, θ is absent from f, therefore we must have

$\frac{\partial f}{\partial \theta'} = c$ (const.) $\Rightarrow \frac{\partial f}{\partial \theta'}[\{h - \phi(r)^{1/2}\{1 + r^2\theta'^2\}^{1/2]} = c$

$$\Rightarrow \{h - \phi(r)\}^{1/2}\frac{r^2}{\left\{r^2 + \left(\frac{dr}{d\theta}\right)^2\right\}^{1/2}} = c$$

$$\Rightarrow c^2\left\{r^2+\left(\frac{dr}{d\theta}\right)^2\right\}=r^4\{h-\phi(r)\}.$$

This is the required equation.

Example 6:

A projectile is launched in a vertical plane with a velocity whose horizontal and vertical components are v_x *and* v_y *respectively. Calculate the value of the integral* $\int_0^{t_0} l\,dt$, *where* $t_0=\frac{n\pi}{\omega}$.

Evaluate this integral for the varied path given by the equations

$$x=v_x t, y=V_y t-\frac{1}{2}gt^2+\varepsilon\sin\omega t,$$

where ε *is a small constant quantity.*

Show that the integral $\int_0^{t_0} L\ dt$ *is greater for the varied path than for the actual path, but the result is in agreement with Hamilton's principle.*

Solution:

The path of the projectile is given by

$$x=V_x t, y=V_y t-\frac{1}{2}gt^2.$$

Also V = mgy, where V denotes the potential,

whence $L=T-V=\frac{1}{2}m(\dot{x}^2+\dot{y}^2)-mgy$

$$=\frac{1}{2}m(y_x^2+(v_y-gt)^2]-mg\left(v_y t-\frac{1}{2}gt^2\right)$$

$$=\frac{1}{2}m[v_x^2+v_y^2)-4gtv_y+2g^2t^2] \quad ...(1)$$

$$\Rightarrow\int_0^{t_0}Ldt=\int_0^{t_0}(T-V)dt=\frac{1}{2}m\int_0^{t_0}[(v_x^2+v_y^2)-4gtv_y+2g^2t^2]dt$$

$$=\frac{1}{2}m\left[(v_x^2+v_y^2)t_0-2t_0^2v_y+\frac{2}{3}g^2t_0^3\right] \quad ...(1)$$

Also the varied path is given by : $x=v_x t, y=v_y t-\frac{1}{2}gt^2+\varepsilon\sin\omega t$

Now after obtaining $\dot{x},\dot{y}$; we have

$$T = \frac{1}{2}m[v_x^2 + (y_v - gt + \varepsilon\omega\cos\omega t)^2] \text{ and}$$

$$V = mg\left(v_y t - \frac{1}{2}gt^2 + \varepsilon\sin\omega t\right) \Rightarrow \int_0^{t_0 = n\pi/\omega} L dt = \int_0^{t_0}(T - V)dt$$

$$\int_0^{t_0}\left[\frac{1}{2}m\{v_x^2 + v_y - gt + \varepsilon\omega\cos\omega t)^2\} - mg(v_y t - \frac{1}{2}gt^2 + \varepsilon\sin\omega t\right]dt$$

$$= \frac{1}{2}m\left[(v_x^2 + v_y^2)t_0 - 2t_0^2 g v_y + \frac{2}{3}g^2 t_0^3 + \frac{1}{2}\varepsilon^2\omega^2 t_0\right]$$

[∵ other integrals vanish)

$$= \frac{1}{2}m[(v_x^2 + v_y^2)t_0 - 2t_0^2 g v_y + \frac{2}{3}g^2 t_0^3] + m\pi n\omega\varepsilon^2 \qquad ...(2)$$

From equations (1) and (2), we see that the integral $\int_0^{t_0} L\, dt$ is greater for the varied path then for the actual path and is minimum when $\varepsilon = 0$. Evidently, the minimum value in this case, the varied *path coincides with actual path.*

Example 7:

A particle moves in a straight line with central acceleration $\omega^2 x$ between two fixed points x_0 and x_1 in the prescribed time $t_1 - t_0$. Show that Hamilton's principal function S is

$$\frac{\omega}{2\sin\omega(t_1 - t_0)}\left[(x_1^2 + x_0^2)\cos\omega(t_1 - t_0 - 2x_1 x_0\right]$$

Solution:

Equation of motion is, $\ddot{x} = \omega^2 x$ ...(1)

Solving (1), the path is $x = A\cos\omega t + B\sin\omega t$ ...(2)

Now, $\dot{x} = \omega[-A\sin\omega t + B\cos\omega t] \Rightarrow T = \frac{1}{2}\dot{x}^2$.

Also, we have potential V = work done against the force

$$= \int_0^{t_0}\omega^2 x dx = \frac{1}{2}\omega^2 x^2 \qquad ...(3)$$

Again, we also have

$A\cos\omega t_0 + B\sin\omega t_0 - x_0 = 0$ $\qquad (t = t_0 ; x = x_0)$

and $A\cos\omega t_1 + B\sin\omega t_1 - x_1 = 0$ $\qquad (t = t_1 ; x = x_1)$

Saving these

$$A = \frac{x_0 \sin\omega t_1 \; 0 \; x_1 \sin\omega t_0}{\sin\omega(t_1 - t_0)}, B = \frac{x_1 \cos\omega t_0 - x_0 \cos\omega t_1}{\sin\omega(t_1 - t_0)}$$

$$\Rightarrow B^2 - A^2 = [x_1^2 \cos 2\omega t_0 + x_0^2 \cos 2\omega t_1 \; 2x_1 x_0$$

$$- \cos\omega\,(t_1 + t_0)]\frac{1}{\sin^2\omega(t_1 - t_0)} \quad ...(4)$$

$\therefore$ S = Hamilton's principal function

$$= \int_{t_0}^{t_1} L dt = \int_{t_0}^{t_1} (T - V) dt \quad (\because L = T - V)$$

$$= \int_{t_0}^{t_1} \left(\frac{1}{2}\dot{x}^2 - \frac{1}{2}\omega^2 x^2\right) dt \text{ [from above]}$$

$$= \frac{1}{2}\omega^2 \int_{t_0}^{t_1} [(-A \sin\omega t + B\cos\omega t)^2 - (A \cos\omega t + B\sin\omega t)^2]dt$$

$$= \frac{1}{2}\omega\left[(B - A)^2 \sin 2\omega t + 2AB \cos 2\omega t\right]_{t_0}^{t_1}$$

$$= \frac{1}{2}\omega\left[(B^2 - A^2)\cos\omega(t_1 + t_0)\sin\omega(t_1 - t_0) - 2AB\sin\omega(t_1 - t_0 \sin\omega(t_1 - t_0)\right]$$

Substituting for $B^2 - A^2$ and 2AB, we obtain

$$S = \frac{\omega}{2\sin\omega(t_1 - t_0)}[(x_1^2 \cos 2\omega t_0 + x_0^2 \cos 2\omega t_1 - 2x_1 x_0$$

$$\cos w(t_1 + t_0)\} \cos w\,(t_1 + t_0) +$$

$$\{x_1^2 \sin 2\omega t_0 + x_0^2 \sin 2\omega t_1 - 2x_1 x_0 \sin\omega(t_1 + t_0)\} \sin\omega(t_1 + t_0)]$$

$$= \frac{\omega}{2\sin\omega(t_1 - t_0)}[x_1^2 \cos\omega(t_1 - t_0) + x_0^2 \cos\omega(t_1 - t_0) - 2x_1 x_0]$$

$$= \frac{\omega}{2\sin\omega(t_1 - t_0)}[x_1^2 + x_0^2 \cos\omega(t_1 - t_0) - 2x_1 x_0]$$

This is the required result.

Example 8(a):

A particle moves in a plane curve, under the central acceleration $\omega^2 r$, between two fixed points (x_0, y_0) and (x_1, y_1) in the prescribed time $t_1 - t_0$; prove that Hamilton's principle function S is

$$= \frac{\omega}{2\sin\omega(t_1 - t_0)}[(x_1^2 + y_1^2 + x_0^2 + y_0^2)\cos\omega(t_1 - t_0) - 2(x_1 x_0 + y_1 y_0)]$$

Solution:

Equations of motion of the particle are given by

$$\dot{x} = -\omega^2 x \text{ and } \ddot{y} = -\omega^2 y \qquad ...(1)$$

$$x = A\cos\omega t + B\sin\omega t;\ y = C\cos\omega t + D\sin\omega t$$

Now, kinetic energy $T = \frac{1}{2}gw^2(\dot{x}^2 + \dot{y}^2)$

$$= \frac{1}{2}\omega^2\left[(-A\sin\omega t + B\cos\omega t)^2 + (-C\sin\omega t + D\cos\omega t)^2\right]$$

Potential $V = \frac{1}{2}\omega^2 r^2 = \frac{1}{2}\omega^2(x^2 + y^2)$

Hence Hamilton's principal function S is given by

$$S = \int_{t_0}^{t_1} L dt = \int_{t_0}^{t_1} (T - V)\ dt \text{ since } L = T - V$$

$$= \int_{t_0}^{t_1} \left[\frac{1}{2}(\dot{x}^2 + \dot{y}^2) - \frac{1}{2}\omega^2(x^2 + y^2)\right] dt \qquad ...(2)$$

But $\int_{t_0}^{t_1} \left(\frac{1}{2}\dot{x}^2 - \frac{1}{2}\omega^2 x^2\right) dt = \frac{\omega}{2\sin\omega(t_1 - t_0)}\left[(x_1^2 - x_0^2)\cos\omega(t_1 - t_0) - 2x_1x_0\right]$

and $\int_{t_0}^{t_1} \left(\frac{1}{2}\dot{y}^2 - \frac{1}{2}\omega^2\dot{y}^2\right) dt = \frac{\omega}{2\sin\omega(t_1 - t_0)}$

$$\times\left[(y_1^2 - y_0^2)\cos\omega(t_1 - t_0) - 2y_1y_0\right]$$

$$\therefore S = \frac{\omega}{2\sin\omega(t_1 - t_0)}$$

$$\left[(x_1^2 - x_0^2 + y_1^0 + y_0^2)\cos\omega(t_1 - t_0) - 2(x_1x_0 + y_1y_0\right].$$

Example 8(b):

Use Hamilton's principle to find the equation of motion of one - dimensional harmonic oscillator.

Solution:

The kinetic energy of harmonic oscillator is given by $T = m\dot{x}^2$ and the potential energy of the harmonic oscillator is given by

$$V = -\int F dx = \int kx dx = \frac{1}{2}kx^2.$$

$\therefore$ The Lagrangian of the system $L = T - V \ V = \frac{1}{2}m\dot{x}^2 - \frac{1}{2}kx^2$.

Using Hamilton's Principle, we have $\delta \int_{t_1}^{t_2} L\, dt = 0$.

$$\Rightarrow \delta \int_{t_1}^{t_2} \left(\frac{1}{2}m\dot{x}^2 - \frac{1}{2}kx^2\right) dt = 0 \Rightarrow \int_{t_1}^{t_2} \delta\left(\frac{1}{2}m\dot{x}^2 - \frac{1}{2}kx^2\right) dt = 0$$

$$\Rightarrow \delta \int_{t_1}^{t_2} (m\dot{x}\delta\dot{x} - kx\delta x)dt = 0 \text{ But } \delta\dot{x} = (d/dt)(\delta x). \qquad \text{...(1)}$$

$$\therefore (1) \Rightarrow \int_{t_1}^{t_2} m\dot{x}\frac{d}{dt}(\delta\dot{x})\, dt - \int_{t_1}^{t_2} kx\, \delta x\, dt = 0$$

$$\text{i.e.,} [m\dot{x}\, \delta x]_{t_1}^{t_2} - \int_{t_1}^{t_2} m\frac{d}{dt}(\dot{x})\, dt - \int_{t_1}^{t_2} kx\, \delta x\, dt = 0 \qquad \text{...(2)}$$

But $[m\dot{x}\, \delta x]_{t_1}^{t_2} = 0$ [$\because$ $\delta x = 0$ at fixed points, i.e, at instants t_1 and t_2]

$$\therefore (2) \Rightarrow \int_{t_1}^{t_2} m\frac{d}{dt}(\dot{x})\, \delta x\, dt - \int_{t_1}^{t_2} kx\, \delta x\, dt = 0 \Rightarrow \int_{t_1}^{t_2} (m\dot{x} + kx)\delta\, dt = 0$$

But δx is arbitrary, hence the above equation is only satisfied if $m\dot{x} + kx = 0$. This is the equation of motion for one dimensional harmonic oscillator.

Example 8(c):

A particle of mass m moves in a force field of potential V. Write

(a) *The Hamiltonian and*

(b) *Hamilton's equations in spherical polar co-ordinates.*

Solution:

K.E. is given by

$$T = \frac{1}{2}m(\dot{r}^2 + r^2\dot{\theta}^2 + r^2\sin^2\theta\dot{\phi}^2) \qquad \text{...(1)}$$

$$\therefore L = T - V = \frac{1}{2}m(\dot{r}^2 + r^2\dot{\theta}^2 + r^2\sin^2\theta\dot{\phi}^2 - V \qquad \text{...(2)}$$

We have $p_r = (\partial L/\partial\dot{r}) = m\dot{r}, p_\theta = (\partial L/\partial\dot{\theta}) = mr^2\dot{\theta}$,

$p_\phi = (\partial L/\partial\dot{\phi}) = mr^2\sin^2\theta\dot{\phi}$

$$\Rightarrow \dot{r} = (p_r/m), \dot{\theta} = (p_\theta/mr^2), \dot{\phi} = p\phi/(mr^2\sin^2\theta). \qquad \text{...(3)}$$

Now Hamiltonian is given by

$$h = \Sigma p_i \dot{q}_i - L = p_r \dot{r} + p_\theta \dot{\theta} + p\phi\dot{\phi} - L$$

$$= \frac{p_r^2}{2m} + \frac{p_\theta^2}{2mr^2} + \frac{p_\theta^2}{2mr^2 \sin^2\theta} + V(r, \theta, \dot{\phi})$$

= total energy of the system.

(b) Hamilton's equations are given by

$$\dot{q}_i = (\partial H / \partial p_i), \dot{p}_i = -(\partial H / \partial q_i)$$

i.e. $\dot{r} = \frac{\partial H}{\partial p_r} = \frac{p_r}{m}$ $\qquad \dot{p}r = -\frac{\partial H}{\partial r} = \frac{p_\theta^2}{mr^3} + \frac{p_\phi^2}{mr^3 \sin^2\theta} - \frac{\partial V}{\partial r}$

$\dot{\theta} = \frac{\partial H}{\partial p_\theta} = \frac{p_\theta}{mr^2}$ $\qquad \dot{p}\theta = -\frac{\partial H}{\partial \theta} = \frac{p_\phi^2 \cos\theta}{mr^2 \sin^3\theta} - \frac{\partial V}{\partial \theta}$

$\dot{\phi} = \frac{\partial H}{\partial p_\phi} = \frac{p_\phi}{mr^2 \sin^r\theta}$ $\qquad \dot{p}\phi = -\frac{\partial H}{\partial \phi} = -\frac{\partial V}{\partial \phi}$

Example 9:

If H is the Hamiltonian prove that if f is any function depending (df/dt) = $((\partial f / \partial t) + [H.f]$

Solution:

We have

$$(df / dt) = (\partial f / \partial t) + \{\partial f / \partial q_i)(dq_i / dt) + (\partial f / \partial p_i)(dp_i / dt)\}$$

$$\Rightarrow (df / dt) = (\partial f / \partial t) + \Sigma\{\partial f / \partial q_i)(\partial H / \partial q_i) - (\partial f / \partial p_i)(\partial H / \partial qi)\}$$

$\{\because$ By Hamilton's equations $\dot{q}i = \partial H / \partial p_i, \dot{p}_i = -(\partial H / \partial q_i)\}$

$\Rightarrow (df / dt)\, (\partial f / \partial t) + [H, f]$ where [H, f] is the Poisson Bracket)

Example 10:

A particle of mass ma moves in a force field of potential V.

(a) Write the Hamiltonian and

(b) Hamilton's equations in artesian co-ordinates.

Solution:

(a) we have

$$T = \frac{1}{2} m(\dot{x}^2 + \dot{y}^2 + \dot{z}^2)$$

$$\Rightarrow L = \frac{1}{2}m(\dot{x}^2 + \dot{y}^2 + \dot{z}^2) - V(x,y,z) \qquad ...(1)$$

$$\Rightarrow \dot{x} = (p_x / m), \;\; \dot{y} = (p_y / m), \dot{z} = (p_z / m).$$

Thus $H = \Sigma p_x \dot{q}_z - \dot{p}_x x + \dot{p}_y y + \dot{p}_z z - \frac{1}{2}m(\dot{x}^2 + \dot{y}^2 + \dot{z}^2) + V(x,y,z)$

$$= p_x (p_x/m) + p_y (p_y/m) + (p_z (p_z/m)$$

$$- \frac{1}{2}m[p_x^2 / m^2) + (p_y^2 / m^2) + (p_z^2 / m^2)] + V(x,y,z)$$

$$= (p_x^2 / 2m) + (p_y^2 / 2m) + (p_z^2 / 2m) + V(x,y,z)$$

= total energy of the system.

(b) Hamilton's equations are:

$\dot{p}_x = -(\partial H / \partial x); \dot{p}_y = -(\partial H / \partial y); \dot{p}_z = -(\partial H / \partial z)$ and

$\dot{x} = (\dot{p}_x = -(\partial H / \partial x); \dot{p}_y \; \dot{x} = (\partial H / \partial p_x); \dot{y} = (\partial H / \partial p_y); \dot{z} = (\partial H / \partial p_z).$

$$\Rightarrow \dot{p}_x = -(\partial V / \partial x), \dot{p}_y = -(\partial V / \partial y), \dot{p}_z = -(\partial V / \partial z)$$

$$\dot{x} = (p_x / m), \dot{y} = (p_y / m), \dot{z} = (p_z / m).$$

Example 11(a):

A particle moves in the xy-plane under the influence of a central force depending only on its distance from the origin.

(a) Set up the Hamiltonian for the system.

(b) Write Hamilton's equations of motion.

Solution:

(a) Let the p potential due to the central force be V(r). Then, we have

$T = \frac{1}{2}mv^2 = \frac{1}{2}m$ [(radial velocity)^2 + (transverse velocity^2)]

$$= \frac{1}{2}m(\dot{r}^2 + r^2\dot{\theta}^2),$$

$$L = T - V = \frac{1}{2}m(\dot{r}^2 + r^2\dot{\theta}^2) - V(r).$$

$$\therefore p_r = (\partial L / \partial \dot{r}) = m\dot{r}, p_\theta = (\partial L / \partial \dot{r}) = mr^2\dot{\theta}$$

$$\Rightarrow \dot{r} = (p_r / m), \dot{\theta} = (p_\theta / mr^2)$$

Thus $H = \Sigma p_i \dot{q}_i - L = p_r \dot{q}_r + p_\theta \dot{q}_\theta - L$

$$= p_r.\dot{r} + p_\theta\dot{\theta} - \frac{1}{2}m(\dot{r}^2 + r^2\dot{\theta}^2) + V(r)$$

$$= p_r(p_r/m) + p_\theta(p_\theta/mr^2)$$

$$-\frac{1}{2}m\{p_r^2/m^2) + r^2.(p_\theta^2/m^2r^4)\} - V(r)\}$$

$= (p_r^2/2m) + (p_\theta^2/2m_r^2) + V(r) =$ total energy of the system.

(b) Hamilton's equations are

$$\dot{p}_i = -(\partial H/\partial q_i), \dot{q}_i = (\partial H/\partial p_i)$$

$$\Rightarrow \dot{r} = (\partial H/\partial p_r) = (p_r/m), \dot{\theta} = (\partial H/\partial p_\theta) = (p_\theta/mr^2)$$

$$\dot{p}_1 = -(\partial H/\partial r) = (p_\theta^2/mr^3) - V(r), \dot{p}_\theta = -(\partial H/\partial\theta) = 0.$$

Example 11(b):

A sphere rolls down a rough includedplane; if x be the distance of the point of contact of the sphere from a fixed point on the plane, find the acceleration.

Solution:

We have $T = \frac{1}{2}m(\dot{x}^2 + k^2\dot{\theta}^2) = \frac{1}{2}m\left(\dot{x}^2 + \frac{2}{5}a^2\dot{\theta}^2\right)\left(\because k^2 = \frac{2}{5}a^2\right)$

$$= \frac{1}{2}m\left(\dot{x}^2 \frac{2}{5}\dot{x}^2\right) = \frac{7}{10}m\dot{x}^2; \qquad ...(1)$$

$$V = -mgx \sin\alpha \qquad ...(2)$$

$$\therefore L = T - V == = \frac{7}{10}m\dot{x}^2 + mgx\sin\alpha \qquad ...(3)$$

Now $p_x = (\partial L/\partial\dot{x}) = \frac{7}{10}m\dot{x} \Rightarrow \dot{x} = (5p_x/7m)$

Thus $H = -L + p_x\dot{x} = -\frac{7}{10}m\dot{x} - mgx\sin\alpha + p_x.(5p_x/7m)$

$$= \frac{5}{14}(p_x^2/m) - mgx\sin\alpha$$

$\therefore$ One of Hamilton's equations gives

$$p_x = -(\partial H/\partial x) = mg\sin\alpha \Rightarrow \frac{7}{5}m\ddot{x} = mg\sin\alpha \Rightarrow \ddot{x} = \frac{5}{7}g\sin\alpha$$

Example 11(c):

If the Hamiltonian H is independent of time explicity, prove that it is

(a) a constant, and

(b) equal to the total energy of the system.

Solution:

(a) $(dH/dt) = \sum_{i=1}^{n}(\partial H/\partial p_i)\dot{p}_i + \sum_{i=1}^{n}(\partial H/\partial q_i)\dot{q}_i$

$\sum_{i=1}^{n} \dot{q}_i \dot{p}_i + (-\dot{p}_i)\dot{q}_i = 0$ $\quad [\because (\partial H/\partial p_i) = \dot{q}_i, (\partial H/\partial q_i) = -\dot{p}_i]$

$\Rightarrow H = \text{constant} = E$ say.

(b) By Euler's theorem on homogeneous functions, we have

$$\sum_{i=1}^{n} \dot{q}_i (\partial T/\partial \dot{q}_i) = 2T. \qquad ...(2)$$

Put

$$p_i = (\partial L/\partial \dot{q}_i) = \{\partial(T - V)/\partial \dot{q}_i\} = (\partial T/\partial \dot{q}_i) - (\partial V/\partial \dot{q}_i) = (\partial T/\partial \dot{q}_i)$$

$\{\because (\partial V/\partial \dot{q}_i) = 0$ as V is independent of $\dot{q}_i\}$

$\therefore (2) \Rightarrow \sum_{i=1}^{n} \dot{q}_i p_i = 2T$

Thus $H = \sum_{i=1}^{n} p_i \dot{q}_i - L = 2T - L = 2T - (T - V) = T + V = E.$

Example 12:

Write the Hamiltonian and equation of motion for a simple pendulum.

Solution:

We have $T = \frac{1}{4} ml^2\dot{\theta}^2$ and $V = mgl\,(1 - \cos\theta)$,

$$\therefore L = T - V = \frac{1}{2} ml^2\dot{\theta}^2 - mgl(1 - \cos\theta) \qquad ...(1)$$

$$\Rightarrow H = \Sigma p_i \dot{q}_i - L = p_\theta \dot{\theta} - L = ml^2\dot{\theta}^2 - \left\{\frac{1}{2} ml^2\theta^2 - mgl(1 - \cos\theta)\right\}$$

$$= \frac{1}{2} ml^2\dot{\theta}^2 + mgl\,(1 - \cos\theta) = T + V = \text{total energy}.$$

Now $p_\theta = (\partial L/\partial\dot{\theta}) = ml^2\dot{\theta}^2 = (p_\theta/ml^2)$

$$\therefore H = \frac{1}{2} ml^2(p_\theta/ml^2)^2 + mgl(1 - \cos\theta) = (p_\theta^2/2ml^2) + mgl(1 - \cos\theta)$$

$\Rightarrow (\partial H/\partial p_\theta), \dot{p}_\theta = -(\partial H/\partial\theta) \Rightarrow \dot{\theta} = (p_\theta/ml^2)$ and $\dot{p}_\theta = -mgl\sin\theta$,

These represent Hamilton's equations for a simple pendulum.

From above, we have $p_\theta = ml^2\dot{\theta}$, i.e. $p_e = ml^2\ddot{\theta}$

$\therefore ml^2\ddot{\theta}, = -mgl\sin\theta \Rightarrow \ddot{\theta} + (g/l)\sin\theta = 0$

This gives the equation of motion of the simple pendulum.

Example 13:

Write the Hamiltonian function and equation of motion for a compound pendulum.

Solution:

We have $L = \frac{1}{2}I\dot{\theta}^2 + mgh\cos\theta \Rightarrow p_\theta = (\partial L/\partial\theta) = I\theta.$

where I mk^2

$\therefore H = \Sigma p_i\dot{q}_i - L = p_\theta\dot{\theta} - L = I\dot{\theta}\dot{\theta} - \frac{1}{2}I\dot{\theta}^2 - mgh\cos\theta$

$= \frac{1}{2}I\dot{\theta}^2 - mgh\cos\theta$

$\Rightarrow \frac{1}{2}H = (p_\theta/I)^2 - mgh\cos\theta = (p_\theta^2/2I)\ 2\ mgh\cos\theta\ \{\because \dot{\theta} = (p_\theta/I)\}$

$\therefore (\partial H/\partial p_\theta) = (p_\theta/I), (\partial H/\partial\theta) = mgh\sin\theta$

Thus the Hamilton's equations for $\dot{\theta}$ and $\dot{p}_\theta$ are given by

$\dot{\theta} = (\partial H/\partial p_\theta), \dot{p}_\theta = -(\partial H/\partial\theta)$

i.e. $\dot{\theta} = (p_\theta/I)$ and $\dot{p}_\theta = -mgh\sin\theta$. But. $p_\theta = I\theta \Rightarrow \dot{p}_\theta = I\ddot{\theta}$.

$\therefore I\ddot{\theta} = -mgh\sin\theta \Rightarrow \ddot{\theta} + \frac{mgh}{I}\sin\theta = 0.$

This is exactly the same as obtained previously using Lagrange's equations.

Example 14:

Use Hamilton's principle to find the equation of motion of one - dimensional harmonic oscillator.

Solution:

The kinetic energy of harmonic oscillator is given by $T = m\dot{x}^2$ and the potential energy of the harmonic oscillator is given by

$$V = -\int F dx = \int kx dx = \frac{1}{2}kx^2.$$

$\therefore$ The Lagrangian of the system $L = T - V \; V = \frac{1}{2}m\dot{x}^2 - \frac{1}{2}kx^2$.

Using Hamilton's Principle, we have $\delta \int_{t_1}^{t_2} L\, dt = 0$.

$$\Rightarrow \delta \int_{t_1}^{t_2} \left(\frac{1}{2}m\dot{x}^2 - \frac{1}{2}kx^2\right) dt = 0 \Rightarrow \int_{t_1}^{t_2} \delta\left(\frac{1}{2}m\dot{x}^2 - \frac{1}{2}kx^2\right) dt = 0$$

$$\Rightarrow \delta \int_{t_1}^{t_2} (m\dot{x}\delta\dot{x} - kx\delta x)dt = 0$$

But $\delta\dot{x} = (d/dt)(\delta x)$. ...(1)

$$\therefore (1) \Rightarrow \int_{t_1}^{t_2} m\dot{x}\frac{d}{dt}(\delta\dot{x})\, dt - \int_{t_1}^{t_2} kx\, \delta x\, dt = 0$$

i.e., $[m\dot{x}\,\delta x]_{t_1}^{t_2} - \int_{t_1}^{t_2} m\frac{d}{dt}(\dot{x})\, dt - \int_{t_1}^{t_2} kx\,\delta x\, dt = 0$ ...(2)

But $[m\dot{x}\,\delta x]_{t_1}^{t_2} = 0$ [$\because$ $\delta x = 0$ at fixed points, i.e, at instants t_1 and t_2]

$$\therefore (2) \Rightarrow \int_{t_1}^{t_2} m\frac{d}{dt}(\dot{x})\,\delta x\, dt - \int_{t_i}^{t_2} kx\,\delta x\, dt = 0 \Rightarrow \int_{t_1}^{t_2} (m\dot{x} + kx)\delta\, dt = 0$$

But δx is arbitrary, hence the above equation is only satisfied if $m\dot{x} + kx = 0$. This is the equation of motion for one dimensional harmonic oscillator.

EXERCISES

1. A particle of unit mass is projected so that its total energy is k, in a field of a force a of which the potential energy is ϕ (r) at a distance **r** from the origin. Find the differential equation of the path.
2. Apply the technique of calculus of variation to obtain Euler Lagrange's equations.
3. State the variational principle of Hamilton.
4. State & prove the principle of least action. 3. Write the Hamiltonian for the motion of a spherical pendulum and obtain Hamilton's equations. Solve them to show that Hamiltonian is a constant.
5. A heavy bead of mass **m** freely movable on a smooth circular wire of radius a which is made to rotate about a vertical diameter with spin ω, prove that the action,

$$A = ma^2 \int_{\theta_1}^{\theta_2} \left[\frac{2H}{ma^2} + \frac{2g}{a} \cos\theta + \omega^2 \sin^2\theta \right]^{1/2} d\theta$$

where **H** is the Hamiltonian and θ is the angle the radius through the bead makes with the downward vertical.

6. Derive the Euler's dynamical equations from Hamilton's equations.
7. If H is the Hamiltonian, prove that if f is any function depending on position, momentum and time, then

   $(df/dt) = (\partial f/\partial t) + [H, f]$ where [H,f] is the poisson's Bracket.

3

Principle of Conservation Energy and Forces

INTRODUCTION

If a system moves under the action of finite forces and if the geometrical equations of the system do-not contain time explicity the change in the kinetic energy of the system in passing from one configuration to any other is equal to the corresponding work done by the forces.

Let (x, y, z) be the coordinates of a particle of mass m at the time t, and let the resolved parts of the external forces parallel to the axes at that time be X, Y, Z.

Then, by D' Alembert's principle, the forces $X - m\ddot{x}, Y - m\ddot{y}, Z - m\ddot{z}$ acting at the point (x, y, z) and the similar forces acting at other particles of the system, form a system of forces in equilibrium.

If dx, dy, dz denote arbitrary small displacements of the particle of mass m at (x, y, z) consistent with geometrical conditions at any time t, then by the principle of virtual work, we have

$$\Sigma[X - m\ddot{x})\,\delta x + (Y - m\ddot{y})\,\delta y + (Z - m\ddot{z})\delta z] = 0 \qquad ...(1)$$

If the geometrical conditions are expressed by the equations which do not contain time explicity, then the conditions which hold at time t, will also hold through the time δ t. Therefore, we can take the arbitrary displacements $\delta x, \delta y, \delta z$ as the actual displacements $\frac{dx}{dt}\delta t, \frac{dy}{dt}\delta t, \frac{dz}{dt}\delta t,$

Replacing δx by $\frac{dx}{dt}\delta t, \delta y$ by $\frac{dz}{dt}\delta t, \delta z$by $\frac{dz}{dt}\cdot\delta t$ in (1), we have

$$\Sigma\left[(X - m\ddot{x})\frac{dx}{dt} + (Y - m\ddot{y})\frac{dy}{dt} + (Z - m\ddot{z})\frac{dz}{dt}\right]\delta t = 0$$

particle, then

$$\text{or } \frac{d}{dt}\Sigma m\frac{dx}{dt} = \Sigma X.$$

But from the given conditions $\Sigma X. = 0$, hence $\frac{d}{dt}\Sigma m\frac{dx}{dt} = 0$

i.e. $\Sigma m\frac{dx}{dt} = \text{constant}$

$$\Rightarrow M\frac{d\bar{x}}{dt} = \text{constant} \qquad \left(\because \bar{x} = \frac{\Sigma mx}{M}\right) \qquad ...(1)$$

where $\bar{x}$ is the x - coordinate of the centre of gravity of body and M its total mass.

From (1), we observe that the total momentum of the body parallel to the axis of x is constant throughout the motion.

MOMENT OF MOMENTUM UNDER FINITE FORCE

If the forces acting on a system be such that they have no moment about a certain fixed straight line, throughout the motion then the angular momentum about this straight line is constant.

Let the co-ordinates of a particles of mass m at time t referred to fixed axes be (x, y, z) where the fixed straight line has been taken as the axis of x. Let X, Y, Z be the resolved parts parallel to the axis of the external forces acting on particle m. The sum of the moments of the external force about the given fixed line is Σ (yZ - zY). By D- Alembert's principle, we have

$$\Sigma m\left(y\frac{d^2z}{dt^2} - z\frac{d^2y}{dt^2}\right) = \Sigma(yZ - zY).$$

But from the given condition that throughout the motion, the sum of the moments of the external forces about the axis of x is zero.

i.e. $\Sigma(yZ - zY) = 0$

$$\therefore \Sigma m\left(y\frac{d^2z}{dt^2} - z\frac{d^2y}{dt^2}\right) = 0 \text{ throughout the motion}$$

i.e. $\frac{d}{dt}\Sigma m\left(y\frac{dz}{dt} - z\frac{dy}{dt}\right) = 0$

i.e. $$\Sigma m\left(y\frac{dz}{dt} - z\frac{dy}{dt}\right) = \text{constant.} \qquad ...(2)$$

From (2), we observe that total moment of moment of momentum of the body about x - axis is constant throughout the motion.

IMPULSIVE FORCES

If the sum of the impulsive forces parallel to a certain fixed line vanishes, the total momentum in that direction remains the same just before and after the application of impulses.

Let us take the fixed line as the axis of x, then

$$\Sigma.m\frac{d^2x}{dt^2} = \Sigma X \ \ i.e. \ \ \frac{d}{dt}\left[\Sigma.m\frac{dx}{dt}\right] = \Sigma X. \qquad ...(1)$$

If τ be the time (small during which the impulsive forces act on the body,

then on integration (1), we get $\left[\Sigma\frac{dx}{dt}\right]_0^\tau = \Sigma\int_0^\tau Xdt = \Sigma x_1$

$$i.e. \ \ \Sigma m(v-u) = \Sigma X_1, \text{ when } X_1 = \int_0^\tau Xdt \qquad ...(2)$$

where u value are the velocities before and after the impulse and X_1 is the impulse of the forces parallel to the x–axis . But if $\sum X_1 = 0$, we have from (2) $\sum m(v-u) = 0$ *i.e.* $\sum mv = \sum mu$),

i.e. total momentum after impulse = the total momentum before impulse

CONSERVATIVE FORCES

If the forces acting on the body are such that Σ (xdx + Y dy + Zdz) is the complete differential of some quantity say V, then the forces are said to be conservative.

Thus we see, If the forces are conservative, then

$$\Sigma(Xdx + Ydy + Zdz) = \int_{t_1}^{l_2}\{\Sigma(Xdx + Ydy + Zdz)\}$$

$$= \int_{t_1}^{l_2} dV = Vt_2 - V_{t_1} \qquad ...(1)$$

which implies that the potential of forces is V. Thus we note that L.H.S. of (1) is independent of the path followed by the body in moving from one position to the other and depends only on the configuration of the body at time t_1 and t_2.

Thus if A is the configuration at time t_1 and B the configuration at time t_2, then have K.E. at time t_2 – K.E. at time t_1 = $V_t - V_t$ *i.e.* K.E. at B – K.E. at A = $V_B - V_A$.

When a body moves under the action of a system of Conservative faces, the sum of its kinetic and potential energies is constant throughout the motion. (Conservation of energy)

In the last article, we have shown that if A and B are the configuration of the body at times t_1 and t_2 then

K.E. at time t_2 *i.e.* at the position B – K.E. at t_1 *i.e.* at the position A

$$= \int_{t_1}^{t_2} dV = VB - V_A \qquad ...(1)$$

The potential energy of the body in any position A is the work done by the forces in moving the body form position A to some standard position C (say)

$\therefore$ Potential energy at time t_1

$$= \Sigma \int_A^C (Xdx + Ydy + Ydz) = \int_A^C dV = V_C - V_A \qquad ...(2)$$

and potential energy at time $t_2 = \Sigma \int_A^C (Xdx + Ydy + Ydz)$

$$= \int_A^C dV = V_C - V_B \qquad ...(3)$$

Combining (2) and (3), we have P.E. at time t_1 – P.E. at time t_1

$$=V_C - V_A - (V_C - V_B) = V_B - V_A \qquad ...(4)$$

Comparing (1) and (4), we have P.E. at time t_1 – P.E. at time t_2

= K.E. at time t_2 – K.E. at time t_1

i.e. K.E. at time t_1 + P.E. at time t_1

= K.E. at time t_2 + P.E. at time t_2.

i.e. sum of the potential energy and kinetic energy is constant at all times.

The kinetic energy of a rigid body, moving in any manner, is at any instant equal to the kinetic energy of the whole mass, supposed to be collected at its centre of inertia and moving with it, together with the kinetic energy of the whose mass relative to its centre of inertia.

Consider a particle of mass m of the body. Let (x, y, z) be the coordinates of this particle at any time t. Let $(\overline{x}, \overline{y}, \overline{z})$ be the co-ordinates of G, the centre of gravity of the body. If (x', y', z') are the co-ordinates of some particle relative to G then, we have $x = \overline{x} + x', y = \overline{y} + y', z = \overline{z} + z'$ and *Smx′ = 0, Smy′ = 0, $\Sigma mz' = 0$

so that $\Sigma m \frac{dx'}{dt} = 0, \Sigma m \frac{dy'}{dt} = 0; \Sigma m \frac{dz'}{dt} = 0$

The total K.E. of the body

$$=\frac{1}{2}\Sigma m\left[\left(\frac{dx}{dt}\right)^2+\left(\frac{dy}{dt}\right)^2+\left(\frac{dz}{dt}\right)^2\right]$$

$$=\frac{1}{2}\Sigma m\left[\left(\frac{d\bar{x}}{dt}+\frac{dx'}{dt}\right)^2+\left(\frac{d\bar{y}}{dt}+\frac{dy'}{dt}\right)^2+\left(\frac{d\bar{z}}{dt}+\frac{dz'}{dt}\right)^2\right]$$

$$=\frac{1}{2}\Sigma m\left[\left(\frac{d\bar{x}}{dt}\right)^2+\left(\frac{d\bar{y}}{dt}\right)^2+\left(\frac{d\bar{z}}{dt}\right)^2\right]+\frac{1}{2}\Sigma m\left[\left(\frac{dx'}{dt}\right)^2\right.$$

$$\left.+\left(\frac{dy'}{dt}\right)^2+\left(\frac{dz'}{dt}\right)^2\right]+\Sigma m\left[\frac{d\bar{x}}{dt}\frac{dx'}{dt'}+\frac{d\bar{y}}{dt}\frac{dy'}{dt}+\frac{d\bar{z}}{dt}\frac{dz'}{dt}\right]$$

$$=\frac{1}{2}\Sigma m\left[\left(\frac{d\bar{x}}{dt}\right)^2+\left(\frac{d\bar{y}}{dt}\right)^2+\left(\frac{d\bar{z}}{dt}\right)^2\right]$$

$$+\frac{1}{2}\Sigma m\left[\left(\frac{dx'}{dt}\right)^2+\left(\frac{dy'}{dt}\right)^2+\left(\frac{dz'}{dt}\right)^2\right]$$

$$+\frac{d\bar{x}}{dt}\Sigma m\frac{dx'}{dt}+\Sigma m\frac{dy'}{dt}\Sigma m\frac{dy'}{dt}+\frac{d\bar{z}}{dt}\Sigma m\frac{dz'}{dt}$$

$$=\frac{1}{2}\left[\left(\frac{d\bar{x}}{dt}\right)^2+\left(\frac{d\bar{y}}{dt}\right)^2+\left(\frac{d\bar{z}}{dt}\right)\right]^2\Sigma m$$

$$+\frac{1}{2}\Sigma m\left[\left(\frac{dx'}{dt}\right)^2+\left(\frac{dy'}{dt}\right)^2+\left(\frac{dz'}{dt}\right)\right]^2$$

$$=\frac{1}{2}\text{ (velocity of centre of gravity G)}^2\,\Sigma m$$

$$+\frac{1}{2}\text{ (velocity of G)}^2\times\Sigma m+\frac{1}{2}\Sigma m$$

$$=\frac{1}{2}Mv^2+\frac{1}{2}\Sigma m\text{ (velocity of m relative to G)}^2$$

Taking G as the origin the co-ordinates of general point are (x′, y′, z′). Then co-ordintes of the C.G. of the body are given as x coordinate of $C.G=\frac{\Sigma mx'}{Zm}$

But x coordinate of G = 0 (since G is the origin) $\therefore$ $\Sigma mx' = 0$ Similarly $\Sigma my' = 0$ $\Sigma mz' = 0$. where M is the total mass of the body and v is the velocity of the centre of gravity G. *To sum up:*

Total K.E. of the body = K.E. of mass M supposed to be collected at G + K. E. of the body relative to G.

CONSERVATION OF ANGULAR MOMENTUM (IMPULSIVE FORCES)

If the sum of the moments of the impulses about a certain line vanishes then the angular momentum about it remains the same before and after the application of the impulse.

Let us take the fixed line as axis of x. Let t be the time (small) during which the impulses act on the body. By taking moments about the fixed axis, we have

$$\Sigma m(y\ddot{z} - z\ddot{y}) = \Sigma(yZ - zY). \quad \text{...(1)}$$

integrating it w.r.t. "t" we have

$$\int_0^\tau \Sigma m(y\ddot{z} - z\ddot{y})dt = \int_0^\tau \Sigma(yz - zY)dt$$

$$\Rightarrow [\Sigma m(y\dot{z} - z\dot{y})]_0^\tau = \Sigma(yZ_1 - zY_1)$$

where $Z_1 = \int_0^\tau Zd$ and $Y_1 = \int_0^\tau Ydt.$ ...(2)

L.H.S. of (2) is angular momentum after the impulse minus angular momentum before impulse.

R.H.S. of (2) is the sum of the moments of the impulses about x – axis. In case the sum of the moments of impulses about x – axis is zero

i.e., $\Sigma\,(yZ_1 - zY_1) = 0$, we have

Angular momentum after impulse = angular momentum before impulse.

MISCELLANEOUS EXAMPLES

Example 1:

An elliptic area, of eccentricity e is rotating with angular velocity ω about one latus rectum; suddenly this latus rectum is loosened and the other fixed. Show that the new angular velocity is $\frac{1-4e^2}{1+4e^2}\omega$.

Solution:

Consider the ellipse rotating with angular velocity ω, about its latus rectum L S L′.

Let this latus- rectum be set free and suddenly the other latus rectum L″S′L‴ be fixed. Let the angular velocity be ω′ just after the fixture. The impulse is at L″S′L‴ . Hence the moment of the impulse about the rectum L¢¢S¢L¢¢¢ remains the same before and after the fixture.

$$i.e.\ m\frac{a^2}{4}\omega' + ma^2e^2\omega' = m\frac{a^2}{4}\omega - ma^2e^2\omega$$

$$\text{i.e } \left(a^2 + 4a^2e^2\right)\omega' = \left(a^2 - 4a^2e^2\right)\omega \ \ i.e.\ \omega' = \frac{1-4e^2}{1+4e^2}\omega.$$

Example 2:

An elliptic lamina is rotating about its centre on a smooth horizontal table. If $\omega_1, \omega_2, \omega_3$ be its angular velocities when the extremity of its major axis its focus and the extremity of its minor axis respectively become fixed prove that

$$\left(\frac{7}{\omega_1}\right) = \left(\frac{6}{\omega_2}\right) + \left(\frac{5}{\omega_3}\right)$$

Solution:

Initially the lamina was rotating with angular velocity ω about its O.

(i) *Let the centre O be set free and the end A of the major axis suddenly fixed.* The impulse is at A, therefore moment of the impulse about the axis through A (perpendicular to the plane of motion) is zero, therefore the moment of momentum about this axis remains the same before and after the fixture. If ω1 ισ the angular velocity just after the fixture, we have

$$\Rightarrow mk^2\omega + o = mk^2\omega_1 + ma\omega_1 \cdot AOm\frac{a^2+b^2}{4}\omega_1 + ma^2\omega_1.$$

$$\text{or } \left(a^2 + b^2\right)\omega = \left[\left(a^2 + b^2\right) + 4a^2\right]\omega_1\ i.e.\ \omega_1 = \frac{a^2b^2}{5a^2 + b^2}\omega \qquad ...(1)$$

(ii) *Let the centre O be set free and the focus S be suddenly fixed.*

Again the impulse will be at S and the moment of the impulse about S is zero. Therefore the moment of momentum about the axis through S will remain the same before and after the fixture

$$i.e.\ m\frac{a^2+b^2}{4}\omega = m\frac{a^2+b^2}{4}\omega_2 + ma^2e^2\omega_2$$

$$\text{or } \left(a^2 + b^2\right)\omega = \left(a^2 + b^2 + 4a^2e^2\right)\omega_2\ i.e.\ \omega^2 = \frac{a^2+b^2}{a^2+b^2+4a^2e^2}\omega$$

$$\text{or } \omega_2 = \frac{a^2+b^2}{a^2+b^2+4(a^2-b^2)}\omega \qquad \left[\because b^2 = a^2\left(1-e^2\right)\right] = \frac{a^2+b^2}{5a^2+3b^2}\omega$$

(iii) *Let the centre be set free and the end B of the minor axis be suddenly fixed.* In this case, we have

$$m\frac{a^2+b^2}{4}\omega = m\frac{a^2+b^2}{4}\omega_3 + mb^2\omega_3$$

or $\left(a^2+b^2\right)\omega = \left(a^2+b^2+4b^2\right)\omega_3$

i.e. $\omega_3 = \frac{a^2+b^2}{a^2+5b^2}\omega$ ...(3)

from (1), (2) and (3), we have

$$\frac{7}{\omega_1}-\frac{6}{\omega_2}-\frac{5}{\omega_3}=\frac{35a^2+7b^2}{\left(a^2+b^2\right)\omega}-\frac{30a^2-18b^2}{\left(a^2+b^2\right)\omega}-\frac{5a^2+25b^2}{\left(a^2+b^2\right)\omega}$$

$$=\frac{\left[35a^2+7b^2-30a^2+18b^2-5a^2-25b^2\right]}{\left(a^2+b^2\right)\omega}=0 \Rightarrow \frac{7}{\omega_1}=\frac{6}{\omega_2}=\frac{5}{\omega_3}.$$

Example 3:

A uniform square lamina, of mass M and side 2a, is moving freely about a diagonal with uniform angular velocity where one of the corners not in that diagonal becomes fixed, show that the new angular velocity is $\frac{1}{7}\omega$ and that the impulse of the force on the fixed point is $\frac{\sqrt{2}}{7}Ma\omega$.

Solution:

Initially the square ABCD is revolving about the diagonal AC with angular velocity ω; this diagonal AC is set free and the corner B is suddenly fixed. The square will begin to turn about the line BE through B parallel to AC. Let ω′ be angular velocity about BE. The impulse is at B, thus the moment of the impulse about BE zero. Therefore the moment of momentum about BE remains the same before and after the fixture,

i.e. $Mk^2\omega' + M(BG)^2\omega' = Mk^2\omega$ ...(1)

But k^2 about AC $=\frac{a^2}{3}\cos^2\frac{\pi}{4}+\frac{a^2}{4}\sin^2\frac{\pi}{4}=\frac{a^2}{3}$

and $BG = a\sqrt{2}$ $\therefore M\frac{a^2}{3}\omega' + M2a^2\omega' = M\frac{a^2}{3}\omega$ *i.e.* $\omega' = \frac{\omega}{7}$.

After the fixture, the velocity of G, the centre of gravity of the square is $a\sqrt{2}\omega'$ at right angles to BG.

$$= M\left(a\sqrt{2}\omega' - O\right) = Ma\sqrt{2}\omega' = Ma\sqrt{2}.\frac{1}{7}\omega = \frac{\sqrt{2}}{7}Ma\omega.$$

Example 4:

A heave ring of radius a is moving in its own plane which is vertical. At a certain instant when its velocity is V horizontally from left to right and the angular velocity is (V/2a) clockwise, the highest point of the ring is suddenly fixed. Prove that the ring will describe a complete revolution about the point of fixing, if $V^2 > 32$ ag

Solution:

Consider a ring of radius am moving in its own plane with velocity V. Let its angular velocity be ω, τηεν $\omega = \frac{V}{2a}$ (given).

Just after fixing at the highest point of A, let the angular velocity be ω'. Τηε velocity of the centre will be = $a\omega'$.

Obviously the impulse is at A, therefore the moment of the impulse about the axis through A (which is horizontal) is zero thus moment of momentum about A will remain unaltered just before and after the fixture at the point

A *i.e.* $mk^2\omega' + ma\omega'.a = mk^2\omega - m.\,a = mk^2\omega - mVa$

or $mk^2\omega' + ma\omega'.a = mk^2\omega - mVa$

or $(a^2\ a^2)\omega' = a^2\left(\frac{V}{2a}\right) - Va i.e. \omega' = -\left(\frac{V}{4a}\right)$

When the ring is turning about A, let the radius through A make an angle θ with the vertical. Also the energy equation gives

$$\frac{1}{2}m\left\{a^2\dot{\theta}^2 + \dot{\theta}^2\right\} - \frac{1}{2}m\left(a^2 + \omega'^2 + \right) = mg\left(a - a\cos\theta\right)$$

or $\dot{\theta}^2 - \omega'^2 = -\left(\frac{g}{a}\right)\left(a - \cos\theta\right)$

or $\dot{\theta}^2 - \omega'^2 - \left(\frac{g}{a}\right)\left(1 - \cos\theta\right) = \left(\frac{V^2}{16a^2}\right) - \frac{g}{a}\left(1 - \cos\theta\right)$

Now the ring will make a complete revolution if θ is greater than or equal to zero when $\theta = \pi$.

When $\theta = \pi$, $\theta = \pi.\,\dot{\theta}^2 = \left(\frac{V^2}{16a^2}\right) - \frac{g}{a}\left(1 - \cos\pi\right) = \frac{2g}{a}$

it is positive only when $\left(\frac{V^2}{16a^2}\right) > \left(\frac{2g}{a}\right)$ i.e. $V^2 > 32ag$.

Example 5(a):

Three equal uniform rods placed in a straight line are freely jointed and move with a velocity v perpendicular to their lengths. If the middle point of the middle rod be suddenly fixed, show that the ends of the other two rods will meet in time (4 π a)/(9v), *where a is the length of each rod.*

Solution:

Let AB, BC, CD be the three rods each of length a and mass m. Freely jointed at B and C, and placed in a straight line. Let these rods move with velocity v in a direction perpendicular to their lengths. Let G_1, the middle point of the BC be suddenly fixed.

Consider the motion of the rod AB, the impulsive action acts at B with the result the moment of the impulsive force at B vanishes. Hence the angular momentum of AB about the end B remains unaltered before and after the fixture.

$$\therefore mv\left(\frac{a}{2}\right) = m(a/2)^2\,\omega + (m/3)(a/2)^2\omega = \left(\frac{ma^2\omega}{3}\right)$$

(where ω is the angular velocity of AB just after the fixture)

$$i.e.\ \ \omega = \left(\frac{3v}{2a}\right)$$

Similarly, considering the motion of the rod CD, we can prove that its angular velocity after the fixture $= \left(\frac{3v}{2a}\right)$

The end A and D of the rods will meet in E and in that position the triangle CBE will be an equilateral triangle so that $\angle ABE = \angle DCE = 120°$. Now $(d\theta/dt) = \omega$. *i.e.* $\theta = \omega t$.

$$\therefore \text{Required time} = \frac{\theta}{\omega} = \frac{120°}{(3v/2a} = \frac{2\pi}{3}\cdot\frac{2a}{3v} = \frac{4\pi a}{9v}.$$

Example 5(b):

If an actant of ellipsoid bounded by three principal planes be rotating about the axis of "a" with angular velocity ω and if this axis suddenly becomes free when the axis of "b" becomes suddenly fixed at the same instant, show that the new angular velocity is $\left(\frac{2abw}{\pi(a^2+c^2}\right)$.

Solution:

Let the axis of "a" be taken as the axis of x. Let $\omega_{x'}$, $\omega_{y'}$, ω_z be the angular velocities about the axes just before the fixture, then we have $\omega_x = \omega$ given $\omega_y = 0$, $\omega_{z'} = 0$.

Let $\omega'_{x'}$, $\omega'_{y'}$, $\omega'_{z'}$ be the angular velocities about the same axis after the fixing of the axis b.*i.e.* y – axis.

$\therefore\ \omega'_{x'} = 0,\ \omega'_{z'} = 0$ and $\omega'_{y'}$ is to be determined.

Now as y – axis is suddenly fixed, the impulse is at y – axis and moment of the impulse about y – axis vanishes, therefore the moment of momentum about y – axis remains unaltered before and after the fixture.

i.e. $B\ (\omega'_{y'} - \omega_y) - D\,(\omega'_z - \omega_z) - (\omega'_x - \omega_x) = 0$

where B, D, F have their usual meaning in moments of inertia.

Now substituting the values of $\omega_{x'}$, $\omega_{y'}$, $\omega_{z'}$ and $\omega'_x\ \omega'_{z'}$ we get $B\ \omega'_{y'} = -F$

ω But $B = M\left[\dfrac{c^2 + a^2}{5}\right]$, therefore, we have

$$M\frac{c^2 + a^2}{5}\omega'y = -\,\omega F = -\,\omega\iiint p.xydxdydz$$

the integration being taken throughout the region bounded by

$$x > 0,\ y > 0, z > 0 \text{ and } \frac{x^2}{a^2} + \frac{y^2}{b^2} + \frac{z^2}{c^2} \le 1,$$

and M is the mass of the octant. Let $dx = \frac{1}{2}aX^{-\frac{1}{2}}dX$

Similarly, $dy = \frac{1}{2}bY^{-1/2}\ \ dZ = \frac{1}{2}cZ^{-1/2}dZ.$

making these substituting in (1), we get

$$M\frac{c^2 + a^2}{5}\omega'y$$

$$= -\,\omega p\iiint aX^{1/2}bY^{1/2}\frac{1}{2}aX^{-1/2}\frac{1}{2}bY^{-1/2}\frac{1}{2}cZ^{-1/2}\ dXdYdZ$$

$$= -\,\frac{a^2b^2c}{8}\omega\rho\iiint X^{\circ}Y^{\circ}Z^{-1/2}dXdYdZ \text{ where } X + Y + Z \le 1$$

$$= -\,\frac{a^2b^2c}{8}\omega\rho\iiint X^{1-1}Y^{1-1}Z^{(1/2)^{-1}}\ dXdYdZ$$

$$= -\frac{a^2b^2c}{8}\frac{1.1.\Gamma(1)\Gamma\left(\frac{1}{2}\right)}{\frac{5}{2\cdot}\frac{3}{2\cdot}\frac{1}{2\cdot}\Gamma\left(\frac{1}{2}\right)}\omega\rho = \frac{6M}{\pi abc\cdot}$$

$$M\frac{(c^2+a^2)}{5}\omega' y = \frac{a^2b^2c}{15}\cdot\omega\frac{6M}{\pi abc} \text{ i.e. } \omega' y = \frac{2abc\omega}{\pi(c^2+a^2)}$$

Example 6:

A cube is turning with angular velocity ω about a diagonal, when suddenly the diagonal is let go and of the edges which does not meet this diagonal is fixed; show that the resulting angular velocity about this edge is $\left(\frac{\omega\sqrt{3}}{12}\right)$

Solution:

Let 2a be the side of the cube. Take OA, OB, OC as the axes of x,y,z, respectively. Then the direction cosines of the diagonal OP are $\frac{1}{\sqrt{3}}, \frac{1}{\sqrt{3}}, \frac{1}{\sqrt{3}}$. Let the cube revolve about the diagonal OP with angular velocity ω. Obviously the resolved parts of this angular velocity about OA, OB, OC will be $\frac{\omega}{\sqrt{3}}, \frac{\omega}{\sqrt{3}}, \frac{\omega}{\sqrt{3}}$ respectively.

Let the diagonal OP be set free and edge LA be suddenly fixed. Let ω' be the angular velocity about the edge. The distance of the edge from the centre is

Since the moment of momentum remains the same before and after the fixture, therefore we have $M\left(\frac{2a^2}{3}\omega' + 2a^2\omega'\right) = m.o + m\frac{2a^2}{3}\frac{\omega}{\sqrt{3}}$

[$\because$ For the cube about any axis through its centre, $k^2 = \frac{2a^2}{3}$]

$$\text{or } \frac{8a^2}{3}\omega' = \frac{2a^2}{3\sqrt{3}}\omega \; i.e.\; \omega' = \frac{\omega}{4\sqrt{3}} = \frac{\sqrt{3}\omega}{12}.$$

Example 7(a):

A uniform cube is spinning freely with angular velocity w about a diagonal through a corner O, when suddenly the diagonal through O′ of one of the faces through O becomes fixed, show that the new angular velocity is $\frac{2\sqrt{3}}{5\sqrt{3}}\omega$.

Solution:

Let the side of cube be 2a, so the length of its diagonal

$= \sqrt{[(4a^2 + 4a^2 + 4a^2)]} = 2a\sqrt{3}$

and the length of the diagonal of a face $= \sqrt{[(4a^2 + 4a^2)]} = 2a\sqrt{2}.$

$\therefore$ inclination of the diagonal of the cube to the diagonal of the face

$$= \cos^{-1}\frac{2a\sqrt{2}}{2a\sqrt{3}} = \cos^{-1}\sqrt{\left(\frac{2}{3}\right)}$$

Initially the cube was revolving about its diagonal with angular velocity ω, resolved part of this angular velocity along diagonal of the face is $\omega\sqrt{\left(\frac{2}{3}\right)}$.

The distance of the centre of gravity of the cube from the diagonal of the face equal to a. Since the diagonal of the face is suddenly fixed the impulse is at this diagonal, so moment of the impulse about this diagonal is zero, therefore the moment of momentum about this diagonal of the face is the same before and after the fixture.

$$\text{i.e } m\left(\frac{2a^2}{3}\omega' + a^2\omega'\right) = m\left[\left(\frac{2a^2}{3}\omega\right)\sqrt{\left(\frac{2}{3}\right)}\right]$$

$$5\omega' = 2\sqrt{\left(\frac{2}{3}\right)}\omega \text{ or } 5\omega' = \left(\frac{2\sqrt{2}}{5\sqrt{3}}\right)\omega.$$

Example 7(b):

Two like rods AB and BC, each of length 2a are freely jointed at B; AB can turn round the end A, and C can move freely on a vertical straight line through A and they are then released. Initially the rods are held in a horizontal line, C being in coincidence with A and they are then released. Show that when the rods are inclined at an angle θ to the horizontal, the angular velocity of either is

$$\sqrt{\left(\frac{3g}{a}\cdot\frac{\sin\theta}{1+3\cos^2\theta}\right)}$$

Solution:

Take the fixed point A as origin, the lines AX and AY as the axes. Let G_1 and G_2 be the centres of gravity of the rods AB and BC respectively. Then Co-ordinates of G_1 are (a cos θ, a sin θ). Hence the energy equation gives

$$\frac{1}{2}m\left[\frac{a^2}{3}\dot{\theta}^2 + (-a\sin\theta\dot{\theta})^2 + (a\cos\theta\dot{\theta})^2\right]$$

$$\frac{1}{2}m\left[\frac{a^2}{3}\dot{\theta}^2 + (-a\sin\theta\dot{\theta})^2\right] = \text{mg a sing } \theta + \text{mg. } 3 \text{ a } \sin\theta$$

Or $\frac{1}{2}m\left[\frac{4}{3}a^2\dot{\theta}^2\right] + \frac{1}{2}m\left[\frac{1}{3}a^2\dot{\theta}^2 + a^2\sin^2\dot{\theta}\theta^2 + 9a^2\cos^2\theta\dot{\theta}^2\right]$

$= 4\text{g a} \sin\theta$

Or $\left[\frac{4}{3}a^2\dot{\theta}^2 + a^2\dot{\theta}^2 + 8a^2\cos^2\theta\dot{\theta}\right] = 8\text{g a}\sin\theta$

Or $\frac{8}{3}(a^2\dot{\theta}^2 + a^2\cos^2\theta\dot{\theta}^2 = 8\text{ga} \sin\theta$

Or $(a^2\dot{\theta}^2 + 3a^2\cos^2\theta\dot{\theta}^2) = 3\text{ga} \sin\theta$

Or $a\dot{\theta}^2(1+3\cos^2\theta) = 3g\sin\theta$ or $\dot{\theta} = \sqrt{\left(\frac{3g}{a}\frac{\sin\theta}{1+3\cos^2\theta}\right)}$.

Example 7(c):

A uniform straight rod of length 2a. has two small rings at its ends which can respectively slide on thin smooth horizontal and vertical wires OX and OY. The rods starts at an angle α, to the horizontal with angular velocity [{3g (1 – sin α)}/2a]$^{1/2}$ and downwards. Show that it will strike the horizontal wire at the end of time

$$2\sqrt{\left(\frac{a}{3g}\right)}\log\left[\left\{\cot\left(\frac{\pi}{8} - \frac{\alpha}{4}\tan\right)\right\}\frac{\pi}{8}\right].$$

Solution:

At any time t, let the rod AB be inclined at an angle θ to the horizontal al. If G be the C.G. of the rod AB, then co-ordinates of G are given by x = a cos θ, y = a sin θ, so that

$$\dot{x} = -a\sin\theta\dot{\theta},\ \dot{y} = a\cos\theta\dot{\theta},$$

$\therefore$ velocity of $G = \sqrt{\{(\dot{x}^2 + \dot{y}^2)\}}$

$= \sqrt{[(-a\sin\theta\dot{\theta})^2 + (a\cos\theta\ddot{\theta}^2] = \dot{\theta}}$

$\therefore$ kinetic energy of the rod at time t

$$= \frac{1}{2}m\left[\frac{a^2}{3}\dot{\theta}^2 + a^2\dot{\theta}^2\right] = \frac{2}{3}ma^2\dot{\theta}^2 \quad ...(1)$$

Initial angular velocity is

$$=\left[\frac{3a}{2a}(1-\sin\alpha)\right]^{1/2} \quad \text{(given)}$$

$\therefore$ Initial kinetic energy $=\frac{2}{3}ma^2\cdot\frac{3g}{2a}(a-\sin\alpha)$

[from (1)] = mga (1 – sin α).

Hence energy equation gives

$$\frac{2}{3}ma^2\dot{\theta}^2 = mga(1-\sin\alpha) = mg(a\sin\alpha) - mg(a\sin\theta)$$

i.e. $\frac{2}{3}a^2\dot{\theta}^2 - ga(1-\sin\alpha) = ga\,(\sin\alpha - \sin\theta)$

i.e. $\frac{2}{3}a^2\dot{\theta}^2 = ga\,(1-\sin\theta)$

or $\dot{\theta} = \frac{d\theta}{dt} = -\sqrt{\left(\frac{3g}{2a}\right)}\sqrt{(1-\sin\theta)}$

(negative sign is taken with radical sign, since θ decreases)

or $dt = -\left(\frac{2a}{3g}\right)^{1/2}\frac{1}{\sqrt{1-\sin\theta}}d\theta.$

Hence the required time

$$= -\left(\frac{2a}{3g}\right)^{1/2}\int_{\alpha}^{0}\frac{1}{\sqrt{1-\sin\theta}}d\theta = \left(\frac{2a}{3g}\right)^{1/2\alpha}\int_{0}^{\alpha}\frac{1}{\sqrt{(1-\sin\theta)}}d\theta$$

$$= \left(\frac{2a}{3g}\right)^{1/2}\int_{a}^{\alpha}\frac{d\theta}{[\{\cos^2(\theta/2)+\sin^2(\theta/2)-2\sin(\theta/2)\cos(\theta/2)]^{1/2}}$$

$$= \left(\frac{2a}{3g}\right)^{1/2}\int_{0}^{\alpha}\frac{d\theta}{\{\cos(\theta/2)-\sin(\theta/2)\}}$$

$$= \left(\frac{2a}{3g}\right)^{1/2}\int_{0}^{\alpha}\frac{d\theta}{\sqrt{2}\left[\frac{1}{\sqrt{2}}\cos(\theta/2)-\frac{1}{\sqrt{2}}\sin(\theta/2)\right]}$$

$$= \left(\frac{2a}{3g}\right)^{1/2}\frac{1}{\sqrt{2}}\int_{0}^{\alpha}\frac{d\theta}{\sin\left(\frac{\pi}{4}-\frac{\theta}{2}\right)} = \left(\frac{a}{3g}\right)^{1/2}\int_{9}^{\alpha}\text{cosec}\left(\frac{\pi}{4}-\frac{\theta}{2}\right)d\theta$$

$$= -2\sqrt{\left(\frac{a}{3g}\right)}\left[\log\tan\left(\frac{\pi}{8}-\frac{\theta}{4}\right)\right]_0^{\alpha}$$

$$= 2\sqrt{\left(\frac{a}{3g}\right)}\left[\log\tan(\pi/8) - \log\tan\left(\frac{\pi}{8} - \frac{\alpha}{4}\right)\right]$$

$$= 2\sqrt{\left(\frac{a}{3g}\right)}\frac{\tan(\pi/8)}{\tan[(\pi/8)-(\alpha/4)]}$$

$$= 2\sqrt{\left(\frac{a}{3g}\right)}\log\left[\tan\frac{\pi}{8}\cot\left(\frac{\pi}{8} - \frac{\alpha}{4}\right)\right].$$

Example 8:

A uniform rod of length 2a, is placed with one end in contact with a smooth horizontal table and is then allowed to fall, if α be its initial inclination to the vertical, show that its angular velocity when it is inclined at an angle θ is

$$\left[\frac{6g}{a\cdot}\frac{\cos\alpha - \cos\theta}{1+3\sin^2\theta}\right]^{1/2}$$ *find also the reaction of the table.*

Solution:

Let AB be the rod such that AB = 2a. Let G be the C.G. of the rod AB. Since there is no horizontal force on the rod AB, therefore its centre of gravity G moves down a vertical line through G Let this vertical line through G cut the horizontal table at the point O, then O is fixed point.

Taking O as the origin and the line OX and OY as the axes, the co-ordinates of G are (0, a cos θ).

i.e. $x_G = 0, y_G = a\cos\theta$

$$\Rightarrow \dot{x}_G = 0.\dot{y}_G = -a\sin\theta\dot{\theta} \Rightarrow v_{G^2} = \dot{x}_{G^2} + \dot{y}_{G^2} = a^2\sin^2\theta\dot{\theta}^2$$

Hence the energy equation gives

$$\frac{1}{2}m\left[v_G^2 + k^2\dot{\theta}^2\right] = \frac{1}{2}m\left[\frac{a^2}{3}\dot{\theta}^2 + a^2\sin^2\theta\dot{\theta}^2\right] = m\,ga\cos\alpha - mga\cos\theta$$

i.e., $$\frac{a^2\dot{\theta}^2}{6}[1+3\sin^2\theta] = ga(\cos\alpha - \cos\theta)$$

i.e. $$\dot{\theta}^2 = \frac{6g}{a}\frac{\cos\alpha - \cos\theta}{1+3\sin^2\theta}$$

i.e. $$\dot{\theta} = \sqrt{\left(\frac{6g}{a}\right)}\sqrt{\left(\frac{\cos\alpha - \cos\theta}{1+3\sin^2\theta}\right)} \quad ...(1)$$

To find the reaction R, take moments about G.

$$R\, a \sin \dot{\theta} = m\frac{a^2}{3}\theta\ddot{\theta} \text{ i.e., } R = \frac{m}{3} a \operatorname{cosec} \theta\ddot{\theta} \qquad ...(2)$$

To find the value of $\ddot{\theta}$, we differentiate $\dot{\theta}^2$ found above

$$2\,\dot{\theta}\ddot{\theta} = \frac{6g}{a}\frac{[(1+3\sin^2\theta)\sin\theta - (\cos\alpha - \cos\theta).6\sin\theta\cos\theta]\dot{\theta}}{(1+3\sin^2\theta)^2}$$

$$\Rightarrow \ddot{\theta} = \frac{3g}{a}\frac{\sin\theta[1+3\sin^2\theta - 6\cos\theta(\cos\alpha - \cos\theta)]}{(1+3\sin^2\theta)^2}$$

$$= \frac{3g\sin\theta}{a(1+3\sin^2\theta)^2}[1+3\sin^2\theta + \cos^2\theta - 6\cos\theta\cos\alpha]$$

$$= \frac{3g\sin\theta}{a(1+3\sin\theta)^2}[1+3(\sin^2\theta+\cos^2\theta)+3\cos^2\theta - 2\cos\theta\cos\alpha]$$

$$= \frac{3g\sin\theta}{a(1+3\sin^2\theta)^2}[4+3(\cos^2\theta - 2\cos\theta\cos\alpha) - 3\cos^2\alpha]$$

$$= \frac{3g\sin\theta}{a(1+3\sin^2\theta)^2}[4-3\cos+3(\cos\theta-\cos\alpha)^2].$$

Substituting this value of $\ddot{\theta}$ in (2), we get

$$R = \frac{ma}{3\sin\dot{\theta}} = \frac{3g\sin\theta}{a(1+3\sin^2\theta)^2}[4-3\cos^2\alpha)+3(\cos\theta-\cos\alpha)^2]$$

$$= \frac{ma}{(1+3\sin^2\theta)^2}[4-3\cos^2\alpha+3(\cos\theta-\cos\alpha)^2]$$

we observe that value of the reaction is positive for all values of θ, hence the lower end of the rod will never leave the horizontal table.

Example 9:

A straight uniform rod, of mass m, is placed at right angles to smooth plane of inclination α with one end in contact with it; the rod is then released. Show that when its inclination to the plane is ϕ, the reaction of the plane will be $mg\dfrac{3(1-\sin\phi)^2+1}{(3\cos^2\phi+1)^2}\cos\alpha$.

Solution:

Let at any time t, the rod AB make an angle ϕ with the inclined plane. Then the equation of motion of the C.G. of the rod

$$AB \quad m\frac{d^2}{dt^2}(a\sin\phi) = R - mg\cos\alpha \qquad \text{...(i)}$$

$$m\frac{a^2}{3}\ddot{\phi} = -R\,a\cos\phi \qquad \text{...(ii)}$$

Substituting the value of R from (ii) in (i),

we get $m\frac{d^2}{dt^2}(a\sin\phi)$

$$= -mg\cos\alpha - \frac{ma}{3}\sec\phi.\ddot{\phi}$$

or $\frac{d}{dt}(a\cos\phi.\dot{\phi}) = -g\cos\alpha - \frac{1}{3}a\sec\phi.\ddot{\phi}$

or $\left(a\cos\phi + \frac{a}{3}\sec\phi\right)\ddot{\phi} - a\sin\phi\dot{\phi}^2 = -g\cos\alpha$

or $a\left(\frac{1}{3} + \cos^2\phi\right)\ddot{\phi} - a\sin\phi\cos\phi\dot{\phi}^2 = -g\cos\alpha\cos\phi$

Multiplying by 2 $\dot{\phi}$ throughout, we get

$$2a\left(\frac{1}{3} + \cos^2\phi\right)\dot{\phi}\ddot{\phi} - 2a\sin\phi.\dot{\phi}^3 = -2g\cos\alpha\cos\phi\dot{\phi}$$

or $\frac{d}{dt}\left[a\left(\frac{1}{3} + \cos^2\phi\right)\dot{\phi}^2\right] = -2g\cos\alpha\cos\phi\dot{\phi}.$

Integrating it we get a $a\left(\frac{1}{3} + \cos^2\phi\right).\dot{\phi}^2 = -2g\cos\alpha\sin\phi + C.$

Initially $\phi = (\pi/2)$ and $\dot{\phi} = 0$ $\therefore C = 2g\cos\alpha.$

Hence $a\left(\frac{1}{2} + \cos^2\phi\right).\dot{\phi}^2 = 2g\cos\alpha(1 - \sin\phi)$

or $\dot{\phi}^2 = \dfrac{2g\cos\alpha(1-\sin\phi)}{a\left(\frac{1}{3} + \cos^2\alpha\right)}$

Substituting the value of $\dot{\phi}^2$ in (iii), we get

$$a\left(\frac{1}{3} + \cos^2\phi\right)\ddot{\phi} - a\sin\phi\cos\phi.\frac{2g\cos\alpha(1-\sin\phi)}{a\left(\frac{1}{3} + \cos^2\phi\right)} = -g\cos\alpha\cos\phi$$

or $a\left(\frac{1}{3}+\cos^2\phi\right)^2\ddot{\phi}-2g\cos\alpha\sin\phi\cos\phi(1-\sin\phi)$

$=-g\cos\alpha\cos\phi\left(\frac{1}{3}+\cos^2\phi\right)$

or $a(1+3\cos^2\phi)^2\ddot{\phi}-3g\cos\alpha\cos\phi[6\sin\phi-6\sin^2\phi-1-3\cos^2\phi)]$

$=3g\cos\alpha\cos\phi[6\sin\phi-3\sin^2\phi-3(\sin^2\phi+\cos^2\phi)-1]$

$=3g\cos\alpha\cos\phi[-3(\sin^2\phi-2\sin\phi+1)-1]$

$=3g\cos\alpha\cos\phi[-3(1-\sin\phi)^2-1]$

$=-3g\cos\alpha\cos\phi[3(1-\sin\phi)^2+1]$

or $\ddot{\phi}=-\dfrac{3g\cos\alpha\cos\phi[3(1-\sin\phi)^2+1]}{a(11+3\cos^2\phi)^2}$

putting this value of $\ddot{\phi}$ in (ii), we get R = m g cos `

$\alpha\dfrac{3(1-\sin\phi)^2+1}{(1+3\cos^2\phi)^2}$.

Example 10(a):

A hemisphere of mass M and radius a is placed with its plane base on a smooth table, and a heavy rod of mass m is constrained to move in a vertical line with one end p on the curved surface of the hemisphere. If at any time t, the radius to p makes an angle θ with the vertical, show that

$$a\dot{\theta}^2[M\cos^2\theta+m\sin^2\theta]=2mg(\cos\alpha-\cos\theta).$$

Solution:

Let pQ (= 2a) be the rod moving in a vertical line, with its one end p in contact with the curved surface of the hemisphere. Let G_1 be the C.G. of the hemisphere and G_2 the C.G. of the rod pQ.

Let the vertical line along which the rod move cut the table at O, then O is a fixed point. Take O as the origin and horizontal line OA as - axis and vertical line through O as y-axis. Then the co - ordinate of G_2 are

(0, a + a cosθ) and co-ordinates of $G_1$are $\left(a\sin\theta,\frac{3a}{8}\right)$

Hence the energy equation gives

$\frac{1}{2}k\left(-a\sin\theta\dot{\theta}\right)^2+\frac{1}{2}M\left(a\cos\theta\dot{\theta}\right)^2=mg\left(a\cos\alpha-a\cos\theta\right)$

or $a\dot{\theta}(M\cos^2\theta + m\sin^2\theta) = 2mg(\cos\alpha - \cos\theta)$.

Example 10(b):

A uniform rod, of length 2a is freely jointed at one end to a small ring, whose mass is equal to that of the rod. The ring is free to slide on smooth wire and initially is at rest and the rod is vertical and below the ring rotating with angular velocity $\sqrt{}(3g/a)$ in a vertical plane passing through the wire. When the rod is inclined at an angle θ to the vertical, show that its angular velocity is $\sqrt{\left(\frac{3g}{a}\frac{1+\cos\theta}{8-3\cos^2\theta}\right)}$ and find the velocity of the ring.

Solution:

Let A′ B′ be the initial position of the rod.

At time t, let q be the angle through which the rod A B has turned and the ring has moves a distance x from its initial position A ′ along the smooth horizontal wire.

Now taking A′ as the origin and the lines, A′A,A′B′ as axes of x and y repetitively , the co-ordinates of G are (X + a sinθ, a cos θ).

$\therefore$ Its velocity components are $\left(\dot{x} + a\cos\theta\dot{\theta}\right)$ and $\left(-a\sin\theta\dot{\theta}\right)$ and velocity of the ring is $\dot{x}$. Thus the horizontal momentum at time t is

$$= m\dot{x} + m\left(\dot{x} + a\cos\theta\dot{\theta}\right)$$

Initial horizontal momentum $= m\left[0 + a\left(\frac{3g}{a}\right)^{1/2}\right] = m\sqrt{(3ag)}$

$$\{\text{since initially } \theta = 0, \theta = (\frac{3g}{a}\}^{1/2} \text{ and } x = 0\}$$

Since there is no horizontal force on the system, therefore the horizontal momentum remains unaltered throughout the motion.

$$\therefore m\ddot{x} + m(\dot{x} + a\cos\theta\dot{\theta}) = m\sqrt{(3ag)}$$

$$\text{or } \ 2\dot{x} + a\cos\theta\dot{\theta}) = \sqrt{(3ag)} \text{ i.e.} \dot{x} = \frac{1}{2}[\sqrt{(3ag)} - a\cos\theta\dot{\theta}] \qquad ...(1)$$

Kinetic energy at time t

$$= \frac{1}{2}m[\dot{x}^2 + \frac{1}{2}m[\dot{x} + a\cos\theta.\dot{\theta})^2 + (-a\sin\theta.\dot{\theta})^2 + \frac{1}{3}a^2\dot{\theta}^2]$$

$$=\frac{1}{2}m[\dot{x}^2+\dot{x}^2+a^2\dot{\theta}^2+2a\dot{x}\dot{\theta}\cos\theta+\frac{1}{3}a^2\dot{\theta}^2]$$

$$=\frac{1}{2}m[2\dot{x}^2+\frac{4}{3}a^2\dot{\theta}^2+2a\dot{x}\dot{\theta}\cos\theta]$$

Initial kinetic energy

$$\frac{1}{2}m\left\{0+\frac{4}{3}a^2\cdot\frac{3g}{a}+0\right\}\left(\text{since initially }\dot{x}=0,\theta=0\text{ and }\dot{\theta}^2=\frac{3g}{a}\right)$$

$$=2mga$$

Thus the energy equation gives

$$\frac{1}{2}m[2\dot{x}^2+\frac{4}{3}a^2\dot{\theta}^2+2a\dot{x}\dot{\theta}\cos\theta]-2mga=-mg(a-a\cos\theta)$$

$$\text{or }\dot{x}^2+\frac{2}{3}a^2\dot{\theta}^2+a\dot{x}\dot{\theta}\cos\theta-2ga+ga\cos\theta)$$

$$\text{or }\dot{x}^2+\frac{2}{3}a^2\dot{\theta}^2+a\dot{x}\dot{\theta}\cos\theta=ga(1+\cos\theta)$$

Substituting the value of $\dot{x}$ from (1), we get

$$\left\{\frac{\sqrt{(3ag)}-a\cos\theta\dot{\theta}}{2}\right\}^2+\frac{2}{3}a^2\dot{\theta}^2+a\dot{\theta}\cos\theta\times$$

$$\times\left\{\frac{\sqrt{(3ag)}-a\cos\theta\dot{\theta}}{2}\right\}=ga(1+\cos\theta)$$

$$\text{or }\frac{1}{4}[3ag+a^2\cos^2\theta\dot{\theta}^2-1\sqrt{(3ag)}.a\cos\theta\dot{\theta}+\frac{2}{3}a^2\dot{\theta}^2$$

$$+\frac{1}{2}a\dot{\theta}\cos\theta\sqrt{(3ag)}-\frac{1}{2}a^2\cos^2\theta\dot{\theta}^2=ga(1+\cos\theta)$$

$$\text{or }\left(\frac{2}{3}a^2-\frac{1}{4}a^2\cos^2\theta\right)\dot{\theta}^2=\frac{ga}{4}+ga\cos\theta$$

$$\text{or }(8a^2-3a^2\cos^2\theta)\dot{\theta}^2=3ga(1+4\cos\theta)$$

$$\dot{\theta}=\sqrt{\left\{\frac{3g(1+4\cos\theta)}{a(8-3\cos^2\theta)}\right\}}$$

Example 11:

A uniform rod, of length 2a, hangs in a horizontal position being supported by vertical strings, each of length 1, attached to its ends, the other extremities *being attached to fixed points. The rod its given an angular velocity ω about a vertical axis through its centre; find its angular velocity when it has turned through an angle, and show that it will rise through a distance $(a^2\omega^2)/(6g)$. prove also that the time of a small oscillation about the position of equilibrium is* $2\pi\sqrt{(1/3g)}$.

Solution:

Let A'B' be the initial position of the rod, supported by two vertical strings DA', CB, such that $DA' = CB' = 1$. The rod is then turned in such a way that its middle point moves in a vertical through O'. Let AB be the position of the rod when it has turned through an angle θ. Let $O'O = x$ and φ is the angle which either string makes with the vertical.

Let the horizontal plane through AB cut CB', DA' in N and M respectively.

Obviously $\angle AMD = \angle BNC = \pi/2$. and $x = OO' = CB' - CN = 1 - l\cos\phi$ where t is the length of the vertical string

$$\therefore \dot{x} = l\sin\phi\dot{\phi}$$

From the right angle Δ CBN we have BN = l sin φ Also from isosceles triangle OBN, we have BN = 2a sin (θ/2)

$$\therefore l\sin\phi = 2a\sin(\theta/2)$$

so that $l\cos\phi\dot{\phi} = a\cos(\theta/2).\dot{\theta}$ *i.e.* $\dot{\phi} = \dfrac{a\cos(\theta/2}{l\cos\phi}\dot{\theta}$

Therefore form (2), we have $\dot{x} = l\sin\phi.\dfrac{a\cos(\theta/2)}{l\cos\phi}\dot{\theta}$

$$= 2a\sin\frac{\theta}{2}a\cos\frac{\theta}{2}\frac{1}{\sqrt{l^2 - 4a^2\sin^2\theta/2}}\dot{\theta}\sin\text{cel}\sin\phi = 2a\sin(\theta/2)$$

$$= \frac{la^2\sin\theta\dot{\theta}}{[l^2 - 4a^2\sin^2(\theta/2)]\cdot}$$ Now energy equation gives

$$\frac{1}{2}m\dot{x}^2 + \frac{1}{2}mK^2\dot{\theta}^2 - \frac{1}{2}k^2\omega^2 = -mgx \qquad ...(3)$$

$$\Rightarrow \frac{1}{2}m\left(\dot{x}^2 + \frac{a^2}{3}\dot{\theta}^2\right) - \frac{1}{2}m\frac{a^2}{3}\omega^2 = -mgx$$

$$\Rightarrow \frac{1}{2}m\left[\frac{l^2a^4\sin^2\theta\dot{\theta}^2}{l^2 - 4a^2\sin^2(\theta/2)]^2} + \frac{a^2\dot{\theta}^2}{3}\right] - \frac{1}{2}m\frac{a^2}{3}\omega^2 = -mgx$$

$$\Rightarrow a^2\dot{\theta}^2\left[\frac{la^2\sin^2\theta}{[l^2-4a^2\sin^2(\theta/2)]^2}\right]-\frac{1}{3}a\omega^2=2gx \qquad ...(4)$$

This equation gives the angular velocity of the rod in any position. The rod will come to instantaneous rest when $\dot{\theta}=0$.

Thus putting $\dot{\theta}=0.$ in (4), we get $2gx=\frac{1}{3}a^2\omega^2$ i.e. $x=\frac{a^2\omega^2}{6g}$

For finding the time of small oscillation, we consider the motion of AB in the vertical direction.

$$\therefore 2T\cos\varphi-mg=mx$$

$$=m\frac{d^2}{dt^2}(1-\text{loc}\phi))=m\frac{d^2}{dt^2}[l(1-\cos\phi)]$$

$$=m\frac{d^2}{dt^2}\left[1.2\sin^2\frac{\phi}{2}\right]=m\frac{d^2}{dt^2}\left[1.2.\frac{[\sin(\phi/2]^2}{(\phi/2)^2}\left(\frac{\phi}{2}\right)\right]$$

$$=m\frac{d^2}{dt^2}\left[\frac{1}{2\cdot}\left(\frac{a\theta}{1}\right)^2\right]=m\frac{d^2}{dt^2}\left[\frac{a^2\theta^2}{2l}\right]$$

$$\left[\sin\text{cel}\sin\phi=2a\sin\frac{\theta}{2}\therefore\frac{l\sin\phi}{\phi}\cdot\phi=2a\frac{\sin\theta/2}{(\theta/2)}\cdot(\theta/2\text{i.e.}l\phi=a\theta\right]$$

$$\therefore 2T\cos\phi-mg=\frac{ma^2}{2l}\frac{d}{dt}(2\theta\dot{\theta})=\frac{2ma^2}{2l}[\theta\ddot{\theta}+\dot{\theta}^2]=\frac{ma^2}{l}[\theta\ddot{\theta}+\dot{\theta}^2]\Big]$$

or $2T-mg\ 0$ [since $\dot{\theta}$ $\ddot{\theta}$ and are small so they could be neglected ϕ being small $\dot{\phi}=1$].

Now taking moments about the vertical axis through O, we have

$$m\frac{a^2}{3}\ddot{\theta}=-2T\sin\phi \text{ [perpendicular from O on BN]}$$

$$=-2T\sin\phi.a\cos\frac{\theta}{2}=-2a\cos\frac{\theta}{2\cdot}T\sin\phi$$

$$=-2T\cos\frac{\theta}{2\cdot}\frac{2a}{1}\sin\frac{\theta}{2}=-\frac{2a^2}{1}T\sin\theta$$

Now since $2T=mg$ i.e $T=\frac{1}{2}mg.$ putting this value above, we get

$$m\frac{a^2}{3}\ddot{\theta}=-\frac{2a}{1\cdot}\frac{2a}{1}\sin\frac{\theta}{2}=-\frac{2a^2}{1}T\sin\theta$$

or $\ddot{\theta} = -\frac{3g}{l}\sin\theta$, when θ is small.

$\therefore \ddot{\theta} = -\frac{3g}{l}\theta$, which is simple harmonic motion. Hence time of small oscillation $= 2\pi\sqrt{\left(\frac{1}{3g}\right)}$

Example 12:

If the earth, supposed to be a uniform sphere, had in a certain period contracted slightly so that its radius was less by (1/n) th than before, show that the length of the day would have shortend by (48/n) hours.

Solution:

Let a be the radius, m be the mass and ω the angular velocity of the earth. Supposed to be a uniform sphere, about its axis ofrotation which is its diameter.

$\therefore$ The length of a day $= l_1 = 2\pi/\omega = 24$ hours.

Now when the earth contracted slightly, let a_1 be its new radius then a_1 be its new radius then $a_1 = a - a/n = a(1 - 1/n)$. ...(1)

If ω_1 is the angular velocity of the earth in this position, then shortened length of a day $= l_2\ 2\pi/\omega_1$.

$$\therefore \frac{\text{Decrease in the length of a day}}{\text{Length of a day}} = \frac{l_1 - l_1}{l1} = 1 - \frac{l_2}{l_1} = 1 - \frac{\omega}{\omega_1}$$

i.e. Decrease in the length of a day $= \left(1 - \frac{\omega}{\omega_1}\right)$

Length of a day $= \left(1 - \frac{\omega}{\omega_1}\right)$. 24 hours. ...(2)

The moment of the external forces (*i.e.* weight o the earth and the reaction on the axis of rotation) about the axis of rotation (*i.e.* diameter) of the earth is zero. Therefore the moment of the momentum about the axis of rotation (*i.e.* diameter) of the earth will remain constant.

$$\therefore mk^2\omega = mk_1{}^2\omega_1$$

or $\frac{1}{5}ma^2\omega = \frac{2}{5}ma^2\ \omega_1$ or $\frac{\omega}{\omega_1} = \left(1 - \frac{1}{n}\right)^2$,

$$= 1 - \frac{2}{n'}\left(\text{Neglecting } \frac{1}{n^{2'}} \text{ etc. being small quantities}\right)$$

Putting in (2), we have

the decrease in the length of a day = $\left\{1-\left(1-\frac{2}{n}\right)\right\}24 = \frac{48}{n}$ hours.

i.e. the length of the day is shortened by (48/n) hours.

Example 13(a):

A sphere of radius b, rolls without slipping down the cycloid x = a (θ + sin θ); y = a (1 – cos θ). It starts from rest with its centre on the horizontal line y = 2a. Show that velocity v of its centre when at its lowest point is given by $v^2 = \frac{10}{7}g(2a-b)$.

Solution:

Initially, let the point A′ of the sphere be at the cusp A. Let C be the centre of this sphere, then the radius CA′ is a fixed line in the sphere.

Let at time t the line CA make an angle ϕ with the vertical and tangent at the point of contact P, make an angle ψ with the horizontal line OX.

Let the arc OP = s and arc OA = c.

∴ are AP = c – s.

But we are given that sphere rolls without slipping.

∴ arc AP = arc A′ P *i.e.* c – s = b (ϕ – ψ)

[since ∠A¢CP = ϕ – ψ]

Differentiating it, we get – s = b (ϕ – ψ) ...(1)

Let the centre C of the sphere move with the velocity v, then

v = velocity of P + velocity of C relative to P.

or $v = \dot{s} - b\dot{\psi} = -b(\dot{\phi}-\dot{\psi}) - b\dot{\psi} = -b\dot{\phi}$ ...(2)

Now energy equation gives $\frac{1}{2}m\left[v^2 + \frac{2b^2\dot{\phi}^2}{5}\right] = mg(2a-b)$

or $\left[v^2 + \frac{2b^2}{5}\cdot\frac{v^2}{b^2}\right] = 2g(2a-b)$ [from (2)]

or $\frac{7}{5}v^2 = 2g(2a-b)$ *i.e.* $v^2 = \frac{10g(2a-b)}{7}$

Example 13(b):

A sphere of radius a, and of radius of gyration k about an axis through its centre, rolls with linear velocity v on a horizontal plane, the direction of motion being perpendicular to a vertical face of a fixed rectangular block are perfectly rough and perfectly elastic. Show that the sphere will surmount the block if $(a^2 - ah + k^2)v^2 > 2gha^2(a^2 + k^2)$.

Solution:

Let O be the centre of the sphere and AB vertical face of the fixed block; AB = h. The sphere strikes the block at A, therefore the impulse is at the point A. Let ω and ω′ be the angular velocities of the sphere just after the impulse. After striking at A, the sphere begins to turn about A so the velocity of the centre O after impulse is at right angles to OA and is equal to aw. Since the impulse is at the point A, therefore the moment of the impulse about A is zero, hence the moment of momentum about A remains unaltered before and after the impulse.

Moment of momentum about A before impulse = $mk^2\omega + mv(a - h)$

[∴ moment of momentum = $mk^2\dot{\theta} + mvp$]

Moment of momentum about A after impulse = $mk^2\omega' + ma\omega'.a$

$\therefore mk^2 + mv(a - h) = mk^2\omega' + ma^2\omega'$ or $kk^2\omega + (a - h) = (k^2 + a^2)\omega'$

$$\text{or } k^2\frac{v}{a} + v(a - h) = (k^2 + a^2)\omega' \qquad [\text{since } \omega = v/a]$$

$$\text{or } (k^2 + a^2)\,\omega' = \frac{v}{a}\left(a^2 - ah + k^2\right) \qquad \text{or } \omega' = \frac{(a^2 - ah + k^2)}{(k^2 + a^2)}$$

Now the sphere will surmount the block if the sphere goes on turning about A till its lowest point L rises a height h, the height of the block. This can happen only if the sphere after impact acquires a kinetic energy, which is sufficient to take the sphere to a height h. It implies that the K.E. of the sphere must be more then that would be required for doing work of taking it upto a height h. But K.E. of the sphere after the impact

$$= \frac{1}{2}m\left(k^2\omega'^2 + a^2\omega'^2\right),$$

$$\therefore \frac{1}{2}m(k^2\omega'^2 + a^2\omega'^2) > mgh \text{ or } (k^2 + a^2)\omega'^2 > 2gh$$

$$\text{or } (k^2 + a^2)\frac{v^2(a^2 - ah + k^2)^2}{a^2(k^2 + a^2)^2} > 2gh \qquad [\text{from (1)}]$$

$$\text{or } (k^2 - ah + k^2)^2 v^2 > 2gha^2(k^2 + a^2)$$

Example 14:

A uniform plane of thickness 2h rests across the top of a fixed circular cylinder of radius a whose axis is horizontal. prove that if it be set in motion, the equation of energy is

$$\frac{1}{2}[k^2 + h^2 + a^2\theta^2]\dot{\theta}^2 + g[a\theta\sin\theta - (a+h)(1-\cos\theta)] = \text{const.}$$

on the assumption that the motion is of pure rolling. Hence show that if a>h, horizontal position is stable and period of small oscillation is the same as for a simple pendulum of length $\frac{k^2+h^2}{a-h}$.

Solution:

The figure is the vertical cross section of the system through the centre of gravity G of the plank. From G drawn GB perpendicular to the edge of the plank, so GB = h. Initially the point B of the plank coincided with the highest point A of the fixed cylinder. Let ω be the initial angular velocity. After time t, the edge of the plank make an angle θ with the horizontal. Since it is given that there is no slipping between the plank and the cylinder, therefore arc AP = BP *i.e.* BP = aθ. With O as origin OX and OY as axes, the co-ordinates of G are given by

$$x = MP - PL + BD = a \sin\theta - a\theta\cos\theta + h\sin\theta$$

$$y = OM + BL + GD = a\cos\theta + a\theta\sin + h\cos\theta$$

$$\therefore \dot{x} = a\cos\theta\dot{\theta} + h\cos\theta\dot{\theta} - a(\cos\theta\dot{\theta} - \theta\sin\theta\dot{\theta})$$

$$= (a\theta\sin\theta + h\cos\theta)\dot{\theta}$$

and $\dot{y} = -a\sin\theta\dot{\theta} - h\sin\theta\dot{\theta} + a(\theta\dot{\theta} - \sin\theta\dot{\theta})$

$$= (a\theta\cos\theta - h\sin\theta)\dot{\theta}$$

$\therefore$ velocity of G at time t = $\sqrt{\dot{x}^2 + \dot{y}^2}$

$$= \theta\sqrt{[(a\theta\sin\theta + h\cos\theta)^2 + (a\theta\cos\theta - h\sin\theta)^2]}$$

$$= \dot{\theta}\sqrt{(a^2\theta^2 + h^2)}$$

K.E. at time $t = \frac{1}{2}m[k^2 + \dot{\theta}^2 + h^2\dot{\theta}^2] = \frac{1}{2}m\dot{\theta}^2[k^2 + a^2\theta^2 + h^2]$

Initial K.E. = $\frac{1}{2}m\omega^2[k^2 + h^2]$

[since in the beginning, $\theta = 0$, and $\dot{\theta} = \omega$]

Hence energy equation gives

$$\frac{1}{2}m[k^2 + a^2\theta^2 + h^2]\dot{\theta}^2 - \frac{1}{2}m(k^2 + h^2)\omega^2$$

$$= mg\,(a+h) - (a\cos\theta + a\theta\sin\theta + h\cos\theta)]$$

or $(k^2 + a^2\theta^2 + h^2)\dot{\theta}^2 - [(k^2 + h^2)\omega^2 = 2g(a+h) - (a+h)\cos\theta - a\theta\sin\theta]$

or $(k^2 + a^2\theta^2 + h^2)\dot{\theta}^2 - 2g[(a+h)(1-\cos\theta) - a\theta\sin\theta]$

$$= (k^2 + h^2)\omega^2 = \text{const.}$$

or $\frac{1}{2}(k^2 + h^2 + a^2\theta^2)\dot{\theta}^2 + g\,[a\theta\sin\theta - [a+h)(1-\cos\theta)]\text{const.}$

Small oscillations: We take q to be small then from the above equation,

we have $\frac{1}{2}(k^2 + h^2)\dot{\theta}^2 + g\left[\theta^2 - (a+h).\frac{2\sin^2(\theta/2)}{(\theta/2)^2}\left(\frac{\theta}{2}\right)^2\right] = \text{const}$

$$\left[\therefore \theta\sin\theta = \theta.\frac{\sin\theta}{\theta}\theta = \theta^2 \text{ approx.}\right]$$

or $\frac{1}{2}(k^2 + h^2)\dot{\theta}^2 + ga\theta^2 - 2ga(a+h).\left(\frac{\sin\frac{1}{2}\theta}{\frac{1}{2}\theta}\right)^2 \frac{\theta^2}{4} = \text{const.} = 0$

or $\frac{1}{2}(k^2 + h^2)\dot{\theta}^2 = g(a+h)\theta^2 - 2ga\theta^2 = -\,g(a-h)\theta^2$

or $\dot{\theta}^2 = -\frac{g(a-h)}{(k^2+h^2)}2\theta\dot{\theta}$ Differentiating it, we get

$$2\dot{\theta}\ddot{\theta}^2 = -\frac{g(a-h)}{(k^2+h^2)}2\theta\dot{\theta}$$

i.e. $\ddot{\theta} = -\frac{g(a-h)}{(k^2+h^2)}\theta$

which is a simple harmonic motion implying that the horizontal position of equilibrium is stable.

The length of the simple equivalent pendulum $= \frac{k^2 + h^2}{a-h}$.

Example 15:

A heavy circular disc is revolving in a horizontal plane about the centre which is fixed. An insect of mass $\frac{1}{n}$ *th that of the disc walks from the centre along a radius and then flies away. Show that the final angular velocity is* $\frac{n}{n+2}$ *times the original angular velocity of the disc.*

Solution:

Let ω be the initial angular velocity of the disc. The insect flies away when it reaches the edge of the disc, let at this time ω′ be the angular velocity of the disc.

The forces acting on the system are either vertical or pass through O; thus their moments vanish about the vertical through O; so moment of, momentum about the vertical through O remains constant the motion

$$\therefore m0.\omega' + m\frac{a^2}{2}\omega' + \frac{m}{n}a^2\omega'$$

$$= m.0.\omega + m\frac{a^2}{2}\omega$$

Example 16(a):

A circular ring of mass M and radius a, lies on a smooth horizontal plane and an insect of mass m, resting on it starts and walks round it with uniform velocity v relative to the ring. Show that the centre of the ring describes a circle with angular velocity $\frac{m}{M+2m}\cdot\frac{v}{a}\cdot$

Solution:

Let O be the centre of the ring and G the common centre of gravity of the ring and the insect resting at A.

If OG = x and GA = y, we have Mx = my

$$\Rightarrow \frac{x}{m} = \frac{y}{m} = \frac{x+y}{m+M} = \frac{a}{m+M}$$

$$\Rightarrow x = \frac{ma}{m+M'}\ y = \frac{Ma}{m+M}$$

where m is the mass of the insect and M the mass of ring.

So there is no horizontal force on the system throughout the motion, the horizontal momentum remains constant throughout the motion, which implies that the velocity of G (common C. G. of ring and the insect) in the horizontal plane will be zero throughout the motion because of its initial velocity being zero. In other words G remains fixed.

Hence the point O will describe a circle of radius $\frac{ma}{m+M}$ about G.

Now let ω be the angular velocity of the ring about G. Since all the forces [weight of the ring, weight of the insect] acting on the system are vertical, therefore their moments vanish about the vertical through G, hence moment of momentum about the vertical through G remains constant throughout the motion.

i.e. $(Mk^2\omega + M x \omega. GO) + (m \omega G A. G A + m v. G A) = 0$

for ring $\leftarrow$ for in sec t $\rightarrow$

$\Rightarrow Ma^2\omega + M\omega x^2 + (m\omega y^2 + mv\, y) = 0$

i.e. $\{Mk^2\omega_1 + Mx\omega_1.GO\} + m\omega_2 GA.GA = 0 + mu.GA$

for tubefor particle [tube] [practical]

$\therefore M(a^2 - x^2)\omega + my^2\omega + mv(a - x) = 0 (\because y = a - x)$

$\Rightarrow \omega[Ma^2 + Mx^2 + My^2] = - mv(a - x)$

$$\Rightarrow \omega[Ma^2 + \frac{Mm^2a^2}{(m+M^2} + \frac{mM^2a^2}{(m+M^2}] = -mv\left[a - \frac{ma}{m+M}\right]$$

$$\Rightarrow \frac{Ma^2}{(m+M)^2}[(m+M)^2 + m^2 + mM]\omega = -\frac{amMv}{(m+M)}$$

$$\Rightarrow \frac{Ma^2}{(m+M)^2}[(m+M)^2 + m(m+M)]\omega = -\frac{amMv}{(m+M)}$$

$$\Rightarrow \frac{a(m+M)(2m+M)\omega}{m+M} = -mv \Rightarrow \omega = -\frac{m}{M+2m}.(v/a)$$

Negative sign shows that disc will turn in a direction which is opposite of the direction of v

Example 16(b):

A particle of mass m, moves within a rough circular tube of mass M lying on a horizontal plane and initially the tube is at rest while the particle has an angular velocity round the tube. Show that by the time the relative motion ceases, the fraction $\frac{M}{M+2m}$ *of the initial kinetic energy has been dissipated by friction.*

Solution:

Let O be the centre of the circular tube; the common C.G. of m and M where m = mass of the particle and M = mass of the tube. If OG = x and GA = y, then M x = my.

$$\Rightarrow \frac{x}{m} = \frac{y}{m} = \frac{x+y}{m+M} = \frac{a}{M+m}$$

$\Rightarrow x = ma/(M + m),\ y = Ma/M + m).$

Let u be the initial of the particle when the tube is at rest and v the velocity with which G, the common C.G. begins to move.

Now there being no force acting on the system in the horizontal plane throughout the motion, the horizontal momentum remains unchanged throughout the motion.

$$\therefore (M + m)v = mu \Rightarrow v = \frac{mu}{M + m}$$

When the system begins to move, let ω_1 and ω_2 be the angular velocities of the tube and the particle about G. Since the forces acting on the system are all vertical, therefore their moments vanish about the vertical axis through G throughout the motion, which implies that the moment of momentum about G remains the same throughout the motion.

i.e. $muy = M\,(a^2\omega_1 + x^2\omega_1) + my^2\,\omega_2$

The relative motion will cease when $\omega_1 = \omega_2$, then from (1), we have.

$$\omega_1 = \frac{muy}{M(a^2 + x^2) + my^2} \qquad \text{...(2)}$$

At this time, the kinetic energy of the system is

$$= \frac{1}{2}(M + m)v^2 + \frac{1}{2}M\,(a^2 + x^2)\omega_1^2 + \frac{1}{2}my^2)\omega_1^2$$

[particle] [Tube] [K.E. of G.]

$$= \frac{1}{2}(M + m)\frac{m^2u^2}{(M + m)^2} + \frac{1}{2}(my^2 + Ma^2) \times \frac{m^2u^2y^2}{(my^2 + Ma^2 + Mx^2)^2}$$

$$= \frac{1}{2}m^2\,u^2\left[\frac{1}{M + m} + \frac{y^2}{(my^2 + Ma^2 + Mx^2)}\right]$$

$$= \frac{1}{2}mu^2 - \frac{1}{2}mu^2\frac{M}{M + 2m}$$

[by putting the values of x and y]

$$\therefore \text{Loss of K.E.} = E\cdot = \frac{1}{2}mu^2 - \left[\frac{1}{2}mu^2 - \frac{1}{2}mu^2\frac{M}{M + 2m}\right]$$

$$= \frac{1}{2}mu^2 \times \frac{M}{M + 2m} = \frac{M}{M + 2m} \text{ of the initial K.E.}$$

Example 16(c):

A small insect moves along a uniform bar; of mass equal to itself and of length 2a the ends of which are constrained to remain on the circumference of a fixed circle whose radius is $(2a/\sqrt{3})$. *If the insect starts from the middle point of the bar and moves along the bar with relative velocity V, show that the bar in time t will turn through an angle.*

$$\frac{1}{\sqrt{3}}\tan^{-1}\left(\frac{Vt}{a}\right)$$

Solution:

Let θ be the angle through which the bar AB turn in time t, the dotted line through A represents the original position. Let P be the position of insect time t.

$$\therefore CP = Vt \text{ and} OC = \left[\left(\frac{2a}{\sqrt{3}}\right)^2 - a^2\right]^{1/2} = \frac{a}{\sqrt{3}}$$

$$\therefore OP = \sqrt{[OC^2 + CP^2)]} = \sqrt{\left[(\frac{1}{3}a^2 + V^2t^2)\right]}$$

Since the moment of external forces [weights, reaction about vertical axis through O vanishes throughout the motion so the moment of momentum about this line remains unchanged.

$$i.e.\ 0 = (OC\dot{\theta}.OC + \frac{1}{3}a^2\dot{\theta}) + m(VOC + OP\dot{\theta}.OP)$$

$$\Rightarrow \frac{1}{3}a^2\dot{\theta} + \left(\frac{1}{3}a^2 + V^2t^2\right)\dot{\theta} + \frac{a}{\sqrt{3}}V = O$$

$$\Rightarrow a^2\dot{\theta} + V^2t^2\ \dot{\theta} + \frac{a}{\sqrt{3}}V = O$$

$$\Rightarrow \dot{\theta} + (a^2 + V^2t^2) + \frac{a}{\sqrt{3}}V = 0 \Rightarrow \theta = -\frac{Va}{\sqrt{3}\left(a^2 + V^2t^2\right)}$$

$$\therefore \theta = \frac{Va}{\sqrt{2}}\int_0^1 \frac{dt}{V^2t^2 + a^2} = -\frac{Va}{\sqrt{3}}\frac{1}{V^2}\int_0^1 \frac{dt}{t^2 + \left(a^2/V^2\right)}$$

$$= -\frac{Va}{\sqrt{3}\cdot}\frac{1}{V^2}\frac{V}{a}\left[\tan^{-1}\frac{Vt}{a}\right]_0^1 = -\frac{1}{\sqrt{3}}\tan^{-1}\left(\frac{Vt}{a}\right)$$

Negative sign shows that the rod turns in a direction opposite to that of the insect.

Example 17:

A uniform circular wire of radius a, lies on a smooth horizontal table and is movable about a fixed point O on its circumference. An insect of mass equal to that of the wire, starts from the other end of the diameter through O and crawls along the wire with a uniform velocity v relative to the wire. Show that at the end of time t, the wire has turned through an angle

$$\left(\frac{vt}{2a}\right)-\left(\frac{1}{\sqrt{3}}\right)\tan^{-1}\left[\left(\frac{1}{\sqrt{3}}\right)\tan\left(\frac{vt}{2a}\right)\right]$$

Solution:

Let in time t the diameter O A have turned through an angle ϕ from its initial position OA_0, let the insect be at P such that

$\angle ACP = \theta \therefore OP = 2a\cos(\theta/2)$.

Let O N be perpendicular on the tangent at P

$\therefore ON = OP\cos(\theta/2) = 2a\cos^2(\theta/2)$

The forces acting on the system are (i) Reaction at the point O (ii) Weight of the insect vertically downwards. (iii) Weight of the wire acting vertically down wards. Since the forces acting on the system are either vertical or pass through O, therefore moment of all forces about the vertical axis through O vanishes throughout the motion. Thus the moment of momentum of the system about this axis remains unchanged throughout the motion. The insect at P has a velocity v along the tangent at P and a velocity OP perpendicular to OP ϕ due to rotation of the circular wire.

So the moment of momentum of the insect about a vertical through

$$O = m(v.ON + OP\dot{\phi}.OP = m\left[2av\cos^2\frac{\theta}{2}.+\left(2a\cos\frac{\theta}{2}\right)^2\dot{\phi}\right]$$

$$= m\left[2av\cos^2\frac{\theta}{2}+4a^2\cos^2\frac{\theta}{2}.\dot{\phi}\right]$$

Also the moment of momentum of the circular wire about the vertical axis though O.

= moment of momentum of the mass m moving with the C.G + moment of momentum about the C.G.

$= m[a\dot{\phi}.a + a^2\dot{\phi}]$

$= 2ma^2\dot{\phi}$ [C.G. describes a circle of radius a with velocity a $\dot{\phi}$]

Hence the total moment of momentum at time t

$$= m\left[2av\cos^2\frac{\theta}{2} + 4a^2\cos^2\frac{\theta}{2}.\dot{\phi}\right] + 2ma^2\dot{\phi}$$

But the moment of momentum of the system remains unchanged and initial the body was at rest, therefore we have

$$m\left[2av\cos^2\frac{\theta}{2} + 4a^2\cos^2\frac{\theta}{2}.\dot{\phi}\right] + 2ma^2\dot{\phi}$$

$$\Rightarrow 2a^2\left(1 + 2\cos^2\frac{\theta}{2}\right)\dot{\phi} = -2av\cos^2\frac{\theta}{2}$$

$$\Rightarrow 2a^2\left(1 + 2\cos^2\frac{\theta}{2}\right)\dot{\phi} = -2av\cos^2\frac{\theta}{2}$$

$$\Rightarrow \dot{\phi} = -\frac{v\cos^2(\theta/2)}{a\{1 + 2\cos^2(\theta/2)\}} \qquad ...(1)$$

Now the arc AP is covered by the insect in time moving with uniform velocity v.

$$\therefore \text{ arc AP} = vt, \text{ also arc AP} = a\theta \;\therefore\; a\theta = vt \Rightarrow \theta = \left(\frac{vt}{a}\right)$$

putting this value of θ in (1), we get

$$= -\left(\frac{\pi}{4}\right) + \left(\frac{72}{3\sqrt{7}}\right)\tan^{-1}\left(\frac{\sqrt{7}}{3}\right) = \left(\frac{24}{\sqrt{7}}\right)\tan^{-1}\left(\frac{\sqrt{7}}{3}\right) - (\pi/4).$$

Hence the angle turned by the diameter OA in time t is given by

$$\theta = \int_0^t \frac{-v\cos^2(vt/2a)}{a[1 + 2\cos^2(vt/2a)}dt = -\frac{v}{a}\int_0^t \frac{\cos^2(vt/2a)}{1 + 2\cos^2(vt/2a)}dt$$

$$= -\frac{v}{a}\int_0^t \frac{1}{2}\left[1 - \frac{1}{1 + 2\cos^2(vt/2a)}\right]dt$$

$$= -\frac{v}{2a}\int_0^t \left[1 - \frac{\sec^2(vt/2a)}{\sec^2(vt/2a) + 2}\right]dt$$

$$= -\frac{v}{2a}\int_0^t \left[1 - \frac{\sec^2(vt/2a)}{3 + \tan^2(vt/2a)}\right]dt$$

$$= -\frac{v}{2a}\int_0^t \left[t - \frac{2a}{v}\frac{1}{\sqrt{3}}\tan^{-1}\left(\frac{1}{\sqrt{3}}\tan\frac{vt}{2a}\right)\right]$$

$$=-\left[\frac{vt}{2a}-\frac{1}{\sqrt{3}}\tan^{-1}\left(\frac{1}{\sqrt{3}}\tan\frac{vt}{2a}\right)\right].$$

Negative sign shows that wire moves in a direction opposite to that of the insect.

Example 18:

A circular disc is moving with an angular velocity ω about an axis through its centre perpendicular to its plane. An insect alights on its edge and crawls along a curve drawn on the disc in the form of lemniscale with uniform relative angular velocity $\frac{1}{8}\omega$, the curve touching the edge of the disc. The mass of the in sect being $\frac{1}{16}$th of that of the disc. Show that the angle turned through by the disc when the insect gets to the centre is

$$\left(\frac{24}{\sqrt{7}}\right)\tan^{-}\left[\frac{\sqrt{7}}{3}\right]-\left(\frac{\pi}{4}\right)$$

Solution:

Let the disc turn through an angle ϕin time t and the distance of insect from 0, the centre of the circle, be than r (OP = r) at an angle θ with the axis of the lemniscate, then equation of leminiscate is

$$r^2 = a^2 \cos 2\theta$$

Moment of momentum before insect alights of the disc $= M\frac{1}{2}a^2\omega$,

and moment of momentum of the system about O after time t

$$=\frac{1}{2}Ma^2\dot{\phi}+\frac{1}{16}M\left(r^2\dot{\phi}+r^2.\frac{1}{8}\omega\right)$$

Since all forces acting on the system are vertical or pass through the centre O, therefore the moment of all the forces about a vertical axis through O. vanishes, thus the moment of momentum about this axis through O remains unchanged throughout the motion.

$$\therefore M.\frac{1}{2}a^2\dot{\phi}+\frac{1}{16}M\left(r^2\dot{\phi}+\frac{1}{8}r^2\omega\right)=\frac{1}{2}Ma^2\omega$$

$$\text{or } Ma^2\dot{\phi}-\frac{1}{8}M\left(\dot{\phi}+\frac{1}{8}\omega\right)a^2\cos 2\theta = Ma^2\omega\left[\because r^2 = a^2\cos 2\theta\right]$$

or $64\phi + \left(8\dot{\phi} + \omega\right)\cos 2\theta = 64\omega$

or $(64 + 8\cos 2\theta)\dot{\phi} = 64\omega - \omega\cos 2\theta$

or $\dot{\phi} = \frac{d\theta}{dt} = \frac{(64-\cos 2\theta)\omega}{64+8\cos 2\theta.}$ or $\frac{d\phi}{d\theta}\cdot\frac{d\theta}{dt} = \frac{(64-\cos 2\theta)\omega}{64+8\cos 2\theta}$

or $\frac{d\phi}{d\theta}\cdot\frac{1}{8}\omega = \frac{(64-\cos 2\theta)\omega}{8(8+\cos 2\theta)}$ $\left[\because \frac{d\theta}{dt} = \frac{1}{8}\omega\right]$

or $d\phi = \frac{64-\cos 2\theta}{8+\cos 2\theta}d\theta.$

If ϕ be the angle turned by the disc when the insect reaches the centre of the circle (*i.e.* when $\theta = \pi/4$), then

$$\phi = \int_0^{\pi/4} \frac{64-\cos 2\theta}{8+\cos 2\theta}d\theta = \int_0^{\pi/4}\left[-1 + \frac{72}{8+\cos 2\theta}\right]d\theta$$

$$= \int_0^{\pi/4}\left[-1 + \frac{72}{8+\left(2\cos^2\theta - 1\right)}\right]d\theta$$

$$= \int_0^{\pi/4}\left[-1 + \frac{72}{(7+2\cos^2\theta}\right]d\theta = \int_0^{\pi/4}\left[-1 + \frac{72\sec^2\theta}{7\sec^2\theta + 2}\right]d\theta$$

$$= \int_0^{\pi/4}\left[-1 + \frac{72\sec^2\theta}{(9+7\tan^2\theta}\right]d\theta = \int_0^{\pi/4}\left[-1 + \frac{72}{7}\frac{\sec^2\theta}{\left(\frac{9}{7}\right)+\tan^2\theta}\right]d\theta$$

$$= \left[-\theta + \frac{72}{3\sqrt{7}}\tan^{-1}\left(\frac{\sqrt{7}}{3}\tan\theta\right)\right]_0^{\pi/4}$$

$$= -\left(\frac{\pi}{4}\right) + \left(\frac{72}{3\sqrt{7}}\right)\tan^{-1}\left(\frac{\sqrt{7}}{3}\right) = \left(\frac{24}{\sqrt{7}}\right)\tan^{-1}\left(\frac{\sqrt{7}}{3}\right) - (\pi/4).$$

Example 19:

A rod OA can turn freely in a horizontal plane about the end O and lies at rest. An insect, whose mass is one third that of the rod alights on the end A and commences crawling along the rid with uniform velocity V, at the same instant the rod is set in rotation about O in such a way that the initial velocity of A is V, when the insect reaches O, prove that the rod has rotated through a right angle and that the angular velocity of the rod is then twice the initial angular velocity.

Solution:

Let OA be the initial position of the rod, when the insect is at A. Let the rod turn through an angle θ time t, also let the insect reach the point p in this time. since the insect moves along the rod will uniform velocity V so the distance moved in time t.

AP = Vt, and OP = 2a –t where 2a is length of the rod.

Since the rod moves about 0 with initial velocity V so its angular velocity in its initial position

$$=\frac{V\cdot 2a}{(a)^2}=\frac{V}{2a} \qquad \left[\therefore \text{angular velocity}=\frac{pV}{r^2}\right]$$

Moment of momentum of the system about 0, when the insect is at A

$$=\frac{1}{3}M.V.2a+M\left(a\frac{V}{2a}a+\frac{1}{3}a^2\frac{V}{2a}\right)=\frac{4}{3}MVa$$

also moment of momentum after time t

$$=\frac{1}{3}M(2a-vt)\theta(2a-Vt)+M\left(a\theta a+\frac{1}{3}a^2\dot{\theta}\right)$$

[formula $mvp + mk^2\dot{\theta}$]

[since now the angular velocity of rod is θ and that of insect due to rotation of the rod is $(2a - Vt)\theta$]

$$=\frac{M}{3}(2a-Vt)^2+\frac{4}{3}Ma^2 \qquad \text{...(ii)}$$

The moment of all the forces about the vertical through 0 vanishes , so the moment of momentum about 0 remains unchanged throughout the motion. therefore we have from (i) and (ii)

$$\frac{M}{3}(2a-Vt)^2\theta+\frac{4}{3}Ma^2\theta=\frac{4}{3}MVa$$

or

$$[(2a-Vt)^2+4a^2]\dot{\theta}=4Va$$

or

$$\frac{d\theta}{dt}=\frac{4Va}{(2a-Vt)^2+4a^2}$$

Now the insect will reach the point 0 (moving with uniform velocity V) in time (2a/V). If θ be the angle described by the rod in this time, then

$$\int_0^{2a/v} \frac{4Va}{\left(2a - Vt^2\right) + 4a^2} dt \text{ Put } 2a\text{–}Vt = z, \therefore -Vdt = dz$$

$$i.e.\ \theta = \int_{2a}^{0} \frac{4Va}{z^2 + 2a^2} \cdot \frac{-dz}{V} = \int_{2a}^{0} \frac{4a}{z^2 + (2a^2)} dz$$

$$= 4a \int_0^{2a} \frac{dz}{z^2 + (2a)^2} = 4a \cdot \frac{1}{2a} \left[\tan^{-1} \frac{z}{2a} \right]_0^{2a} = \left(\frac{\pi}{4} - 0 \right) = (\pi/2).$$

Hence the rod has turned through a right angle in this time the angular velocity of the rod $\dot{\theta}$ when the insect reaches 0 is obtained by putting $t = \left(\frac{2a}{V}\right)$ in (iii), i.e., $\dot{\theta} = \left(\frac{V}{a}\right) = 2\left(\frac{V}{2a}\right)$ = twice the initial angular velocity

Example 20:

A circular plate rotates about an axis through its centre perpendicular to its plane with angular velocity ω. This axis is set free and a point, in the circumference of the plate is fixed, show that the resulting angular velocity is $\frac{1}{3}\omega$.

Solution:

In this case centre C is set free and a point A on the circumference is suddenly fixed. Let ω′ be the angular velocity just after the fixture. The impulse is at A, so the moment of the impulse about A will be zero. Hence moment of momentum about the axis through A remains the same before and after the fixture. *i.e.*

$$m\left[a^2\omega' + \frac{a^2}{2}\omega'\right] = \frac{ma^2}{2}\omega \text{ or } \omega' = \frac{1}{3}\omega.$$

Example 21:

A circular plate is turning in its own plane about a point A on its circumference. Suddenly A is freed and a point B on the circumference is fixed. Show that the plate will be reduced to rest if the arc AB is one third of the circumference.

Solution:

Consider a circular plate with centre C turning in its own plane about a point A on its circumference. A is set free and B is suddenly fixed. Let w¢ be angular velocity just after the fixture. Let . Since the impulse is at B, so the moment of the impulse about B is zero, hence moment and momentum about B will remain the same before and after fixture

$$i.e.\quad m\left[\frac{a^2}{2}\omega + a^2\omega'\right] = \left[\frac{a^2}{2}\omega + a\omega a\cos\theta\right]$$

$$\text{or}\quad \omega' = \frac{1}{3}(1 + 2\cos\theta)\omega.$$

The plate will reduced to rest if $\omega' = 0$

$$i.e.\ \text{if}\ 1 + 2\cos\theta = 0\ \text{or}\ \cos\theta = -\frac{1}{2}\ \text{i.e.}\ \theta = \frac{2\pi}{3}$$

or $a\,\theta = \frac{1}{3}(2\pi a)$ *i.e.* are AB is $\frac{1}{3}$ of the circumference.

Example 22(a):

A uniform circular disc is spinning with angular velocity ω about a diameter, when a point P on its rim is suddenly fixed. If the radius vector to P makes an angle α with this diameter, show that the angular velocities after the fixing, about the tangent and normal at P are $\frac{1}{5}\omega \sin\alpha$ and $\omega\cos\alpha$.

Solution:

Consider a circular plate with centre C, rotating with angular velocity ω about the diameter ACB. The resolved part of ω about the normal at P *i.e.* about PC = $\omega\cos\alpha$.

The resolved part of ω about the tangent at P = $\omega\sin\alpha$.

Now the diameter AB is set free and the point P is suddenly fixed. Let ω_1 and ω_2 be the angular velocities about the tangent and normal at P just after the fixture.

The impulse is at P, therefore moment of the impulse about P is zero.

Hence the moment of momentum about the tangent at P remains unchanged after the fixture.

i.e. $m\left(\frac{a^2}{4}\right)\omega_1 + ma^2\omega_1 = m\left(\frac{a^2}{4}\right)\omega\ \sin a$. *i.e.* $\omega_1 = \left(\frac{1}{5}\right)\omega\sin\alpha$. Similarly the moment of momentum about the normal must be the same before and after the fixture.

$$m\left(\frac{a^2}{4}\right)\omega_2 = m\left(\frac{a^2}{4}\right)\omega\cos a\ \ i.e.\ \ \omega_2 = \omega\cos\alpha.$$

Example 22(b):

An equilateral triangular lamina is rotating in its plane with uniform angular velocity about an axis through one vertex. If this vertex is released and one of the other vertices fixed. Show that the new angular velocity is 1/5 of its former value.

Solution:

Let the vertex C be released and the vertex B be suddenly fixed. Let ω and ω' be the angular velocities just before and after the fixture. If m is the mass of the triangular lamina, then it is equivalent to three particles each of mass m/3 placed at the mid - points of the sides.

The moment of momentum about axis through B remains unchanged *i.e.*

$$\left(\frac{m}{3}\right)a^2\omega + 0 + 0 = \left(\frac{m}{3}\right)a^2\omega' + \frac{m}{3}a^2\omega' + \left(\frac{m}{3}\right)a^2\omega'$$

i.e. $5\omega' = \omega$ i.e $\omega' = \frac{1}{5}\omega.$

Example 22(c):

A rod of length 2a is moving about one end with uniform angular velocity upon a smooth horizontal plane. Suddenly this end is set free and a point distant b from this end is fixed; find the motion, considering the case when $b > = < \frac{4a}{3}$.

Solution:

Let AB be the rod, moving with uniform angular velocity ω about the end A. The end A is set free and a point O of the rod is suddenly fixed, where AO = b.

Let the angular velocity, of the rod after the fixture be ω'. The impulse is at O, thus moment of impulse about O will be zero, hence the moment of momentum about the axis through O will remain unchanged before and after the fixture *i.e.*

i.e. $m\frac{a^2}{3}\omega' + m(a-b)^2\omega' = m\frac{a^2}{3}\omega + ma(a-b)\omega$

or $\left[\frac{a^2}{3} + (a-b)^2\right]\omega' = \left[\frac{a}{3} + (a-b)\right]\omega$

or $[a^2 + 3(a-b)^2\,\omega' = a(4a-3b)\omega$ *i.e.* $\omega' = \frac{a(4a-3b)}{a^2 + 3(a-b)^2}\omega.$

(i) If $b < \frac{4a}{3}$ then $\omega' = \frac{3\left(\frac{4}{3}a - b\right)}{a^2 + 3(a-b)^2}$ is positive which implies that the rod will rotate in the same direction after the fixture as before.

(ii) If $b = \frac{4a}{3}$, then $\omega' = 0$, the rod is reduced the rest after the fixture.

(iii) If $b < \frac{4a}{3}$, then ω' is negative, which implies that the rod will move in opposite sense after the fixture.

Example 23:

A lamina in the form of an ellipse is rotating in its own plane about one of its foci with angular velocity ω. This focus is set free and other focus, at the same time is fixed, show that the ellipse now rotates about it with angular velocity $\omega\frac{2-5e^2}{2+3e^2}$

Solution:

Consider an ellipse, rotating with angular velocity ω about the focus S. If O is the midpoint of the major axis then OS = OS′ = ae, where length of the major axis AA′ = 2a. The focus S is set free and suddenly the other focus S ′ fixed.

Let the angular velocity after the fixture be ω¢ the impulse is at S ′, therefore the moment of the impulse about S′ is zero. Hence moment of momentum about this axis remains unaltered before and after the fixture

$$i.e.\ m\frac{a^2+b^2}{4}\omega' + ma^2e^2\omega' = m\frac{a^2+b^2}{4}\omega - ma^2e^2\omega$$

$$\text{or}\quad \left(\frac{a^2+b^2}{4} + a^2e^2\right)\omega' = \left(\frac{a^2+b^2}{4} - a^2e^2\right)\omega$$

$$\text{or}\quad \left(a^2+b^2+4a^2e^2\right)\omega' = \left(a^2+b^2-4a^2e^2\right)\omega$$

$$\text{or}\quad \omega' = \frac{a^2+b^2-4a^2e^2}{a^2+b^2+4a^2e^2}\cdot\omega$$

$$\text{or}\quad \omega' = \frac{a^2+a^2(1-e^2)-4a^2e^2}{a^2+a^2(1-e^2)+4a^2e^2}\omega \qquad \{\because b^2 = a^2(1-e^2)\} = \frac{2-5e^2}{2+3e^2}\omega.$$

EXERCISES

1. An elliptic lamina is rotating about a diameter and a point p on the edge is suddenly fixed. Prove that the velocity of the centre immediately afterwards is 1/5 of the velocity of p before it was fixed.

2. A uniform circular disc. of radius a is rolling without slipping along a smooth horizontal plane with velocity V when the highest point becomes suddenly fixed. Prove that the disc will make a complete revolution round the point if $V^2 > 24ag$.

3. Three uniform rods AB, BC, CD freely jointed at B and C lie on a smooth horizontal plane. They are originally moving as a rigid body with velocity V parallel to CB, A and B being on the same side of BC and the angles B and C being right angles, A is suddenly fixed, prove that if 2a, 2b, 2c are the lengths of rods and 3m, m, 2m their respective masses, the initial angular velocities $\omega_1 \omega_3$ of AB, CD are given by $5a\omega_1 = 20\,w = 3V$.

4. A bead of mass M can slide on a smooth straight horizontal wire and a particle of mass m is attached to the body by a light string of length l, the particle is held in contact with the wire with the string taught and is then let fall. Prove that when the string is inclined to the wire at an angle θ, the bead will have slipped a distance along the wire $\dfrac{mgl(1-\cos\theta)}{M+m}$ and

   that the angular velocity ω of the string will be given by the equation

   $(M + m\cos^2\theta)\, l\omega^2 = \omega\,(M + m)\, g \sin\theta$

5. A uniform rod moves in a vertical plane, its ends being in contact with the interior of fixed smooth sphere; when it is inclined at an angle θ to the horizontal, show that the square of its angular velocity is $\dfrac{6cg}{3c^2+a^2}[\cos\theta - \cos\alpha]$, where a is the initial value of θ, 2a is the length of the rod, and c is the distance of its middle point from the centre of the sphere.

6. A horizontal lamina of mass M can rotate freely about an axis through O, some point of it and perpendicular to it. An insect of mass m alights on A, some point of it at distance a from O and crawls along a circle having OA as diameter and returns to the point of starting Initially if the insect and the lamina were at rest, Prove that the lamina has moved

through an angle relative to the ground given by $\pi\sqrt{\{Mk^2/(Mk^2.ma^2)\}]}$ where k is radius of gyration of the lamina about its axis.

7. A uniform square plate A B C D of mass M and side 2a, lies on a smooth horizontal plane; it is struck at A by particle of mass m moving with velocity V in the direction of AB, the particle remaining attached to plate. Determine the subsequent motion of the system and show that its angular velocity is $\frac{m}{M+4m}\frac{3V}{2a}\cdot$

8. The ends of a straight uniform rod of mass m and length 2a slide on a smooth fixed horizontal circle of radius and an insect of mass m crawls along it at a constant rate V relative to the rod. If θ is the angle made by the rod at time t with a fixed direction, show that

$\dot{\theta}\left(2c^2-\frac{5}{3}a^2+V^2t^2\right)=\text{const}$ and if $c=\frac{2a}{\sqrt{3'}}$ the angle turned through by the rod in time t is $\frac{1}{\sqrt{3}}\tan^{-1}\left(\frac{Vt}{a}\right)$.

4

Eulerian Angles and Equations

INTRODUCTION

Ir order to describe the rotation of a rigid body about a point (say O) we use three angular co-ordinates θ, ϕ ψ which are called *Euler's angles* described below. Let a rigid body turn about the point O. Let OA, OB, OC (axes fixed in the body) be the principal axes of the body at the point O which move along with the body. Initially; OA, OB, OC coincide with Ox, Oy, Oz. Construct a spherical surface of radius unity with centre at O. Let the axes OX, Oy, Oz (fixed in space) meet the spherical surface in points (x, y, z). After a time t, when the rigid body is rotating let the axes OA, OB, OC (fixed in the body) meet the spherical surface in points A, B, C. Let $\angle COz = \theta$, angle between the plane zOC and zOx = ϕ and angle between the planes zPC and COA = ψ then these angles are known as Eulerian angles. To obtain the values of ϕ and ψ more precisely we proceed as follows:

Draw a line ON perpendicular to the plane zOC and obviously the normals to the planes COA and zOx are the lines OB and Oy respectively. But angles between two planes = angle between their normals; so we have

ϕ = angle between the planes zOC and zOx = angle between the normals ON OY and ψ = angle between the planes zOC and COA = angle between the normals ON and OB.

EULER'S GEOMETRICAL EQUATIONS

Let ω_1, ω_2, ω_3 be the component angular velocities about the principal axes OA, OB, OC. Obviously $\angle zOC = \theta$ and ON is normal to this plane so the angular velocity θ is about this line. Further the line ON being perpendicular to both Oz and OC lies in the xOy plane and as well as in AOB plane. Angle ϕ is in xOy plane, therefore angular velocity ϕ is about Oz since oz is $\perp$ to this plane. Similarly the angle ψ is in the plane AOB, therefore angular velocity $\dot{\psi}$ is about

OC which is normal to this plane. Now, we proceed to determine the d.cs of the lines ON, OC, OZ w.r.t. the principal axes OA, OB, OC.

Dcs of ON

The line ON ($\perp$ to OC) lies in the plane A OB, therefore it makes an angle $\frac{\pi}{2} - \psi$ with the line OA and ψ with the line OB. The angle between ON and $OC = \frac{\pi}{2}$. Hence the d.c.s. of the line ON are

$$\cos\left(\frac{\pi}{2} - \psi\right), \cos\psi, \cos\frac{\pi}{2} \text{ i.e. } \sin\psi, \cos\psi, 0$$

Dcs of OC

Angle between OC and $OA = \frac{\pi}{2}$, angle between OC and $OB = \frac{\pi}{2}$

Hence the dcs. of OC are $\cos = \frac{\pi}{2}, = \frac{\pi}{2}$ cos 0 *i.e.* 0,0, 1.

Dcs of Oz

Draw a line OM perpendicular to OC (in the plane zOC). Since the line OM lies in zOC plane, it is normal to the ON further the line OM is normal to OC so it also lies in the plane AOB such that $\angle MON = (\pi/2)$. the line ON makes angle ψ with OB, therefore the line OM will make an angle y with the line OA.

Hence the dcs. of the line OM are cos (p – ψ), $\cos\left(\frac{\pi}{2} - \psi\right), \cos\frac{\pi}{2}$

i.e. $-\cos\psi, \sin\psi, 0$. Also the dcs of Oz w.r.t the coplanar lines OC and OM are $\cos\theta$, $\sin\theta$. Now let the line Oz make angles α, β, γ with OA, OB, OC; then we have $\cos\alpha = -\sin\theta$, $\cos\psi$, $\cos\beta = \sin\theta\sin\psi$, $\cos\gamma = \cos\theta$.

$$\text{Hence, components of } \dot{\theta} \text{ (about ON) are } \begin{cases} \dot{\theta}\sin\psi \text{ about OA} \\ \dot{\theta}\cos\psi \text{ about OB} \\ \dot{\theta}\times 0 \text{ about OC} \end{cases}$$

$$\text{Components of } \dot{\phi} \text{ (about Oz) are } \begin{cases} -\dot{\phi}\sin\theta\cos\psi \text{ OA} \\ \dot{\phi}\cos\theta\sin\psi \text{ about OB} \\ \dot{\phi}\times\theta \text{ about OC} \end{cases}$$

$$\text{and components of } \dot{\psi} \text{ (about OC) are } \begin{cases} \dot{\psi}.0 \text{ i.e. } 0 \text{ about OC} \\ \dot{\psi}.0 \text{ i.e. } 0 \text{ about OA} \\ \dot{\psi}.1 \text{ i.e. } \dot{\psi} \text{ about OB} \end{cases}$$

Now : let $\omega_1, \omega_2, \omega_3$ be the angular velocities about OA,OB, and OC respectively. Then we have

$$\left.\begin{aligned} \omega_1 &= \dot{\theta}\sin\psi - \phi\sin\theta\cos\psi \quad &(1)\\ \omega_2 &= \dot{\theta}\sin\psi + \phi\sin\theta\sin\psi \quad &(2)\\ \omega_3 &= \dot{\phi}\theta + \psi \quad &(3) \end{aligned}\right\}$$

These are called Euler geometrical equations.

Invariable Line

Let the rigid turn about O under no forces. Suppose OA, OB OC are the principal axes of the body at O: $\omega_1, \omega_2, \omega_3$ the angular velocities about OA, OB, and OC respectively then component angular momentums about these axes are $A\omega_1, B\omega_2, C\omega_3$ where A,B,C are the moments of inertias of the body about the principal axes. *The resultant line of the angular momentum is known as invariable line and the resultant angular momentum is given by* $H = (A^2\omega_1^2 + B^2\omega_2^2 + C^2\omega_3^2)$. But here the body moves under no forces, therefore the sum of the moments of the external forces about O is zero implying there by that the angular momentum about any straight line through O is constant throughout the me motion *i.e.* H = constant or $A^2\omega_1^2 + B^2\omega_2^2 + C^2\omega_3^2 =$ constant.

The direction ratios of the invariable line are therefore $A\omega_1$, $B\omega_2$, $C\omega_3$. *This also shows that the direction ratios of the invariable line change w.r.t. the principal axes OA, OB, OC because* ω_1, ω_2, ω_3 *change as t changes.*

EQUATIONS OF MOTION OF A TOP

(*Deduced from Lagrange's equations*) : For motion of a top, we have

$$B = A, \dot{\psi} = 0 \text{ and } \omega_1 = -\dot{\phi}\sin\theta, \omega_2 = \dot{\theta}, \omega_3 = n = \dot{\phi}\cos\theta$$

∴ The kinetic energy is

$$\therefore T = \frac{1}{2}m[A(\omega_2^2 + \omega_2^2) + C\omega_3^2] \qquad (\because B = A)$$

$$= \frac{1}{2}m[A(\dot{\phi}^2\sin^2\theta + \dot{\theta}^2) + C\dot{\phi}^2\cos^2\theta]$$

And the work function W is $W = -mgh\cos\theta$ where h is distance of C.G. form the vertex of O. and θ is inclination of OC to the vertical OZ.

Here the generalised co-ordinates are θ and ϕ.

Lagrange's θ- equation is $\frac{d}{dt}\left(\frac{\partial T}{\partial \dot{\theta}}\right) - \frac{\partial T}{\partial \theta} = \frac{\partial W}{\partial \theta}$ which gives

$$\frac{d}{dt}(mA\dot{\theta}) - \{mA\dot{\phi}^2 \sin\theta\cos\theta - C\dot{\phi}^2\cos\theta\sin\theta\} = mgh \sin\theta$$

or $A\ddot{\theta} - A\dot{\phi}^2 \sin\theta\cos\theta + Cn\dot{\phi}\sin\theta = mgh \sin\theta$ ...(2)

And Lagrange's ϕ-equation is $\frac{d}{dt}\left(\frac{\partial T}{\partial \dot{\phi}}\right) - \frac{\partial T}{\partial \phi} = \frac{\partial W}{\partial \phi}$

which gives $\frac{d}{dt}(mA\dot{\phi}\sin^2\theta + C\dot{\phi}\cos^2\theta) = 0$

or $A\dot{\phi}\sin^2\theta + Cn\cos\theta = D$ (const.) ...(2)

Equation (1) and (2) are same equations of the motion of a top.

LOCUS OF THE INVARIABLE LINE

Let the rigid body turn about O under no forces. Suppose OA, OB, OC are the principal axes of the body at O; ω_1, ω_2, ω_3 the angular velocities about OA, OB and OC respectively then the component angular momentum about these axes are A ω_1, $B\omega_2$, $X\omega_3$ where A, B, C are the m, i's of the body about the principal axes.

$$\therefore [(A^2\omega_1^2 + B^2\omega_2^2 + C^2\omega_3^2)/(A^2\omega_1^2 + B\omega_2^2 + C\omega_3^2)] = (H^2/2T)$$

$$\Rightarrow A^2\omega_1^2\left(1 - \frac{H^2}{2AT}\right) + B\omega_2^2\left(1 - \frac{H^2}{2BT}\right) + C^2\omega_3^2\left(1 - \frac{H^2}{2CT}\right) = 0. \qquad ...(3)$$

Further, the equations of the invartiable line are

$(x/A\omega_1) = (y/B\omega_2) = (z/C\omega_3) = r$ (say)

[$\because$ Invariable line passes through O and has direction ratios $A\omega_1$, $B\omega_1$, $C\omega_1$].

Substituting in the previous equation, we get

$$(x^2/r^2)\left(1 - \frac{H^2}{2AT}\right) + (y^2/r^2)\left(1 - \frac{H^2}{2BT}\right) + (z^2/r^2)\left(1 - \frac{H^2}{2CT}\right) = 0$$

$$\Rightarrow x^2\left(1 - \frac{H^2}{2AT}\right) + y^2\left(1 - \frac{H^2}{2BT}\right) + z^2\left(1 - \frac{H^2}{2CT}\right) = 0 \; (\because r \neq 0) \qquad ...(4)$$

Equation (4) evidently represents a cone with vertex at the origin. The quantities H, A, B, C, T involved in equation (4) are constants, hence the above cone fixed in the body and is known as Invariable cone.

EQUATIONS OF MOTION OF A TOP

(Deduced from the principle of Energy and momentum) : Since for motion of top B = A and $\dot{\psi} = 0$

$$\therefore \omega_1 = -\dot{\phi}\sin\theta,\ \omega_2 = \dot{\theta},\ \omega_3 = n = \dot{\phi}\cos\theta$$

$$\therefore \text{K.E.} = \frac{1}{2}[A(\omega_1^2 + \omega_2^2) + C\omega_3^2] = \frac{1}{2}[A\,\dot{\phi}^2 \sin^2\theta + \dot{\theta}^2) + Cn^2]$$

$$(\because B = A)$$

Again potential energy = mgh cos θ where *h* is the distance of C.G. from the vertex O and θ inclination of OC to the vertical. By the principle of energy, the sum of the kinetic and the potential energy remains constant throughout motion. Thus by the principle of energy we have

$$\frac{1}{2}[A(\dot{\phi}^2 \sin^2\theta + \dot{\theta}^2) + Cn^2] + mgh\cos\theta = \text{const.}$$

$$\text{or } A(\dot{\theta}^2 + \dot{\phi}^2 \sin^2\theta) + mgh\cos\theta = E \quad ...(1)$$

Since the external force (weight) is parallel to OZ and the reactions at O cut OZ; so the sum of their moments about OZ is zero. Thus by the principle of momentum, the moment of momentum about OZ is constant throughout the motion.

Again $A\,\omega_1$, $A\omega_2$, $C\omega_3$ are the moments of momentum of the top about OA, OB, OC respectively; and the direction cosines of OZ about these axes are given by – sin θ, 0 cos θ.

Resolving these momentums about OZ, we have moment of momentum about OZ $-A\,\omega_1 \sin\theta + C\omega_3 \cos\theta$.

Therefore by the principle of momentum, we have

$$-A\,\omega_1 \sin\theta + C\omega_3 \cos\theta = \text{const.}$$

$$\text{or } A\sin\theta\ (-\dot{\phi}\sin\theta) + Cn\cos\theta = D\ (\text{constant}). \quad ...(2)$$

Equations (1) and (2) ae same equations of motion of a top.

STEADY MOTION

The motion of a top is said to be steady if it spins about is vertex O in such a way that its axis OC makes the same angle with the vertical OZ throughout the motion.

For steady motion θ = α (constant), *i.e.* the angular velocity about the vertical OZ is constant.

In this case the axis of the top describes a cone about the vertical with uniform angular velocity. Thus precessional velocity ϕ is constant.

In case the motion steady.

Put $\theta = \alpha, \ddot{\theta} = 0$ and $\dot{\phi} = \omega$ in equation

$$A\ddot{\theta} - A\dot{\phi}^2 \sin\theta\cos\theta + Cn\dot{\phi}\sin\theta - mgh\sin\theta = 0$$

Thus, we have ω^2 A cos α – Cnω + mgh= 0 ...(1)

or $\omega \dfrac{Cn \pm \sqrt{(C^2n^2 - 4A\ mgh\cos\theta}}{2A\cos\theta}$

i.e. there are in general two values of the precessional velocity ω; the two precessional velocities, call them ω_1, ω_2 ω_3; are real and distinct if

$$C^2n^2 > 4A\ mgh\cos\alpha$$

This is the necessary condition of steady motion.

Cor. If $\alpha = \frac{1}{2}\pi$, then from (1), $\omega = \left(\frac{mgh}{Cn}\right)$

Thus if the top is given an angular velocity n about OC when OC is horizontal and OC be then given an angular velocity (mgh/Cn) about OZ, then OC will continue to revolve uniformly in a horizontal plane round the vertical OZ.

If the top is set in motion in the usual manner. Then n is very great, thus

$$\omega = \frac{1}{2A\cos\alpha}\left[Cn \pm Cn\left\{1 - \frac{4Amgh\cos\alpha}{C^2n^2}\right\}^{1/2}\right]$$

$$= \frac{1}{2A\cos\alpha}\left[Cn \pm Cn\left\{1 - \frac{1}{2}.\frac{4Amgh\cos\alpha}{C^2n^2}\right\}\right]$$

$$= \frac{Cn}{2A\cos\alpha}\left[1 \pm \left\{1 - \frac{2Amgh\cos\alpha}{C^2n^2}\right\}\right]$$

$$= \frac{Cn}{A\cos\alpha} \text{ or } \frac{mgh}{Cn}$$

(by taking positive and negative signs respectively)

Thus $\omega = \dfrac{Cn}{A\cos\alpha}$ or $\dfrac{mgh}{Cn}$ approx. When n is very great. Obviously the first value of ω is large and second is very small, when n is very great.

INSTANTANEOUS AXIS OF ROTATION

Let the rigid body turn about the fixed point O and at time t, let $\omega_1, \omega_2, \omega_3$ be the component angular velocity about the principal axes OA,OB,OC (fixed in the body), then d.cs of the resultant axes of rotation are $(\omega_1/\omega), (\omega_2/\omega), (\omega_3/\omega)$ where $\omega = \sqrt{(\omega_1^2 + \omega_2^2 + \omega_3^2)}$. But $\omega_1, \omega_2, \omega_3$ have been expressed (in the previous art) in terms of Eulerian angles θ, ϕ, ψ (which change from time to time). Consequently, the resultant axis of rotation also changes implying that at different instants, there are different axis of rotation. is known as instantaneous axis of rotation or simple the instantaneous axis of rotation or simple the instantaneous axis.

LIMITS OF θ

A top is set spinning with angular velocity n about its axis. Initially he axis being instantaneously at rest and inclined at an angle i to the vertical. To show that in the subsequent motion, the angle θ between the axis and the vertical oscillates between

$$i \le \theta \le \cos^{-1}\{p - \sqrt{[(p^2 - 2p\cos i + 1)]}\} \text{ where } p = \left(\frac{C^2n^2}{4Amgh}\right),$$

and the letters have their usual meanings.

By equation (3) and (4), the two equations of motion of top are

$$A\dot{\phi}\sin^2\theta + Cn\cos\theta = D \text{ and } A(\dot{\theta}^2 + \dot{\phi}^2\sin^2\theta) + 2mgh\cos\theta = E$$

where D and E are constants.

Initially $\theta = i, \theta = \dot{0}, \dot{\psi} = 0,$

$\therefore$ D = Cn cos i and E = 2 mgh cos i.

Thus the above equations reduce to

$$A\dot{\phi}\sin^2\theta + Cn\cos\theta = Cn\cos i \quad ...(1)$$

$$\text{and } A(\dot{\theta}^2 + \dot{\phi}^2\sin^2\theta) + 2mgh\cos\theta = 2mgh\cos i \quad ...(2)$$

Now eliminating ϕ. between (1) and (2), we get

$$A\dot{\theta}^2\sin^2\theta + C^2n^2(\cos i - \cos\theta)^2 + 2Amgh(\cos\theta - \cos i)\sin^2\theta = 0$$

$$\text{or } A^2\dot{\theta}^2\sin^2\theta = -2Amgh(\cos\theta - \cos i)\sin^2\theta - C^2n^2(\cos\theta - \cos i)^2$$

$$= -2Amgh(\cos\theta - \cos i)\sin\theta - 4Amgh\,p(\cos\theta - \cos i)^2$$

$$= 2Amgh(\cos\theta - \cos i)[-\sin^2\theta - 2p(\cos\theta - \cos i)]$$

$$= 2Amgh(\cos i)\,[\cos^2\theta - 2p\cos\theta - 1 + 2p\cos i]$$

$$= 2Amgh(\cos\theta - \cos i)\,[(\cos\theta - p)^2 - (p^2 - 2p\cos i + 1])$$

$$= 2Amgh(\cos\theta - \cos i)\,\{\cos\theta - (p - \sqrt{[p^2 - 2p\cos i + 1]})\}$$

$$\times \{\cos\theta - (p + \sqrt{[p^2 - 2p\cos i + 1]})\} \qquad ...(3)$$

Now $\dot{\theta}$ will vanish when $\theta = i$ or θ_1 or θ_2

where $\cos\theta_1 = p - \sqrt{[(p^2 - 2p\cos i + 1)]}$

and $\cos\theta_2 = p + \sqrt{[(p^2 - 2p\cos i + 1)]}$

The second expression is greater than 1, *hence* θ_2 *is imaginary.*

But $\cos\theta_1 < \cos i$. $\quad \therefore \theta_1 > i$=

Again from (2) θ^2 is negative if $\cos\theta > \cos$ *i.e.* $\theta < i$.

Hence possible values of θ are i and θ_1 where

$\cos\theta_1 = p - \sqrt{(p^2 - 2p\cos i + 1)}$

Again from (3) θ^2 is negative if

$\cos\theta > p - \sqrt{(p^2 - 2p\cos i + 1)}$ *i.e.* $\theta > \theta_1$

Thus, we conclude that top is never at a less inclination than i or at a greater inclination than θ_1, *i.e.* the motion is included between these two limits.

EULER'S EQUATIONS FROM LAGRANGE'S EQUATIONS

Let the rigid body turn about some fixed point O under no forces. Suppose OA, OB, OC are the principal axes of the body at O; ω_1, ω_2, ω_3, the angular velocities about OA, OB OC respectively then the kinetic energy T is given by

$$T = \frac{1}{2}(a\omega_1^2 + B\omega_2^2 + C\omega_3^2)$$

where A, B, C are the moments of inertias of the body about OA, OB, OC respectively and Euler geometrical relations are

$$\omega_1 = \dot{\theta}\sin\phi - \dot{\phi}\sin\theta\cos\psi, \omega_2 = \dot{\theta}\cos\psi + \dot{\phi}\sin\theta\sin\psi,$$

$$\omega_3 = \dot{\phi}\cos\theta + \dot{\psi}.$$

Hence T can be thought to be the function of the generalised co - ordinates θ, ϕ, ψ and the generalised velocities $\dot{\theta}$, $\dot{\phi}$, $\dot{\psi}$,

$$\because \frac{\partial T}{\partial \dot{\psi}} = \frac{\partial T}{\partial \omega_1}\frac{\partial \omega_1}{\partial \dot{\psi}} + \frac{\partial T}{\partial \omega_2}\frac{\partial \omega_2}{\partial \dot{\psi}} + \frac{\partial T}{\partial \omega_3}\frac{\partial \omega_3}{\partial \dot{\psi}}$$

$$= \frac{\partial T}{\partial \omega_3}\frac{\partial \omega_3}{\partial \dot{\psi}} \; (\because \omega_1, \omega_2 \text{ do not involve } \dot{\psi})$$

and $\dfrac{\partial T}{\partial \psi} = \dfrac{\partial T}{\partial \omega_1}\dfrac{\partial \omega_1}{\partial \psi} + \dfrac{\partial T}{\partial \omega_2}\dfrac{\partial \omega}{\partial \psi} + \dfrac{\partial T}{\partial \omega_3}\dfrac{\partial \omega_3}{\partial \psi}$

$$= A\omega_1 \frac{\partial \omega_1}{\partial \psi} + B\omega_2 \frac{\partial \omega_2}{\partial \psi} C\omega_3.0 \qquad \text{[using above results]}$$

$$= A\omega_1 \;(\dot{\theta}\cos\psi + \dot{\phi}\sin\theta\sin\psi) + B\omega_2(-\dot{\theta}\sin\psi + \dot{\phi}\sin\theta\cos\psi)$$

$= A\ \omega_1\omega_2 - B\omega_1\omega_2$ (using above results $= (A - B)\ \omega_1\omega_2$

Further, let W be the work function of the system and N the moment of the external forces about axis OC, then we readily obtain.

$$\frac{\partial W}{\partial \dot{\psi}} = N$$

Now, Lagrange's ψ-equation gives

$$\frac{d}{dt}\left(\frac{\partial T}{\partial \dot{\psi}}\right) - \frac{\partial T}{\partial \psi} = \frac{\partial W}{\partial \psi} \text{i.e.} \frac{d}{dt}(C\dot{\omega}_3) - (A - B)\omega_1\omega_2 = N$$

$$\Rightarrow C\frac{d\omega_3}{dt} - (A - B)\omega_1\omega_2 = N$$

This is Euler's third equations; and similarly other two equations can also be deduced.

MOTION OF A SPINNING TOP

Definition : An interesting example of rigid body motion occurs when a symmetrical rigid body having one point on the axis of symmetry fixed in space is set spinning in a gravitational field. A top is a *rigid body which is a solid of revolution about the axis of symmetry.* The adjoining figure shows such a top rotating about a fixed point O on its axis of symmetry., Thus we idealise the ordinary top by supposing it to terminate in a sharp point (or vertex) and to spin on a floor rough enough to prevent slipping.

EQUATION OF MOTION OF A TOP

(Derived from Euler's Equation) : Let us assume that a top spins with its vertex O fixed in contact with a floor rough enough to prevent slipping Fig. 10.01, Let OC be the axis of the top and G its centre of gravity such that OC = h.

Let OX, OY, OZ be the fixed axes and OA, OB, OC the principal axes at O.

Let C, A be the moments of inertia about OC and OA, since the top is symmetrical about OC, so B = A.

Let at zero time OC be in the plane ZOX and at time t it makes an angle θ to OZ and the plane ZOC makes the angle ϕ with the fixed plane ZOX. Let ω_1, ω_2, ω_3 be the angular velocities of the top about OA, OB and OC respectively and L, M, N be the moments of the external forces about OA, OB and OC respectively.

Again the external forces acting on the top are its weight mg acting vertically downwards through G, the C.G. of the top in a direction parallel to OZ and the reactions at point O (0, 0, 0).

The direction cosines of OZ with respect to OA, OB and OC are. $-\sin\theta\cos\psi$, $\sin\theta\sin\psi$, $\cos\theta$

or $-\sin\theta, 0, \cos\theta$ ($\because$ In case of top $\psi = 0$)

Thus the weight – mg acting at G (0, 0, h) has the resolved parts mg $\sin\theta$; 0, $-mg\cos\theta$, parallel to OA, OB and OC respectively.

By $L = \Sigma (y_1 Z_1 - z_1 Y_1)$ etc.,

we have $L = 0, M = hmg\sin\theta, N = 0$

Therefore Euler's equations give

$$A\dot{\omega}_1 - (B-C)\omega_2\omega_3 = L = 0;$$

$$B\dot{\omega}_2 - (C-A)\omega_3\omega_1 = M = hmg\sin\theta$$

$$C\dot{\omega}_3 - (A-B)\omega_1\omega_2 = N = 0$$

Again Euler's geometrical relations are

$$\omega_1 = \dot{\theta}\sin\psi - \dot{\phi}\sin\theta\cos\psi, \omega_2 = \dot{\theta}\cos\psi + \dot{\phi}\sin\theta\sin\psi,$$

$$\omega_3 = \dot{\psi} + \dot{\phi}\cos\theta$$

which reduce to $\omega_1 = -\dot{\phi}\sin\theta.\ \omega_3 = \dot{\phi}\cos\theta, \omega_2 = \dot{\theta}$ since for top $\dot{\psi} = 0$.

Again B = A, thus the Euler's third equation gives

$\dot{\omega}_3 = 0$ or $\omega_3 = n$ (constant). or $\phi\cos\theta = n$

Putting the values of ω_1, ω_2, ω_3.

Euler's first equation $A\omega_1 - (B-C)\omega_2\omega_3 = 0$

deduces to $A\frac{d}{dt}(-\dot{\phi}\sin\theta) - (A-C)\dot{\theta}\dot{\phi}\cos\theta = 0$ as B = A

or $-A\phi^2\sin\theta - 2A\dot{\theta}\dot{\phi}\cos\theta + C\dot{\theta}\dot{\phi}\cos\theta = 0$

or $-A\phi\sin\theta - 2A\dot{\theta}\dot{\phi}\cos\theta + Cn\dot{\theta} = 0$

And Euler's second equation $B\dot{\omega}_2 - (C-A)\omega_3\omega_1 = mgh\sin\theta$

reduces to $A\frac{d}{dt}(\dot{\theta}) + (C-A)\dot{\phi}^2 \sin\theta\cos\theta = mgh\sin\theta$ as B = A

or $A\ddot{\theta} - A\dot{\phi}^2\sin\theta + C\dot{\phi}^2\sin\theta\cos\theta = mgh\sin\theta$

$A\ddot{\theta} - A\dot{\phi}^2\sin\theta + Cn\dot{\phi}\sin\theta = mgh\sin\theta$

Equations (1) and (2) are the *equations of motion of the top* and are of second order.

Multiplying (1) by sin θ and integrating, we have

$$A\dot{\phi}\sin^2\theta + Cn\cos\theta = D\ (\cos nt.) \quad ...(3)$$

Again multiply (1) by $2\dot{\phi}$ sin θ and (2) by $2\dot{\theta}$ and subtracting, we get

$$2A\ddot{\theta}\dot{\theta} + 2A\dot{\theta}\dot{\phi}^2 \sin\theta\cos\theta + 2A\dot{\phi}\ddot{\phi}\sin^2\theta = 2mgh\sin\theta$$

Integrating, we get

$$A(\dot{\theta}^2 + \dot{\phi}^2\sin^2\theta) + 2mgh\cos\theta = E(\text{const.}) \quad ...(4)$$

Equations (3) and (4) are also the equations of motion of the top but are of first order. Equation (4) is called equation of Energy. The motion due to change in θ is called *mutation* and due to change in ϕ is called precision. The general motion of the top about its fixed vertex O is a combination of these two.

MOVING AXES AND THE FIXED AXES

Let a body turn about some fixed point O and let OA, OB, OC be the principal axes fixed in the body (which move with the body). Further let Ox, Oy, Oz the axes fixed in the space. After a time t, let the components of the angular velocity of the body; about Ox, Oy, Oz be θ_x, θ_y and θ_z; about OA, B and OC be ω_1, ω_2 and ω_3. Again let h_1, h_2, h_3 be the components of any vector h (say, angular momentum) along OA, OB, OC; h_x, h_y, h_z be the components of the same vector h along Ox, Oy, Oz then if α, β, γ, are the inclinations of OA, OB, OC to the axis Ox, we readily obtain

$$h_x = h_1\cos\alpha + h_2\cos\beta + h_3\cos\gamma$$

$$\Rightarrow h_1 = \dot{h}_1\cos\alpha + \dot{h}_2\cos\beta + \dot{h}_3\cos\gamma \text{ 0p6}$$

$$+ h_1(-\sin\alpha.\dot{\alpha}) + h_2(-\sin\beta.\dot{\beta}) + h_3(-\sin\gamma\dot{\gamma}) \quad ...(1)$$

Let the axes fixed in the body coincide with the axes fixed in space at time t, then we have

$$\alpha = 0, \beta = \frac{\pi}{2} = \gamma \Rightarrow \dot{h}_1 - h_2\dot{\beta} - h_3\dot{\gamma}$$

Again we know that β is the angle which the line OB makes with Ox, therefore β becomes the rate at which OB is receding from Ox, also the same rate is θζ ιμπλψινγ that $\dot{\beta} = \theta_z$ (they are in the same sense) ...(3)

Again γ denotes the rate at which OC is receding from Ox, also the same rate is denoted by θψ βυτ in opposite sense and hence $\dot{\gamma} = -\theta_y$.

Making these substitutions in the above result, we get

$$\dot{h}_x = \dot{h}_1 - h_2\theta_z + h_3\theta_y \qquad ...(4)$$

By symmetry, we immediately get

$$\dot{h}_y = \dot{h}_2 - h_3\theta_x + h_1\theta_z \qquad ...(5)$$

$$\dot{h}_z = \dot{h}_3 - h_1\theta_y + h_2\theta_x \qquad ...(6)$$

when OA, OB, OC coincide with Ox, Oy, Oz, then we also have

$$\theta_x = \omega_1, \theta_y = \omega_2, \theta_z = \omega_3$$

$$\left.\begin{aligned} \Rightarrow \dot{h}_x &= \dot{h}_1 - h_2\omega_3 + h_3\omega_2 \\ \dot{h}_y &= \dot{h}_2 - h_3\omega_1 + h_1\omega_3 \\ \dot{h}_z &= \dot{h}_3 - h_1\omega_2 + h_2\omega_1 \end{aligned}\right] \qquad ...(a)$$

This establishes the law, how the component of a vector along an axis fixed in space can be expressed in terms of the components of the same vector along the principal axes and the angular velocities $\omega_1, \omega_2, \omega_2$ about them.

To find the velocity of a point when the rigid body turns about O.

Let Ox, Oy, Oz the axes fixed in space. Consider a point P (x,y,z) in the body. Let $\omega_x, \omega_y, \omega_z$ be the components of the angular velocity of P about Ox, Oy and Oz respectively. First of all, we consider the rotation of P about Oy. Due to this rotation, velocity of P becomes ωψ NΠ ιν the direct ; in PA (to NP). This lies in the plane PMN and hence parallel to the plane zOx. Draw a line PB parallel to OX and let it make an angle θ with the line PA. Again PN is perpendicular to PA and PM Now components of the velocity wy NP are.

$\omega_y NP\cos\theta$ alongPB *i.e.* || to O x *i.e.*

$\omega_y KP(PM/PN)$ || to Ox *i.e.* || to Ox

and $\omega_y NP\sin\theta$ along PM *i.e.* || to –z axis

i.e.$-\omega_y NP.(MN/NP)$ || ro Oz *i.e.*$-\omega_y x$ || to Oz

Similarly due to rot action about Oz we have the velocity components as:

$\omega_z x$ || Oy and $-\ \omega_z x$ || to Ox

Finally, due to rotation about Ox, we have the velocity components as :

$\omega_x y$ || Oz and $-\omega_x z$ || to Oy

Hence due to three rotation, components of velocity of P are

$$(\omega_x y - \omega_z z) \parallel Ox, (\omega_z x - \omega_x z) \parallel \text{to } Oy$$

and $$(\omega_x y - \omega_y x) \parallel \text{to } Oz$$

and $$\dot{z} = (\omega_x y - \omega_y x)$$

$$\Rightarrow u + (\omega_y z - \omega_z y), \dot{y} = v + (\omega_z x - \omega_y z)\dot{z} = w + (\omega_x y - \omega_y x).$$

where $(\dot{x}, \dot{y}, \dot{z})$ are the components of the velocity at P parallel to the axes respectively.

Note. If the point O is not at rest but itself moving with a velocity whose components are u,v,w, (|| to the axes), then components of velocity at P will be

$$\dot{x} = u + (\omega_y z - \omega_z y), \dot{y} = v + (\omega_z x - \omega_y z), \dot{z} = w + (\omega_x y - \omega_y x).$$

DYNAMICAL EQUATIONS OF EULER'S

Let a rigid body turn about some fixed point O under the action of external sources such that L,M,N be the moments of the external forces, inertia about the principal axes fixed in the body then to obtain the equations of motion.

Let the body turn about some fixed point O and let OA, OB, OC be the principal axes fixed in the body (which move with the body).

Further, let Ox, Oy, Oz be the axes fixed in apace. After a time t, let the components of the angular velocity of the body, about Ox, Oy, and Oz by θ_1, θ_2, θ_3; about OA, OB and OC be $\omega_1, \omega_2, \omega_3$ .Again let h_1, h_2, h_3 be the components of the vector h, (angular momentum) along OA, OB, OC, h_x, h_y, h_z be the components of the same vector h along Ox, Oy, Oz. then we have

$$\dot{h}_x = \dot{h}_1 - h_2\theta_z + h_3\theta_y, \dot{h}_y = \dot{h}_2 - h_3\theta_x + h_1\theta_z, \text{ and}$$

$$\dot{h}_z = \dot{h}_3 - h_1\theta_y + h_2\theta_x,$$

$$\left.\begin{aligned} \dot{h}_x &= \dot{h}_1 - h_2\omega_3 + h_3\omega_2 \\ \Rightarrow \dot{h}_y &= \dot{h}_2 - h_3\omega_1 + h_1\omega_3 \\ \dot{h}_z &= \dot{h}_3 - h_1\omega_2 + h_2\omega_1 \end{aligned}\right]$$

(when OA, OB, OC concide with Ox, Oy, Oz) ...(1)

Again, if x,y,z are the co-ordinates of a particle m of the rigid body, relative to OA, OB, OC then relative to these axes, we have

$$\dot{x} = z\omega_2 - y\omega_3, \dot{y} = x\omega_3 - z\omega_3, \dot{z} = y\omega_1 - x\omega_2.$$

Now h_1 = components of angular momentum (h) about OA

= angular momentum about OA

$= \Sigma m(y\dot{z} - z\dot{y}) = \Sigma m[y(y\omega_1 - x\omega_2) - z(x\omega_3 - z\omega_1)]$

$= \omega_1 \Sigma m(y^2 + z^2) - \omega_2 \Sigma mxy - \omega_3 \Sigma mzx = A\omega_1 - F\omega_2 - E\omega_3$ ...(2)

By symmetry, we then have

$h_2 = B\omega_2 - D\omega_3 - F\omega_1$...(3) and $h_3 = C\omega_3 - E\omega_1 - D\omega_2$ ...(4)

where A, B, C are the moments and D,E, F are the products of inertia about OA, OB, OC.

Here OA, OB, OC are the principal axes; therefore we have D = E = F = 0.

Now from above we have

$h_1 = A\omega_1, h_2, = B\omega_2$ and $h_3 = C\omega_3$

$\Rightarrow \dot{h}_1 = A\dot{\omega}_1, \dot{h}_2 = B\dot{\omega}_2$ and $\dot{h}_3 = c\dot{\omega}_3$

$$\left.\begin{array}{l}\dot{h}_x = A\dot{\omega}_1 - B\omega_2\omega_3 + C\omega_2\omega_3 \\ \text{whence } h_y = B\dot{\omega}_2 - C\omega_3\omega_1 + A\omega_1\omega_3 \\ \dot{h}_z = C\dot{\omega}_3 - A\omega_1\omega_2 + B\omega_2\omega_1\end{array}\right] \quad ..(A)$$

[when OA, OP, OC coincide with Ox, Oy, Oz]

But $\dot{h}_x$ = angular momentum about OA

= moment of momentum about OA = L (given)

[when OA coincides with Ox]

and similarly $\dot{h}_y = M, \dot{h}_2 = N$

$\therefore A\dot{\omega}_1 - (B - C)\omega_2\omega_3 = L; B\dot{\omega}_2 - (C - A)\omega_3\omega_1 = M;$

$\therefore C\dot{\omega}_3 - (A - B)\omega_1\omega_2 = N.$ ...(B)

These are called as *Euler's dynamical equations.*

Special Case. *When the body moves under on forces.*

then we have L = 0 = M = N

$$\left\{\begin{array}{ll}\therefore (B) \Rightarrow A\dot{\omega}_1 - (B - C)\omega_2\omega_3 = 0 & ...(X) \\ B\dot{\omega}_2 - (C - A)\omega_3\omega_1 = 0 & ...(Y) \\ C\dot{\omega}_3 - (A - B)\omega_1\omega_2 = 0 & ...(Z)\end{array}\right\}$$

Euler's dynamical equations under no forces.

The special case of motion about O under no forces is of importance. In this case multiplying the three equations (X), (Y), (Z) by the ω_1, ω_2, ω_3 and adding, we get

$$A\omega_1\dot{\omega}_1 + B\omega_2\dot{\omega}_2 + C\omega_3\dot{\omega}_3 = 0 \Rightarrow A\omega_1^2 + B\omega_2^2 + C\omega_3^2 = \text{const.}$$

But it is proved in the forthcoming.

$$\text{K.E.} = T = \frac{1}{2}(A\omega_1^2 + B\omega_2^2 + C\omega_3^2)$$

$\therefore$ Above result shows that the kinetic energy remains constant during the motion. It is obvious since no work is being done on the body, there being on external forces. Another integral of equations (X), (Y), (Z) can be obtained then by multiplying these by $A\omega_1$, $B\omega_2$, $C\omega_3$ and adding to give

$$A^2\omega_1\dot{\omega}_1 + B^2\dot{\omega}_2\omega_2 + C^2\omega_3\dot{\omega}_3 = 0 \Rightarrow A^2\omega_1^2 + B^2\omega_2^2 + C^2\omega_3^2 = \text{const.}$$

But we know that

$$h_1 = A\omega_1, h_2 = B\omega_2, h_3 = C\omega_3.$$

i.e. $h = (A\omega_1, B\omega_2, C\omega_3)$ and $/h/ = \sqrt{(A^2\omega_1^2 + B^2\,\omega_2^2 + C^2\,\omega_3^2}$

Hence the above equation implies the constancy of the magnitude of the angular momentum h during the motion. Physically this is attributable to the zero moment of forces about O.

STABLE MOTION

When axis is vertical : A top is executing steady motion with angular velocity n about its axis which is vertical; to show that the motion is stable.

If the body be disturbed in such a way that θ is very small whilst ψ is unaltered initially, so the general equations of motion of the top are

$$A\ddot{\theta} - A\dot{\phi}^2 \sin\theta\cos\theta + Cn\dot{\phi}\sin\theta = mgh\sin\theta$$

and $A\dot{\phi}^2 \sin^2\theta\cos\theta + Cn\cos\theta = D$

Since motion is steady with axis vertical; thus we have

$$\theta = 0, \ddot{\theta} = 0, \dot{\phi} = n, \therefore D = Cn$$

Also $C^2n^2 > 4A\,mgh$ is the condition for steady motion when $\alpha = 0$. Let the motion be slightly disturbed. The disturbed motion is a general motion of the top; hence its equations are

$$A\,\ddot{\theta} - A\dot{\phi}\sin\theta\cos\theta + Cn\dot{\phi}\sin\theta = mgh\sin\theta \qquad ...(1)$$

$$\text{and} \quad A\dot{\phi}\sin^2\theta + Cn\cos\theta = Cn \qquad ...(2)$$

Eliminating ϕ between (1) and (2), we have

$$A^2\ddot{\theta} - \frac{C^2n^2}{(1+\cos\theta)^2}\sin\theta\cos\theta + \frac{C^2n^2}{(1+\cos\theta)}\sin\theta = \text{Amgh}\sin\theta$$

Since θ is small, therefore, writing θ for sin θ and 1 for cos θ the above equation reduces to

$$A^2\ddot{\theta} - \frac{C^2n^2\theta}{4} + \frac{C^2n^2\theta}{2} = \text{Amgh}\,\theta \text{ or } \ddot{\theta} = -\left(\frac{C^2n^2 - 4\text{Amgh}}{4A^2}\right)\theta,$$

(the coefficient of q is negative, $\because C^2n^2 > 4\text{Amgh}$)

The above equation is simple harmonic motion.

i.e. the motion of the axis of the figure is simple Harmonic about the vertical OZ from which θ is measured. In other words, axis of the top, if disturbed slightly from its vertical position in steady motion, will tend to come back to its original position, *i.e.* the steady motion, in which axis of the figure is vertical, is stable.

The period of oscillation

$$= 2\pi\sqrt{\left(\frac{4A^2}{B^2n^2 - 4\text{Amgh}}\right)} = \frac{4\pi A}{\sqrt{(C^2n^2 - 4\text{Amgh})}} \text{ (where axis is not vertical)}$$

Stable Motion

A top is executing steady motion, with its axis inclined at a constant angle α to the vertical, and precessional velocity ω; to show that motion is stable. Here, we shall show that, the axis of the top, when disturbed slightly, from its position in steady motion, will tend to come back to the original position.

Equations of motion of the top are

$$A\ddot{\theta} - A\dot{\phi}^2\sin\theta\cos\theta + Cn\dot{\phi}\sin\theta = mgh\sin\theta$$

and $A\dot{\phi}\sin^2\theta + Cn\cos\theta = D$

Since the motion, originally, is steady with $\theta = \alpha, \dot{\phi} = \omega$.

Thus the above equation reduces to

$Cn\omega - A\omega^2\cos\alpha = mgh$ and $D = A\omega\sin^2\alpha + Cn\cos\alpha$.

Let the steady motion be slightly disturbed, so that motion is then general one and so its equations are

$$A\ddot{\theta} - A\dot{\phi}^2\sin\theta\cos\theta + Cn\dot{\phi}\sin\theta = mgh\sin\theta \quad ...(1)$$

$$\text{and } A\dot{\phi}\sin^2\theta + Cn\cos\theta = A\omega\sin^2\alpha + Cn\cos\alpha \quad ...(2)$$

Eliminating $\dot{\phi}$ between (1) and (2), we get

$$A^2\ddot{\theta}\sin^3\theta - \{A\omega\sin^2\alpha + Cn(\cos\alpha - \cos\theta)\}^2\cos\theta + Cn\sin^2\theta\times$$

$$\{A\omega\sin^2\alpha + Cn(\cos\alpha - \cos\theta)\} = Amgh\sin^4\theta$$

Putting $\theta = \alpha + X$ Where X is small, the above equation reduces to

$$A^2\ddot{X}\sin^3(\alpha + X) - [A\omega\sin^2\alpha + Cn\{\cos\alpha - \cos(\alpha + X)\}]^2\cos(\alpha + X)$$

$$+ Cn\sin^2(\alpha + X)[A\omega\sin^2\alpha + Cn\{\cos\alpha - \cos(\alpha + X)\}]$$

$$= Amgh\ \sin^4(\alpha + X)$$

Since X is small, neglecting quantities of second and higher orders of smallness, we have

$$A^2\ddot{X}(\sin\alpha + X\cos\alpha)^3 - (A\omega\sin^2\alpha + Cn\{\cos\alpha - (\cos\alpha -$$

$$X\sin\alpha)\}]^2(\cos\alpha - X\sin\alpha) + Cn(\sin\alpha + X\cos\alpha)^2[A\omega\sin^2\alpha$$

$$+Cn\{\cos\alpha - (\cos\alpha - X\sin\alpha)\}] = Amgh(\sin\alpha + X\cos\alpha)^4$$

or $A^2\ddot{X}\sin^3\alpha - (A\omega\sin^2\alpha + CnX\sin\alpha)$

$$= A\ mgh(\sin^4\alpha + 4X\cos\alpha\sin^3\alpha)$$

• or $A^2\ddot{X}\sin^3\alpha - (A^2\omega^2\sin^4\alpha + 2XA\ Cn\omega\sin^3\alpha)(\cos\alpha - X\sin\alpha)$

$$+ Cn(\sin^2 + 2X\sin\alpha\cos\alpha)(A\omega\sin^2\alpha + CnX\sin\alpha)$$

$$= A\ mgh(\sin^4\alpha + 4X\cos\alpha\sin^3\alpha)$$

or $A^2\ddot{X}\sin^3\alpha - (A^2\omega^2\sin^4\alpha + 2X\ ACn\omega\sin^3\alpha + XA^2\omega^2\sin^5\alpha)$

$$+ Cn(\sin^2\alpha + 2X\sin\alpha\cos\alpha)A\omega\sin^2\alpha + XC^2n^2\sin^3\alpha$$

$$= A\ mgh(\sin^4\alpha + 4X\cos\alpha\sin^3\alpha)$$

or $A^2\ddot{X} - A^2\omega^2\sin\alpha\cos\alpha + Cn\omega A\sin\alpha + XA^2\omega^2\sin^2\alpha + XC^2n^2$

$$= Amgh(\sin\alpha + 4X\cos\alpha)$$

or $A^2X - A^2\omega^2\sin\alpha\cos\alpha + (A\omega^2\cos\alpha + mgh)A\sin\alpha + XA^2\omega^2\sin^2\alpha$

$$+ X(A\omega^2\cos\alpha + mgh)^2(1/\omega^2) = Amgh(\sin\alpha + 4X\cos\alpha)$$

or $A^2\ddot{X} + X[A^2\omega^4(\sin^2\alpha + \cos^2\alpha) + 2A\ mgh\omega^2\cos\alpha + m^2g^2h^2]$

$$\times[1/\omega^2]\,4XA\ mgh\cos\alpha$$

or $A^2\ddot{X} + X[A^2\omega^4 - 2A\ mgh\ \omega^2\cos\alpha + m^2g^2h^2](1/\omega^2) = 0$

$$\text{or } \ddot{X} = -\frac{A^2\omega^4 - 2Amgh\omega^2\cos\alpha + m^2g^2h^2}{A^2\omega^2}X \qquad \text{[S.H.M]}$$

(the coefficient of X is obviously negative)

This shows that the axis of the top, if disturbed slightly from its position in steady motion will tend to come back to the original (predisturbed), position, *i.e.* the steady motion, with axis inclined at α to the vertical, is stable.

$$\text{The period of oscillation} = 2\pi\frac{A\omega}{(A^2\omega^4 - 2Amgh\,\omega^2\cos\alpha + m^2g^2)^{(1/2)}}$$

In case the top is set in motion in usual manner, then n is very large and two values of the precessional velocity are approximately given by

$$\omega = \frac{Cn}{A\cos\alpha}.\frac{mgh}{Cn}$$

When n is large, the first value is large and the second small. For first value of ω, the period of oscillation is given by

$$\frac{2\pi A\omega}{A\omega^2}\text{ (approx.) } = \frac{2\pi\cos\omega}{Cn}$$

And for second value of ω, the period of oscillation is given by

$$\frac{2\pi A\omega}{mgh}\text{ (approx.) } = \frac{2\pi A}{Cn}$$

KINETIC ENERGY

Suppose a rigid body terms about some fixed point. O. If $\omega_1, \omega_2, \omega_3$ are the angular velocities and A,B, C the moments of inertia about the principal axes OA, OB, OC; then the kinetic energy of the body is

$$T = \frac{1}{2}(A\omega_1^2 + B\omega_2^2 + C\omega_3^2).$$

Proof:

A time t, let (x,y,z) be the co-ordinates, of a particle m of the body, referred to the principal axes OA, OB, OC; and $\omega_1, \omega_2, \omega_3$ the angular velocities about these axes. Then the components of velocity of a particle of mass m of the body are given by

$$\dot{x} = (\omega_2 z - \omega_3 y),\ \dot{y} = (\omega_3 x - \omega_1 z),\ \dot{z} = (\omega_1 y - \omega_2 x)$$

$\therefore$ Kinetic energy of the body

$$T = \frac{1}{2}\Sigma m(\dot{x}^2 + \dot{y}^2 + \dot{z}^2)$$

$$= \frac{1}{2}\Sigma m[(\omega_2 z - \omega_3 y)^2 + (\omega_3 x - \omega_3 z)^2 + (\omega_1 y - \omega_2 x)^2]$$

$$= \frac{1}{2}\Sigma m[(\omega_1^2(z^2 + y^2) + \omega_2^2(z^2 + x^2) + \omega_3^2(x^2 + y^2)$$
$$- 2\omega_2\omega_3 yz - 2\omega_3\omega_1 zx - 2\omega_1\omega_2 xy]$$

$$= \frac{1}{2}[A\omega_1^2 + B\omega_2^2 + C\omega_3^2 - 2D\omega_2\omega_3 - 2E\omega_3\omega_1 - 2F\omega_1\omega^2],$$

where A, B, C are the moments and D, E, F are the product of inertias about OA, OB, OC.

But D = E = F = 0 [$\because$ OA, OB, OC are the principal axes]

$$\therefore T = \frac{1}{2}(a\omega_1^2 + B\omega_2^2 + C\omega_3^2).$$

EULER'S EQUATIONS

(for impulsive forces).

Let the rigid body turn about a fixed point O and let OA, OB, OC (axes fixed in the body) be the principal axes which move along with the body. Let the body be given an impulse so that L, M, N are the moments of the impulse about OA, OB, OC respectively. Further, let ω_1, ω_2 ω_3 and ω_1', ω_2' ω_3' be the component angular velocities just before and just after the action of the impulse. Then we have, Change in angular momentum about OA

= moment of the impulse about OA

$\Rightarrow$ (angular momentum after the action of impulse

– angular momentum before the action of impulse) about OA = L

$\Rightarrow A\omega_1' - A\omega_1 = L(\because h_1 = A\omega_1) \Rightarrow A(\omega_1' - \omega_1) = L$

and similarly $B(\omega_2' - \omega_2) = M$; $C(\omega_3' - \omega_3) = N$.

These are Euler's Equations of motion under impulsive forces.

MISCELLANEOUS EXAMPLES

Example 1:

On the assumption that the earth had once been rotating bout an axis through its centre of gravity not coincident with a principal axis and been subject to the acting of retarding couple whose magnitude at each instant was proportional to the angular velocity at that time and where axis was the instantaneous axis of rotation, find the component angular velocities about the principal axes and show that the instantaneous axis would approach ultimately coincide with the least axis.

Solution:

Euler's dynamical equations are

$$A\dot{\omega}_1 - (B - C)\omega_2\omega_3 = L,\ B\dot{\omega}_2 - (C - A)\omega_3\omega_1 = M,$$

$$C\dot{\omega}_3 - (A - B)\omega_1\omega_2 = N$$

But here $L = -\ \lambda\omega_2, M = -\lambda\omega_2, N = -\lambda\omega_3$

and A = B, therefore the above equations take the form as

$$A\dot{\omega}_1 = (A - C)\omega_2\omega_3 = -\lambda\omega_1 \qquad ...(1)$$

$$A\dot{\omega}_2 = (C - A)\omega_3\omega_1 = -\lambda\omega_2 \qquad ...(2)$$

$$C\dot{\omega}_3 = -\lambda\omega_3 \qquad ...(3)$$

Solving the differential equation (3), we get

$$\omega_3 = pe^{-\lambda t/c} \text{ where p is a const.} \qquad ...(4)$$

$$A\dot{\omega}_1 = (A - C)pe^{-\lambda t/c}\omega_2 - \lambda\omega_1 \qquad ...(5)$$

$$A\dot{\omega}_2 = (C - A)pe^{-\lambda t/c}\omega_1 - \lambda\omega_2 \qquad ...(6)$$

Multiplying (6) by i and adding to (5), we get

$$A(\dot{\omega}_1 + i\omega_2) = (A - C)pe^{-\lambda t/c}(\omega_3 + i^3\omega_1) - \rho(\omega_1 + i\omega_2)$$

$$= (A - C)pe^{-\lambda t/c}(i^4\omega_2 + i^3\omega_1) - \lambda(\omega_1 + i\omega_2)$$

$$= i^3(A - C)pe^{-\lambda t/c}(\omega_1 + i\omega_3) - \lambda(\omega_1 + i\omega_2)$$

Putting $\omega_1 + i\omega_2 = u$, we get $\dot{u} = -\left[\dfrac{A - C}{A} ipe^{-\lambda t/c} + \dfrac{\lambda}{A}\right]u.$

$$i.e.\ \frac{du}{dt} = -\left[\frac{A - C}{A} ipe^{-\lambda t/c} + \frac{\lambda}{A}\right]u.\ -\left[\frac{-C(A - C)}{\lambda A} ipe^{-\lambda t/c} + \frac{\lambda}{A}\right]u.$$

Integrating it, we get $u = \alpha\ e$

where α is const.

$$= \alpha e^{-(\lambda/A)t} e[iC(A - C)/\lambda A]pe^{-(\lambda t/c)}$$

$$\Rightarrow \omega_1 + i\omega_2 = \alpha_e^{-\lambda t/a}\left[\cos\left\{\frac{(A - C)C}{\lambda A} pe^{-\lambda t/c}\right\} + i\sin\left\{\frac{C(A - C)}{\lambda A} pe^{-\lambda t/c}\right\}\right]$$

Separating the real and imaginary parts, we get

$$\omega_1 = \alpha e^{-\lambda/A}\cos\left\{\frac{C(A - C)}{\lambda A} pe^{-\lambda t/C}\right\}, \text{ and}$$

$$\omega_2 = \alpha e^{-\lambda/A}\sin\left\{\frac{C(A - C)}{\lambda A} pe^{-\lambda t/C}\right\}.$$

Example 2:

If the earth be regarded as a solid of revolution, whose principal moments of inertia at the centre of gravity are A, A, C; show that its axis of rotation describes a cone of very small angle about the axis of the figure in period $\frac{A}{C-A}$ *sidereal days.*

Solution:

Euler's equations of motion under on forces are

$$A\dot{\omega}_1 - (B-C)\omega_2\omega_3 = 0,\ B\dot{\omega}_2 - (C-A)\omega_3\omega_1 = 0$$

$$C\dot{\omega}_3 - (A-B)\omega_1\omega_2 = 0.$$

But here principal moments of inertias are given as A, A and C, therefore the equations becomes

$$A\dot{\omega}_1 - (B-C)\omega_2\omega_3 = 0 \quad \text{...(1)}$$

$$A\dot{\omega}_2 - (C-A)\omega_3\omega_1 = 0 \quad \text{...(2)}$$

$$C\dot{\omega}_3 = 0 \quad \text{...(3)}$$

Solving (3), we get ω_3 = const. ...(4)

Now if the earth rotates with angular velocity ω about an axis through centre of gravity and making an angle α with the axis of the figure, then initially, we have $\omega_1 = \omega \sin\alpha$, $\omega_2 = 0$, $\omega_3 = \omega\cos\alpha$

Thus the equation (4) becomes $\omega_3 = \omega\cos\alpha$,

Further dividing (1) by (2), we get

$\omega_1\dot{\omega}_1 + \omega_2\dot{\omega}_2 = 0 \quad \Rightarrow \omega_1^2 + \omega_2^2 = a$, where 'a' is const.

But initially $\omega_1 = \omega\sin\alpha$, $\omega_2 = 0 \ \Rightarrow \omega_1^2 + \omega_2^2 = \omega^2\sin^2\alpha$

Now substituting the values of ω_2, ω_3 in (1) and differentiating, we get

$$A\dot{\omega}_1 - (A-C)\omega\cos a\omega_2 = 0$$

i.e. $\ddot{\omega}_1 = \dfrac{A-C}{A}\omega\cos\alpha\,\dot{\omega}_2$

$$= \frac{A-C}{A}\omega\cos\alpha.\frac{C-A}{A}\omega\cos\alpha\omega \quad \text{[using (2)]}$$

$$\therefore \ddot{\omega}_1 = -\left(\frac{C-A}{A}\omega\cos\alpha\right)^2\omega_1 \quad \text{...(7)}$$

Now proportional direction cosines of the instantaneous axis are

$\omega_1, \omega_2, \omega_3.$

$\therefore$ angle between the instantaneous axis and the axis of the figure.

$$= \cos^{-1}\left\{\frac{\omega_3}{\sqrt{(\omega_1^2+\omega_2^2+\omega_3^2)}}\right\} = \cos^{-1}\left\{\frac{\omega\cos\alpha}{\sqrt{(\omega^2\sin^2\alpha+\omega^2\cos^2\alpha}}\right\}$$

$$= \cos^{-1}(\cos\alpha) = \alpha$$

It shows that instantaneous axis describes about the axis of the figure a right cone of semivertical angle α. The period of rotation is

$$= \frac{2\pi}{\frac{C-A}{A}\omega\cos\alpha} = \frac{2\pi}{\frac{(C-A)\omega'}{A}}, \text{ since a is small.}$$

$$= \frac{2\pi A}{(C-A)\omega} = \frac{A}{C-A}, \qquad \left[\because \text{sidereal day} = \frac{2\pi}{\omega}\right]$$

Example 3(a):

The principal moments of inertia of a body at the centre of mass are A, 3A, 6A. The body is so rotated that its angular velocities about the axes are 3n, 2n, n respectively. If in the subsequent motion under no forces $\omega_1, \omega_2, \omega_3$ denote the angular velocities about the principal axes at time t, prove that $\omega_1 = 3\omega_3 = \frac{9n}{\sqrt{5}}$ sech u and ω_3 = 3 n tanh n where $u = 3nt + \frac{1}{2}\log 5$.

Solution:

Since the body moves under no forces, therefore Euler's equations of motion are

$$A\dot{\omega}_1 - (B-C)\omega_2\omega_3 = 0, B\dot{\omega}_2 - (C-A)\omega_3\omega_1 = 0, \text{ and}$$

$$C\dot{\omega}_3 - (A-B)\omega_1\omega_2 = 0.$$

But here the principal moments are given as A, 3A and 6A, therefore the above equation becomes

$$A\dot{\omega}_1 - (3A-6A)\omega_2\omega_3 = 0 \Rightarrow \dot{\omega}_1 = -3\omega_2\omega_3 \qquad ...(1)$$

$$3A\dot{\omega}_2 - (6A-A)\omega_3\omega_1 = 0 \Rightarrow 3\dot{\omega}_2 = -5\omega_3\omega_1 \qquad ...(2)$$

$$6A\dot{\omega}_3 - (A-3A)\omega_1\omega_2 = 0 \Rightarrow 3\dot{\omega}_3 = -\omega_1\omega_2. \qquad ...(3)$$

Now dividing (1) and (3), we get

$$\omega_1\dot{\omega}_1 - 9\omega_3\dot{\omega}_3 = 0$$

Intergrating $\omega_1^2 - 9\omega_3^2 = a$ (const.)

But initially $\omega_1 = 3n, \omega_3 = n$

Thus we have $\omega_1^2 - 9\omega_3^2 = 0 \Rightarrow \omega_1^2 = 9\omega_3^2 \Rightarrow \omega_1 = 3\omega_3$. ...(4)

Again dividing (2) by (3), we get

$$\omega_2\dot{\omega}_2 + 5\omega_3\dot{\omega}_3 = 0 \Rightarrow \omega_2^2 + 5\omega_3^2 = b \text{ (const.)}$$

Initially $\omega_2 = 2n, \omega_3 = n \ \therefore \ b = 9n^2$

Hence $\omega_2^2 + 5\omega_3^2 = 9n^2$. ...(5)

Now substituting $\omega_1 = 3\omega_3$ in (2), we have

$$3\dot{\omega}_2 = 15\omega_3^2 = 15\left[\frac{9n^2 - \omega_2^2}{5}\right] \text{ [using (5)] } \therefore \dot{\omega}_2 = (9n^2 - \omega_2^2)$$

i.e. $dt = \dfrac{d\omega_2}{9n^2 - \omega_2} \quad \therefore t = \dfrac{1}{3n}\tanh^{-1}\left(\dfrac{\omega_2}{3n}\right) + k.$

But when $t = 0, \ \omega_2 = 2n \ \therefore k = -\dfrac{1}{3}\tanh^{-1}\left(\dfrac{2}{3}\right)$

Hence $T = \dfrac{1}{3n}\tanh^{-1}\left(\dfrac{\omega_2}{3n}\right) - \dfrac{1}{3}\lg\sqrt{5}$ *

$$\therefore 3nt + lon\sqrt{5} = \tanh^{-1}\left(\frac{\omega_2}{3n}\right)$$

or $\omega_2 = 3n\tanh[3nt + \log\sqrt{5}] = 3n \tanh u$.

from (5), we have $5\omega_3^2 = 9n^2 - \omega_2^2 = 9n^2 - 9n^2\tanh^2 u = 9n^2 \operatorname{sech}^2 u$

$$\therefore \omega_3 = \frac{3n}{\sqrt{5}}\operatorname{sech} u \Rightarrow \omega_1 = 3\omega_3 = \frac{9n}{\sqrt{5}}\operatorname{sech} u$$

Example 3(b):

A rectangular parallelopiped whose edges are a, 2a, 3a, can turn about its centre and is set rotating about a line perpendicular to the mean axis and making an angle $\cos^{-1}\left(\dfrac{5}{8}\right)$ *with the least axis. Prove that ultimately the body will rotate about the mean axis.*

Solution:

If ω is the initial angular velocity, then we have

$$\omega_1 = \frac{5}{8}\omega, \omega_2 = 0, \omega_3 = \frac{\sqrt{(39)}}{8}, \text{ since } \omega_1^2 + \omega_2^2 + \omega_3^2 = \omega^2.$$

Also $A = \frac{b^2+c^2}{3} = \frac{(2a)^2+(3a)^2}{3} = \frac{13a^2}{3}$

$B = \frac{c^2+a^2}{3} = \frac{(3a)^2+(a)^2}{3}; C = \frac{a^2+b^2}{3} = \frac{a^2+(2a)^2}{3} = \frac{5a^2}{3}.$

Now Euler's equations of motion under no forces are

$A\dot{\omega}_1 - (B-C)\omega_2\omega_3 = 0;\;\; B\dot{\omega}_2 - (C-A)\omega_3\omega_1 = 0;$

$C\dot{\omega}_3 - (A-B)\omega_1\omega_2 = 0.$

Putting in the above equations, the values of A,B, C, we get

$$\frac{13}{3}a^2\dot{\omega}_1 - \left[\frac{10a^2}{3} - \frac{5a^2}{3}\right]\omega_2\omega_3 = 0 \Rightarrow 13\dot{\omega}_1 = 5\omega_2\omega_3 \qquad ...(1)$$

$$\frac{10a^2}{3}\dot{\omega}_2 - \left[\frac{5a^2}{3} - \frac{13a^2}{3}\right]\omega_3\omega_1 = 0 \Rightarrow 5\dot{\omega}_2 = -4\omega_3\omega_1 \qquad ...(2)$$

$$\frac{5a^2}{3}\dot{\omega}_3 - \left[\frac{13a^2}{3} - \frac{10a^2}{3}\right]\omega_1\omega_2 = 0 \Rightarrow 5\dot{\omega}_3 = 3\omega_1\omega_3 \qquad ...(3)$$

Dividing (1) by (2) and cross multiplying, we get

$$52\omega_1\dot{\omega}_1 + 25\dot{\omega}_2 = 0$$

which on integration gives $52\omega_1^2 + 25\omega_2^2 = k$ (const.)

But initially $\omega_1 = \frac{5}{8}\omega \;\; \omega_3 = 0 \quad \therefore k = \frac{325}{16}\omega^2.$

Hence $52\omega_1^2 + 25\omega_2^2 = \frac{325}{16}\omega^2.$

Again dividing (1) by (3) and cross-multiplying, we have

$$39\omega_1\dot{\omega}_1 - 27\omega_3\dot{\omega}_3 = 0 \Rightarrow 39\omega_1^2 - 27\omega_3^2 = d \text{ (const.)}$$

But initially $\omega_1 = \frac{5}{8}\omega, \omega_3 = \frac{\sqrt{(39)}}{8}\omega \;\; \therefore \; d = 0$

$$\therefore 39\omega^2 - 2\omega_3^2 = 0 \Rightarrow \omega_3 = \frac{\sqrt{(39)}}{5}\omega_1.$$

Using (5) in (2), we have

$$\dot{\omega}_2 = -\frac{4}{5}.\frac{\sqrt{(39)}}{5}\omega_1^2 = -\frac{4\sqrt{(39)}}{25}\omega_1^2$$

$$= -\frac{4\sqrt{(39)}}{25\times 52}\left[\frac{325}{16}\omega_\pi - 25\omega_2^2\right]$$

$$= -\frac{\sqrt{3}}{25\sqrt{(13)}}\left[\frac{325}{16}\omega^2 - 25\omega_2^2\right]$$

$$\therefore \frac{\sqrt{3t}}{\sqrt{(13)}} = -\int\frac{d\omega_2}{\frac{13}{16}\omega^2 - \omega_2^2} = -\frac{4}{\omega\sqrt{(13)}}\tanh^{-1}\frac{4\omega_2}{\omega\sqrt{(13)}}$$

$$\therefore \omega_2 = -\frac{\omega\sqrt{(13)}}{4}\tanh^{-1}\left[\left(\frac{\sqrt{3}}{4}\right)\omega t\right]$$

But as $t \to \infty$, $\tanh\left[{}^4\left(\frac{\sqrt{3}}{4}\right)\omega t\right] \to 1 \therefore \omega_2 = -\frac{\omega\sqrt{(13)}}{4}$

Making use of (4) and (5), we get $\omega_1 = 0$ and $\omega_3 = 0$.

So the body ultimately rotates about the mean axis.

Example 4:

A lamina rotating with uniform angular velocity an about an axis through its centre of gravity perpendicular to its plane has an additional angular velocity λn impressed upon it about the axis of least moment, (A <B<C) where $\lambda^2 = (B + A) / (B - A)$. Prove that at time t, its angular velocities are λn sech nt, λn tanh nt, n sech nt.

Also show that it will ultimately revolve about the axis of mean moment.

Solution:

Euler's equations of motion under no forces are:

$$A\dot{\omega}_1 - (B - C)\omega_2\omega_3 = 0;$$

$$B\dot{\omega}_2 - (C - A)\omega_3\omega_1 = 0;$$

$$C\dot{\omega}_3 - (A - B)\omega_1\omega_2 = 0.$$

In this case C = A + B. Putting A + B for C in the above equations, we get

$$*\tanh^{-1}(x/a) = \frac{1}{2}\text{lon}\frac{a+x}{a-x} \Rightarrow \tanh^{-1}(2/3) = \frac{1}{2}\text{long}\left(\frac{3+2}{3-2}\right) = \log\sqrt{(5)}$$

$$A\dot{\omega}_1 + A\omega_2\omega_3 = 0 \Rightarrow \dot{\omega}_1 = -\omega_2\omega_3 \quad ...(1)$$

$$B\dot{\omega}_2 - B\omega_3\omega_1 = 0 \Rightarrow \dot{\omega}_2 = \omega_3\omega_1 \quad ...(2)$$

$$(A + B)\dot{\omega}_3 - (A - B)\omega_1\omega_3 = 0 \Rightarrow \dot{\omega}_3 = -\frac{1}{\lambda^2}\omega_1\omega_2$$

$$\left(\because \lambda^2 = \frac{B + A}{B - A}\right) \quad ...(3)$$

Now dividing (1) by (2) and then cross-multiplying, we get

$\omega_1\dot{\omega}_1 + \omega_2\dot{\omega}_2 = 0 \Rightarrow \omega_1^2 + \omega_2^2 = a,$ where a is constant of integration.

But initially $\omega_1 = \lambda n,, \omega_2 = 0, \omega_3 \ \therefore\ a = \lambda^2 n^2$

Hence $\omega_1^2 + \omega_2^2 = \lambda^2 n^2$ ...(4)

Again dividing (1) by (3) and then cross-multiplying, we have

$\omega_1\dot{\omega}_1 - \lambda^2\dot{\omega}_3\omega_3 = 0 \Rightarrow \omega_1^2 - \lambda^2\omega_3^2 = b,$ where b is a constant of integration.

Initially, we have $\omega_1 = \lambda n,\ \omega_2 = 0,\ \omega_3 = n \ \therefore\ b = 0$

Hence $\omega_1^2 - \lambda^2\omega_3^2 = 0 \Rightarrow \omega_1^2 = \lambda^2\omega_2^2 \Rightarrow \omega_1 = \lambda\dot{\omega}_3$

Further substituting the value of ω_3 from (5) in (2), we get

$$\dot{\omega}_2 = \frac{1}{\lambda}\omega_1^2 = \frac{1}{\lambda}(\lambda^2 n^2 - \omega_2^2) \quad \text{[using (4)]}$$

i.e. $$\frac{d\omega_2}{dt} = \frac{1}{\lambda}(\lambda^2 n^2 - \omega_2^2) \therefore t = \int \frac{\lambda d\omega_2}{\lambda^2 n^2 - \omega_2^2} = \frac{\lambda}{\lambda n}\tanh^{-1}\left(\frac{\omega_2}{\lambda n}\right) + C,$$

where C is a constant.

But when $t = 0,\ \omega_2 = 0,\ \therefore\ C = 0$

So $$t = \frac{1}{n}\tanh^{-1}\left(\frac{\omega_2}{\lambda n}\right) \Rightarrow \omega_2 = \lambda n \tanh(nt)$$ ...(6)

Substituting this value of ω2 in (4), we get

$$\omega_1^2 = \lambda^2 n^2 \tanh^2(nt) = [\lambda nswxh\,(nt)]^2 \Rightarrow \omega_1 = \lambda n\, \text{sech}\,(nt)$$ ...(7)

From (5) we have $\omega_2 = n$ sech (nt)

Further when $t \to \propto$, we have

$$\omega_1 = \lambda n \lim_{t\to\propto} \text{sech}\,(nt) = \lambda n \lim_{t\to\propto} \frac{2}{e^{nt} - e^{-nt}} = 0$$

$$\omega_3 = \lambda n \lim_{t\to\propto} \tanh nt = \lambda n \lim_{t\to\propto} \frac{e^{nt} - e^{-nt}}{e^{nt} + e^{-nt}} = \lambda n \lim_{t\to\propto} \frac{1 - e^{-2nt}}{1 + e^{-ent}} = \lambda n$$

and $\omega_3 = \frac{1}{\lambda}\omega_1 = 0.$ Thus ultimate values of angular velocities are 0, λn, 0. It show that, ultimately, the lamina revolves about the axis of mean moment.

Example 5:

A uniform elliptic disc is free to move about a focus and is set rotation with initial angular velocity Ω about an axis perpendicular to the corresponding latus rectum and making an angle θ with the plane of the disc. If cos 2θ = (A/ B) *when A,B are the moments of inertia of the disc about the major axis and latus rectum respectively, prove that after* time t, the component angular velocity of the disc about the major axis will be

$$\Omega\sqrt{\left(\frac{A+B}{2B}\right)}\operatorname{sech}\left[\Omega t\left(\frac{B-A}{2B}\right)\right].$$

Solution:

The initial axis of rotation is perpendicular to the latus rectum and makes an angle θ with the major axis; will have direction cosines cos θ, cos (90 – θ), cos 90 *i.e.* cosθ, sin θ, 0. Corresponding to the set of principal axes, lactus rectum and a line through the focus and perpendicular to the plane of the disc. Since initial angular velocity is Ω therefore initially, we have

$$\omega_1 = \Omega\cos\theta, \omega_2 = 0, \omega_3 = \Omega\sin\theta$$

Further for lamina. We have C = A + B, therefore Euler's equations under no forces become

$$A\dot{\omega}_1 + A\omega_2\omega_3 = 0 \Rightarrow \dot{\omega}_1 = -\omega_2\omega_3 \quad ...(1)$$

$$B\dot{\omega}_2 - B\omega_3\omega_1 = 0 \Rightarrow \dot{\omega}_2 = -\omega_3\omega_1 \quad ...(2)$$

$$(A+B)\dot{\omega}_3 - (A-B)\omega_1\omega_2 = 0 \Rightarrow \dot{\omega}_3 = -\frac{B-A}{A+B}\omega_1\omega_2 \quad ...(3)$$

Dividing (1) by (2), we get $\omega_1\dot{\omega}_1 + \omega_2\dot{\omega}_2 + \omega_2\dot{\omega}_2 = 0 \Rightarrow \omega_1^2 + \omega_2^2 = a$

where a is a constant of integration. Since initially

$\omega_1 = \Omega\cos\theta, \omega_2 = 0,$ therefore $\Omega^2\cos^2\theta.$

Hence $\omega_1^2 + \omega_2^2 = \Omega^2\cos^2\theta$ ...(4)

Again dividing (1) by (3), we get

$$\omega_1\dot{\omega}_1 - \frac{B+A}{B-A}\omega_3\dot{\omega}_3 = \Rightarrow \omega_1^2 - \frac{B+A}{B-A}\omega_3^2 = b$$

where b is a constant of integration. Initially

$$\omega_1 = \Omega\cos\theta, \omega_3 = \Omega\sin\theta$$

$$\therefore b = \Omega^2\left[\cos^2\theta - \frac{B+A}{B-A}\sin^2\theta\right] = \Omega^2\frac{(B-A)\cos^2\theta - (A+B)\sin^2\theta}{B-A}$$

$$= \frac{B(\cos^2\theta - \sin^2\theta - A}{B-A} = \frac{B\cos^2\theta - A}{B-A} = 0 \qquad [\because \cos 2\theta = (A/B)]$$

Hence $\omega^2 = \frac{B+A}{B-A}\omega_3^2$

Substituting the value of ω_3 from (5) in (2), we get

$$\dot{\omega}_2 = \sqrt{\left(\frac{B-A}{B+A}\right)}\omega_1^2 = \sqrt{\left(\frac{B-A}{B+A}\right)}\left[\Omega^2\cos^2\theta - \omega_2^2\right] \qquad \text{[using (4)]}$$

$$\therefore t = \sqrt{\left(\frac{B+A}{B-A}\right)}\int\frac{d\omega_2}{\Omega^2\cos^2\theta - \omega_2^2}$$

$$= \sqrt{\left(\frac{B+A}{B-A}\right)}\frac{1}{\Omega\cos\theta}\tanh^{-1}\left\{\frac{\omega_2}{\Omega\cos\theta}\right\}$$

the constant of integration vanish since $\omega_2 = 0$ when $t = 0$

$$= \frac{1}{\Omega}\sqrt{\left(\frac{B+A}{B-A}\right)}\sqrt{\frac{2B}{B+A}}\tanh^{-1}\left\{\frac{\omega_2}{\Omega}\left(\frac{2B}{B+A}\right)\right\}*$$

$$\therefore \omega_2 = \Omega\sqrt{\left(\frac{B+A}{2B}\right)}\tanh^{-1}\left[\Omega t\sqrt{\frac{B-A}{2B}}\right]$$

Substituting this value of ω_2 in (4) we get

$$\omega_1 = \Omega\sqrt{\left(\frac{B+A}{2B}\right)}\operatorname{sech}\left\{\Omega t\sqrt{\frac{B-A}{2B}}\right\}$$

Example 6:

If initially the axis of the top is horizontal and it is set spinning with angular velocity ω *in* a horizontal plane, prove that the axis will start to rise if Cn ω> *mgh, and that when Cn* $\omega = 2m\,gh$*, the axis will rise to an angular distance.* $\cos^{-1}$ (A ω / Cn), *provided that* $A\omega < Cn$*, and will be there at an instantaneous rest.*

Solution:

We have equations of motion of the top as

$$A\ddot{\theta} + Cn\,\dot{\phi}\sin\theta - A\dot{\phi}^2\sin\theta\cos\theta = mgh\sin\theta \qquad ...(1)$$

$$A\,\dot{\phi}\sin^2\theta + Cn\cos\theta = D = A\omega \qquad ...(2)$$

Initially $\dot{\theta} = \frac{1}{2}\pi$ and $\dot{\phi} = \omega$, $\therefore D = A\omega$

Thus initially from (1) $A\dot{\theta} = -(cn\,\omega - mgh)$

which is negative if Cn ω> mgh. In other words, the axis will start to rise if Cn ω> mgh.

Now, eliminating ϕ between (1) and (2) we get

$$A\ddot{\theta} + Cn\frac{A\omega - Cn\cos\theta}{A\sin^3\theta}\sin\theta - A\left(\frac{A\omega - Cn\cos\theta}{A\sin^2\theta}\right)^2\sin\theta\cos\theta$$

$$= mgh\sin\theta$$

$$\text{or } A\ddot{\theta} + Cn\frac{A\omega - Cn\cos\theta}{A\sin^3\theta} - \frac{B\omega - Cn\cos\theta}{A\sin^2\theta}\cos\theta = \frac{Cn\omega\sin\theta}{2}$$

(putting mgh $= \frac{1}{2}$ Cn ω)

$$A\ddot{\theta} = -\frac{A^2\omega^2 + C^2n^2}{A}\frac{\cos\theta}{\sin^3\theta} + Cn\,\omega\left\{\frac{1}{2}\sin\theta - \left(\frac{1}{\sin\theta} + \frac{2\cos\theta}{\sin^3\theta}\right)\right\}$$

Integrating it, $A\dot{\theta}^2 = -\dfrac{A^2\omega^2 + C^2n^2}{A\sin^2\theta} + Cn\,\omega\left(-\cos\theta + \dfrac{2\cos\theta}{\sin^2\theta}\right) + C'$

where C′ is constant.

But initially $\theta = \frac{1}{2}\pi, \dot{\theta} = 0, \quad \therefore C' = \dfrac{A^2\omega^2 + C^2n^2}{A}$

$$\therefore A\dot{\theta}^2 = -\frac{A^2\omega^2 + C^2n^2}{A\sin^2\theta} + Cn\omega\frac{(-\sin^2\theta\cos\theta + 2\cos\theta)}{\sin^2\theta} + \frac{A^2\omega^2 + C^2n^2}{A}$$

$$\text{or } A^2\dot{\theta}^2\sin^2\theta = -(A^2\omega^2 + C^2n^2)\cos^2\theta + Cn\,\omega A(-\sin^2\theta + 2)\cos\theta$$

$$= -(A^2\omega^2 + C^2n^2)\cos^2\theta + Cn\,\omega A(1 + \cos^2\theta)\cos\theta$$

$$= \cos\theta\,(A\omega - Cn\cos\theta)\,(Cn - A\omega\cos\theta)$$

For instantaneous res $\dot{\theta} = 0$

in which cos cosq (A ω – Cn cos θ) (Cn – A ω cos θ) = 0

i.e. when cos θ = 0 or $\theta = \frac{1}{2}\pi$ which is the initial positions.

and when A ω – Cn cos θ = 0 or cos $\theta = \dfrac{A\omega}{Cn}$

i.e. $\theta = \cos^{-1}\dfrac{A\omega}{Cn}$ provided Aω < Cn

and when Cn – Aω cos θ or $\theta = \cos^{-1}\left(\dfrac{Cn}{A\omega}\right)$, which is impossible when

Aω < Cn.

Example 7:

A top consists of a thin uniform spherical shell of radius a and centre C and is free to move about a fixed point O of its surface. The radius OC makes an angle θ with upward down vertical Oz and φ is the angle between the plane zOC and a fixed vertical plane zOx.

Initially $\theta = \frac{1}{2}\pi, \dot{\phi} = \left(\frac{2n}{5}\right)$, where $n = \sqrt{\left(\frac{15g}{2a}\right)}$

Show that at time t, $\cos\theta = \tanh^2 \left\{\left(\frac{3g}{10a}\right)\right\}$.

Solution:

We have equations of motion of the top as

$$A\,(\dot{\theta}^2 + \dot{\phi}^2 \sin^2\theta) + 2mgh\cos\theta = E$$

and $A\,\dot{\phi}^2 \sin^2\theta + Cn\cos\theta = D$

where D,E are constants to be determined by initial conditions.

Initially $\theta = \frac{1}{2}\pi,\ \theta = 0,\ \dot{\phi} = \frac{2n}{5}$.

Putting these in the above equations, we get

$$E = A\left(\dot{\theta} + \frac{4n^2}{25}\right) = (4n^2/25)A \text{ and } D = A\frac{2n}{5} = \frac{2n}{5}A$$

Thus the above equations can be written as

$$A(\dot{\theta}^2 + \dot{\phi}^2 \sin^2\theta) + 2mgh\cos\theta = \left(\frac{4n^2}{25}\right)A \qquad \text{...(1)}$$

and $A\,\dot{\phi}\sin^2\theta + Cn\cos\theta = (2n/5)A$ ...(2)

Again C = moment of inertia about OC

= moment of inertia of a hollow sphere about a diameter

= $(2ma^2/3)$

A = moment of inertia about OA

= moment of inertia of a hollow sphere about a line at ⊥ OC

= $(2ma^2/3) + ma^2 = (5ma^2/3)$

and h = distance of C.G. from O = a

$$\text{or} \quad u = \tanh\frac{\lambda t}{2} \Rightarrow u^2 = \frac{\sinh^2 \frac{1}{2}\lambda t}{\cosh^2 \frac{1}{2}\lambda t} = \frac{\cosh \lambda t - 1}{\cosh \lambda t + 1} = \frac{1 - \text{sech}\,\lambda t}{1 + \text{sech}\,\lambda t}$$

$$\therefore 1 + u^2 = \frac{2}{1 + \text{sech}\,\lambda t} \quad i.e. \; 2\cos\theta = \frac{2}{1 + \text{sech}\,\lambda t}$$

or $\sec\theta = 1 + \text{sech}\,\lambda t$, or $\sec\theta = 1 + \text{sech}\sqrt{(mgh/A)}\,t$.

Example 8:

A rigid body possesses an axis of symmetry OC and moves about O under a retarding couple $\lambda C w_3$ about OC; λ being a constant : A,A, C being the principal moments of inertia at O, and ω_1, ω_2, ω_3 being the angular velocity components about the principal axes, the third axis being OC, Initially the body is given an angular velocity ω about a line inclined at α to OC. Prove that ultimately the instantaneous axis will be perpendicular to OC.

Solution:

Initially we have $\omega_1 = 0$, $\omega_2, = \omega \sin\alpha$, $\omega_3, = \omega\cos\alpha$ Euler's general equations of motion are

$$A\dot{\omega}_1 - (B - C)\omega_2\omega_3 = L;$$

$$B\dot{\omega}_2 - (C - A)\omega_3\omega_1 = M;$$

$$C\dot{\omega}_3 - (A - B)\omega_1\omega_2 = N;$$

But $L = 0$, $M = 0$, $N = -\lambda C\omega_3$ and $B = A$

Therefore the above equations becomes

$$A\dot{\omega}_1 - (B - C)\omega_2\omega_3 = 0;$$

$$B\dot{\omega}_2 - (C - A)\omega_3\omega_1 = 0;$$

$$C\dot{\omega}_3 = -\lambda C\omega_3 \Rightarrow \dot{\omega}_1 = -\lambda\omega_3.$$

i.e. $\dfrac{d\omega_3}{dt} = -\lambda\omega_3 \Rightarrow \omega_3 = ke^{-\lambda t}$, where k is constant.

But when $t = 0$, $\omega_3 = \omega\cos\alpha \therefore k = \omega\cos\alpha$,

Hence $\omega_3 = \omega\cos\alpha e^{-\lambda t}$ ...(1)

Now when $t \to \infty, \omega_3 = \omega\cos\alpha \lim_{t\to\infty} e^{-\lambda} = \cos\alpha \lim_{t\to\infty} \dfrac{i}{\lambda t} = 0$

It shows that ultimately instantaneous axis is perpendicular to OC.

Example 9:

A body having an axis of symmetry OC, moves about a fixed O under no forces, except a constant retarding complex-kC about the axis OC If A,A, C are the moments of inertia about ω_1, ω_2, ω_3 *the angular velocities about the principal axes OA, OB, OC, show that at time t,*

$$\omega_1 = \Omega \cos \lambda t\left(\Omega - \frac{1}{2}kt\right), \omega_2 = -\omega \sin \lambda t\left(\Omega - \frac{1}{2}tk\right), \omega_3 = \Omega - kt$$

where $\lambda = \dfrac{A-C}{A}$, *the initial values of* $\omega_1, \omega_2, \omega_3$ *being* Ω, 0, Ω *respectively.*

Solution:

In this case $L = M = 0$, $N = -kC$ and $B \equiv A$

Putting these values in Euler's dynamical equations, we get

$$A\dot{\omega}_1 \; (A - C)\omega_2\omega_3 = 0 \Rightarrow \dot{\omega}_1 = \lambda\omega_2\omega_3, \text{ since } \lambda = \frac{A-C}{A} \qquad ...(1)$$

$$A\dot{\omega}_2 - (C - A)\omega_3\omega_1 = 0 \Rightarrow \dot{\omega}_2 = -\lambda\omega_3\omega_1, \qquad ...(2)$$

$C\dot{\omega}_3 = -kC \Rightarrow \dot{\omega}_3 = -k \Rightarrow \omega_3 = -kt + a$, where a is a constant of integration. Initially we have $\omega_3 = \Omega \therefore a = \Omega$.

Hence $\omega_3 = \Omega - kt$ ...(3)

Substituting this value of ω_3 in (1) and (2), we have

$$\omega_1 = \Omega\,(\Omega - kt)\omega_2 \qquad ...(4)$$

and $\omega_2 = -\lambda(\Omega - kt)\omega_1$ ...(5)

Multiplying (5) by in and adding to (4), we get

$$\dot{\omega}_1 + i\dot{\omega}_2 = \lambda(\Omega - kt)[\omega_2 - i\omega_1] = -i\lambda(\Omega - kt)(\omega_1 + i\omega_2)]$$

Putting $\omega_1 + i\omega_2 = u$ so that $\dot{\omega}_1 + i\dot{\omega}_2 = \dot{u}$, we have

$$\dot{u} = -i\lambda(\Omega - kt)u \Rightarrow \frac{du}{u} = -i\lambda(\Omega - kt)dt$$

integrating, we get $\log \dot{u} = -i\lambda(\Omega t - \frac{1}{2}kt^2) + b$ where b is a constant of integration.

But when $t \to 0$, $u = \omega_1 + i\omega_2 = \Omega + 0 = \Omega \therefore b = \log\Omega$

$$\therefore \log u - \log\Omega = -i\lambda(\Omega t - \frac{1}{2}kt^2) \Rightarrow \log\frac{u}{\Omega} = -i\lambda(\Omega t - \frac{1}{2}kt^2)$$

$$\Rightarrow u = \Omega e - i\lambda(\Omega t - \frac{1}{2}kt^2)$$

$$\Rightarrow \omega_1 + i\omega_2 = \Omega[\cos\{\lambda(\Omega t - \frac{1}{2}kt^2\} - t\sin\{\lambda(\Omega t - \frac{1}{2}kt^2)\}]$$

$$\Rightarrow \omega_1 = \Omega\cos\{\lambda t(\Omega - \frac{1}{2}kt)\} \text{ and } \Rightarrow \omega_2 = -\Omega\sin\{\lambda t(\Omega - \frac{1}{2}kt)\}$$

Hence we get the required values of $\omega_1, \omega_2, \omega_3$.

$$*\cos\ 2\theta = \frac{A}{B} \Rightarrow 2\cos^2\theta - 1 = \frac{A}{B} \Rightarrow \cos^2\theta\frac{B+A}{2B} \Rightarrow \cos\theta = \sqrt{\left(\frac{B+A}{2B}\right)}.$$

Example 10:

Show that in free motion of the body with an axis of symmetry C about its centre of gravity if n denotes the spin about the axis C and ψ *denotes the Euler's third angle, then* $A\psi sym = (A - C)n$.

Solution:

Here B = A.

$\therefore$ Euler's equation of motion under no forces are

$$A\dot{\omega}_1 - (A - C)\omega_2\omega_3 = 0 \Rightarrow \dot{\omega}_1 = -\frac{(A-C)}{A}\omega_2\omega_3 \qquad ...(1)$$

$$A\dot{\omega}_2 - (C - A)\omega_3\omega_1 = 0 \Rightarrow \dot{\omega}_2 = -\frac{(A-C)}{A}\omega_3\omega_1 \qquad ...(2)$$

$$C\dot{\omega}_3 = 0 \Rightarrow \dot{\omega}_3 = 0 \Rightarrow \omega_3 = \text{const} = n \qquad ...(3)$$

Solving (1) and (2), we get

$$\omega_1\dot{\omega}_1 + \omega_2\dot{\omega}_2 = 0 \Rightarrow \omega_1^2 + \omega_2^2 = a, \text{ where a is const.}$$

Now the direction cosines of the invariable line are proportional to A ω_1, Aω_2, Cω_3 and the direction cosines of the axis of symmetry are 0, 0, 1. If θ is the angle between them

$$\therefore \cos\theta = \frac{C\omega_3}{\sqrt{[(A\omega_1)^2 + (A\omega_2)^2 + (C\omega_3)^2]}} \qquad \text{(Note B = A)}$$

$$= \frac{Cn}{\sqrt{[A^2(\omega_1^2 + \omega_2^2) + C^2n^2]}} \qquad ...(4)$$

It shows that $\cos\theta$ = const. $\Rightarrow \theta$ = const $\Rightarrow \dot{\theta} = 0$

Now substituting $\dot{\theta} = 0$ in Euler's geometrical equations, we get

$$\omega_1 = -\dot{\phi}\sin\dot{\theta}\cos\psi \quad ...(5)$$

$$\omega_2 = \dot{\phi}\sin\dot{\theta}\sin\psi \quad ...(6)$$

$$\omega_3 = \dot{\psi} + \dot{\phi}\cos\theta$$

Squaring and adding (5) and (6), we get $\omega_1^2 + \omega_2^2 = \dot{\phi}^2 \sin^2\theta$

Substituting this value of $\omega_1^2 + \omega_2^2$ in (4), we get

$$\cos^2\theta = \frac{C^2n^2}{A^2\dot{\phi}^2\sin^2\theta + C^2n^2}$$

$$\Rightarrow (A^2\dot{\phi}^2\sin^2\theta + C^2n^2)\cos^2\theta = C^2n^2 = C^2n^2\sin^2\theta$$

$$\Rightarrow A^2\dot{\phi}^2\sin^2\theta\cos^2\theta$$

$$\Rightarrow \dot{\phi}^2\cos^2\theta = (C^2n^2/A^2) \Rightarrow (\omega_3 - \dot{\psi}) = Cn/A \quad \text{[using (7)]}$$

$$\Rightarrow A(n-\dot{\psi}) = \Rightarrow (A-C)n = A\dot{\psi}$$

Example 11:

Show that the vertical pressure of the top on the point of support is equal to its weight when the inclination of its axis to the vertical is given by the least root of the equation

$$3\ \text{Amgh}\cos^2\theta - \cos\theta\,(C^2n^2 + 2\ \text{Amgbh}) + (C^2n^2a - \text{Amgh}) = 0$$

where a and b are constants depending on the initial circumstances of the motion

Solution:

We have equations of motion of the top as

$$-A\ddot{\phi}\sin\theta - 2A\dot{\phi}\,\dot{\theta}\cos\theta + Cn\dot{\theta} = 0 \quad ...(1)$$

$$A\ddot{\theta} + Cn\,\dot{\phi}\sin\theta - A\dot{\phi}^2\sin\theta\cos\theta = Mgh\sin\theta \quad ...(2)$$

$$A\dot{\phi}\sin^2\theta + Cn\cos\theta = D \quad ...(3)$$

and $A(\dot{\theta}^2 + \dot{\phi}^2\sin^2\theta) + 2Mgh\cos\theta = E$ ...(4)

Let the vertical component of the reaction at the vertex O be Y, thus for vertical motion of the top, we have

$$M\frac{d^2}{dt^2}(h\cos\theta) = Y - Mg$$

In case $Y = Mg$, then $\frac{d^2}{dt^2}(\cos\theta) = 0$ ($\because$ h is constant)

or $\sin\theta\ddot{\theta} + \cos\theta\dot{\theta}^2 = 0.$

Now, putting for $\ddot{\theta}$ and $\dot{\theta}^2$ from (2) and (4)

and writing E = 2 Mgh b, we get

$$\sin\theta\,(A\dot{\phi}^2\sin\theta\cos\theta - Cn\dot{\phi}\sin\theta + Mgh\sin\theta)$$

$$+\cos\theta\,(-A\dot{\phi}^2\sin^2\theta + 2M\,ghb + 2Mgh\,b\cos\theta) = 0$$

or $\sin\theta\,(-Cn\dot{\phi}\sin\theta + Mgh\sin\theta)$

$+\cos\theta\,(2M\,ghb - 2M\,gh\,b\cos\theta) = 0$

or $2M\,gh\,(\sin^2\theta + 2b\cos\theta - 2\cos^2\theta) = Cn\dot{\phi}\sin^2\theta.$

In this put for ϕ from (3) and putting D = Cna

We obtain $M\,gh\,(\sin^2\theta + 2b\cos\theta - 2\cos^2\theta) = \frac{Cn}{A}Cn\,(a - \cos\theta)$

or $AM\,gh\,(1 + 2b\cos\theta - 3\cos^2\theta) = C^2n^2\,(a - \cos\theta)$

or $3AM\,gh\cos^2\theta - \cos\theta\,C^2n^2 + 2AM\,gh\,b) + (C^2n^2a - AM\,gh) = 0$

Example 12:

A rectangular lamina ABCD in which BC = $\sqrt{2AB}$ *can turn freely about the middle point O of AD. Initially it is rotating with angular velocity* Ω *about a line through O perpendicular to AD and making an angle 30° with the plane of the rectangle. Show that after time t, the components of the angular velocity of the rectangle about the principal axes at O are*

$$\frac{\sqrt{3}}{2}\operatorname{sech}\left(\frac{1}{2}\Omega t\right), \frac{\sqrt{3}}{2}\Omega\tanh\left(\frac{1}{2}\Omega\tanh\right)\frac{1}{2}\Omega\operatorname{sech}\left(\frac{1}{2}\Omega t\right)$$

Solution:

If G is the C, G. of the lamina ABCD, and OP, the line normal to its plane then OG, OA, OP form a set of principal axes at O. Since it is rotating about the line OR which is perpendicular to OA, therefore OR lies in the plane GOP.

Since $\angle GOR = 30°$,

$\angle POR = 60°$

and $\angle AOR = 90°$

Direction cosines of OR are

cos 30°, cos90°, cos60° *i.e.* $\frac{\sqrt{3}}{2}, 1, \frac{1}{2}.$

Now if we take AB = 2a then BC = $2\sqrt{2a.}$

A = moment of inertia about $OG = \frac{1}{2}M\left(\frac{1}{2}BC\right)^2 = \frac{2Ma^2}{3}$

B = moment of inertia about

$OA = \frac{1}{3}M\left[\left(\frac{1}{2}AB\right)^2 + \left(\frac{1}{3}AB\right)^2 \frac{2Ma^2}{3}\right] = \frac{4Ma^2}{3}$

C = moment of inertia about OP

$= \frac{1}{3}M\left\{\frac{1}{2}AB)^2 + \left(\frac{1}{2}BC\right)^2\right\} + MOG^2 = \frac{1}{3}M(3a^2) + Ma^2 = 2Ma^2$

Euler's equations of motion under no forces are

$A\dot{\omega}_1 - (B - C)\omega_2\omega_3 = 0;$

$B\dot{\omega}_2 - (C - A)\omega_3\omega_1 = 0;$ ...(1)

$C\dot{\omega}_2 - (A - B)\omega_1\omega_2 = 0.$

Substituting the values A, B, C in (1), we get

$\frac{2Ma^2}{3}\dot{\omega}_1 + \frac{2Ma^2}{3}\omega_2\omega_3 = 0 \Rightarrow \dot{\omega}_1 = -\omega_2\omega_3$ ...(2)

$\frac{4Ma^2}{3}\dot{\omega}_2 = \frac{4Ma^2}{3}\omega_3\omega_1 = 0 \Rightarrow \dot{\omega}_2 = \omega_3\omega_1$ ...(3)

$2Ma^2\dot{\omega}_3 + \frac{2Ma^2}{3}\omega_1\omega_2 = 0 \Rightarrow \dot{\omega}_3 = -\frac{1}{3}\omega_1\omega_2.$ ...(4)

Solving (2) and (3), we get $\omega_1\omega_1 = \omega_2\dot{\omega}_2 = 0 \Rightarrow \omega_1^2 + \omega_2^2 = a,$

where a is a constant of integration.

But initially $\omega_1 = \frac{\sqrt{3}}{2}\Omega, \omega_2 = 0$

and $\omega_3 = \frac{1}{3}\Omega_1 \therefore a = \frac{3}{4}\Omega^2.$

Hence $\omega_1^2 + \omega_2^2 = \frac{3}{4}\Omega^2.$ ...(5)

Again solving (2) and (4), we get

$\omega_1\dot{\omega}_1 - 3\omega_3\dot{\omega}_3 = 0 \Rightarrow \omega_1^2 - 3\omega_3^2 = b.$

where b is constant of integration.

Initially $\omega_1 = \frac{1}{3}\sqrt{3\Omega}$ and $\omega_3 = \frac{1}{2}\Omega$ $\therefore$ b = 0.

Hence $\omega_1^2 - 3\omega_3^2 = 0 \Rightarrow \omega_1\sqrt{3\omega_3}$. ...(6)

Substituting the value for ω_3 from (6) in (2), we have

$$\dot{\omega}_2 = \frac{1}{\sqrt{3}}\omega_1 = \frac{1}{\sqrt{3}}\left(\frac{3}{4}\Omega^2 - \omega_2^2\right) \qquad \text{[using (5)]}$$

i.e. $= \frac{d\omega_2}{dt} = \frac{1}{\sqrt{3}}\left(\frac{3}{4}\Omega^2 - \omega_2^2\right) \Rightarrow dt = \sqrt{3}.\frac{d\omega_2}{\frac{1}{4}\Omega^2 - \omega_2^2}$

$\Rightarrow t = \sqrt{3}.\frac{2}{\sqrt{3\Omega}}\tanh^{-1}\left(\frac{2\omega_2}{\sqrt{3\Omega}}\right) + C$, where C is constant.

Now $t = 0,\ \omega_2 = 0,\ \therefore\ C = 0$.

Hence $t = \frac{2}{\Omega}\tanh^{-1}\frac{2\omega_2}{\sqrt{3\Omega}} \Rightarrow \omega_2 = \frac{\sqrt{3}}{2}\Omega\tanh\left(\frac{1}{2}\Omega t\right)$

Putting this value of ω_2 in (5), we get $\omega_1 = \frac{\sqrt{3}}{2}\Omega\,\mathrm{sech}\left(\frac{1}{2}\Omega t\right)$

And from (6), we get $\omega_2 = \frac{1}{2}\mathrm{sech}\left(\frac{1}{2}\Omega t\right)$.

Thus angular velocities at time t are evaluated.

Example 13:

A uniform right circular cone of vertical angle 2α, moves under no forces except at its vertex, which is fixed. It is set rotating about a generator. Show that its axis describes in a space a right cone of angle 2β where

$$\tan\beta = \frac{1}{2}\tan\alpha + 2\cos\alpha$$

Solution:

Let OP be the generator and OZ the axis of the cone, whose semi-vertical angle is α. Let. OX and. OY be any two perpendicular lines perpendicular to OZ. Then OX, OY, OZ form a set of principal axes at O. The direction cosines of OP are cos $(90 - \alpha)$, cos 90, cos α *i.e.* sin α, 0, cos α. If the initial angular velocity is ω, then initially, we have $\omega_1 = \omega\sin\alpha, \omega_2 = 0, \omega_3 = \omega\cos\alpha$

Eular's equations of the motion under no forces are

$A\dot{\omega}_1 - (B - C)\omega_2\omega_3 = 0;$

$$B\dot{\omega}_2 - (C - A)\omega_3\omega_1 = 0;$$

$$C\dot{\omega}_3 - (A - B)\omega_1\omega_2 = 0. \quad ...(1)$$

But here A = B, therefore (1) becomes

$$A\dot{\omega}_1 = (A - C)\omega_2\omega_3 \Rightarrow \dot{\omega}_1 = \frac{(A - C}{A}\omega_2\omega_3 \quad ...(2)$$

$$A\dot{\omega}_2 = -(A - C)\omega_3\omega_1 = 0 \Rightarrow \omega_2 = -\frac{(A - C}{A}\omega_3\omega_1 \quad ...(3)$$

$$C\dot{\omega}_3 = 0 \Rightarrow \dot{\omega}_3 = \text{const} = \omega\cos\alpha \quad ...(4)$$

Hence $\omega_2^2 + \omega_2^2 = \omega^2\sin^2\alpha$ ...(5)

Now the direction cosines of the invariable line are proportional to $A\omega_1, B\omega_2, C\omega_3$.

If the invariable line makes axis (OZ) of the with the figure, an angle θ then $\cos\theta = \frac{C\omega_3}{\sqrt{(A^2\omega_1^2 + B^2\omega_2^2 + C^2\omega_3^2)}}$

But in this case B = A

$$\therefore \cos\theta = \frac{C\omega_3}{\sqrt{(A^2\omega_1^2 + \omega_2^2) + C^2\omega_3^2]}} = \frac{C\omega\cos\alpha}{\sqrt{[A^2\omega^2\sin^2\alpha + C^2\omega^2\cos^2\alpha]}}$$

$$= \frac{C}{\sqrt{[a^2\tan^2\alpha\ C^2]}} \text{ [multiplying num. and denom. by sec } \alpha]$$

$$\Rightarrow \tan\theta = \frac{A\tan\alpha}{C}$$

$$\text{Now } A = B = \frac{3M}{80}(4a^2 + h^2) + M\left(\frac{3h}{4}\right)^2 = \frac{3M}{20}(a^2 + 4h^2)\ [h = a\cot\alpha]$$

Substituting these values of A and C, we get

$$\tan\theta = \frac{1}{2a^2}(a^2 + 4h^2)\ \tan\alpha = \frac{1}{2}\left(1 + 4\frac{h^2}{a^2}\right)\tan\alpha$$

$$= \frac{1}{2}(1 + 4\cot^2\alpha)\tan\alpha = \frac{1}{2}\tan\alpha + 2\cot\alpha$$

It shows that value of tan θ is constant, implying thereby that θ remains constant throughout the motion. So the axis of the cone of semi vertical angle β such that

$$\tan\beta = \frac{1}{2}\tan\alpha + 2\cot\alpha \text{ as } \theta = \beta.$$

Example 14:

Show that in order that it may be possible for a uniform cone, of height h and vertical angle 2α, to spin a top on a rough horizontal plane it is necessary that

$$n^2 > \frac{5g(4+\tan^2\alpha)\sin\alpha}{h\tan^4\alpha}$$

Solution:

The spinning is possible so long as the slant side just touches the horizontal plane. The motion then will be steady with the axis OC of the top making angle $\frac{1}{2}\pi - \alpha$ with the vertical OZ.

The condition for steady motion is

$$C^2n^2 > 4A\,mgh\cos\alpha \qquad ...(1)$$

where A is M.I. about OA and C is M.I about OC (the axis on the top). In this case A = Moment of inertia about OA

= moment of inertia about a line through O ⊥ to OC

$$= \frac{3m}{20}(a^2+4h^2) = \frac{3m}{20}(h^2\tan^2\alpha + 4h^2) \qquad \left(\because \tan\alpha = \frac{a}{h}\right)$$

$$= \frac{3mh^2}{10}(4+\tan^2\alpha)$$

C = moment of inertia about OC (the axis of the top)

= moment of inertia about the axis of the cone

$$= \frac{3ma^2}{10} = \frac{3mh^2}{10}\tan^2\alpha$$

The distance of the C.G. of the top from the vertex is $\frac{3h}{4}$ and the axis of the top is making an angle $\frac{1}{2}\pi - \alpha$ with the vertical.

Putting these values in the condition for steady motion, we have

$$\left(\frac{3mh^2\tan^2\alpha}{10}\right)^2 n^2 > 4\frac{3mh^2}{20}(4+\tan^2\alpha)mg\frac{3h}{4}\cos\left(\frac{1}{2}\pi - \alpha\right)$$

$$\text{or } n^2 > \frac{5g(4+\tan^2\alpha)\sin\alpha}{h\tan^4\alpha}$$

Example 15:

A uniaxial body is supported at its centre of mass and is rotating initially with angular velocity ω about an axis perpendicular to the axis of symmetry.

Prove that if a couple of constant moment N is applied about the axis of symmetry, the instantaneous axis will describe a cone whose equation referred to axis fixed in the body, of which that of z-axis coincide with the axis of symmetry, is

$$2AN\,(x^2 + y^2)^2 \tan^{-1}(y/x) = C\,(C - A)\omega^2 z^2$$

Solution:

Euler's equations of motion in general form are given by

$$A\frac{d\omega_1}{dt} - (B - C)\omega_2\omega_3 = L;$$

$$B\frac{d\omega_2}{dt} - (C - A)\omega_3\omega_1 = M;$$

$$C\frac{d\omega_3}{dt} - (A - B)\omega_1\omega_2 = N.$$

Here, we have L = 0, M = 0, N = N and B = A.

$$\Rightarrow A\frac{d\omega_1}{dt} - (A - C)\omega_1\omega_2 = 0, \text{ or } \dot{\omega}_1 = \lambda\omega_2\omega_3, \qquad \text{...(1)}$$

$$\left[\text{Putting } \lambda = \frac{A - C}{A}\right]$$

$$\Rightarrow A\frac{d\omega_2}{dt} - (C - A)\omega_3\omega_1 = 0 \text{ or } \dot{\omega}_2 = 0\,\lambda\omega_3\omega_1, \qquad \text{...(2)}$$

$$\text{and} \Rightarrow C\frac{d\omega_3}{dt} = N, \text{ or } C\dot{\omega}_3 = N. \qquad \text{...(3)}$$

Dividing (1) by (2), we obtain $\frac{\dot{\omega}_1}{\dot{\omega}_2} = -\frac{\omega_2}{\omega_1}$, or $\omega_1\dot{\omega}_1 + \omega_2\dot{\omega}_2 = 0$

$\Rightarrow \omega_1^2 + \omega_2^2 = a$, where a is constant.

Initially $\omega_1 = \omega$, $\omega_2 = 0$,

$$\therefore a = \omega^2 \Rightarrow \omega_1^2 + \omega_2^2 \qquad \text{...(4)}$$

Equation (3) $\Rightarrow C\omega_3 = N$.

$\Rightarrow Cw_3 = Nt + b$ where b is constant

Initially t = 0, $w_3 = 0$, $\Rightarrow b = 0$.

$$\therefore C\omega_3 = N\,t \Rightarrow \omega_3 = (N/C)\,t. \qquad \text{...(5)}$$

Then $\therefore$ (1) and (2) reduce to $\dot{\omega}_1 = \frac{N}{C}\lambda t\,\omega_3$ and $\dot{\omega}_2 = -\frac{N}{C}\lambda t\omega_3$

Write $u = \omega_1 + i\omega_2$

Whence differentiating,

$$\frac{du}{dt} = \dot{\omega}_1 + i\dot{\omega}_2 = \frac{N}{C}\lambda t(\omega_2 - i\omega_1), \text{ by the above results.}$$

$$= -i\frac{N}{C}\lambda t(\omega_1 + i\omega_2) = -i\frac{N}{C}\lambda tu, \text{ by the above results.}$$

Now integrating, $\log u = -i\dfrac{N\lambda}{C}\dfrac{t^2}{2} + k$ where k is constant.

But at $t = 0$, $u = \omega_1 + i\omega_2 = \omega$, $\quad \therefore k = \log \omega$

$$\Rightarrow \log\frac{u}{\omega} = -i\frac{\lambda N}{2C}t^2 \Rightarrow u = we^{-i\lambda Nt^2/2C}$$

$$\Rightarrow \omega_1 + i\omega_2 = \omega\left[\cos\left(\frac{\lambda N}{2C}\right)t^2 - i\sin\left(\frac{\lambda N}{2C}\right)t^2\right]$$

Whence equating real and imaginary parts, we obtain

$$\omega_1 = \omega\cos\left(\frac{\lambda N}{2C}\right)t^2, \ \omega_2 = -\omega\sin\left(\frac{\lambda N}{2C}\right)t^2,$$

Putting $\lambda = [(A - C)/A]$, we get

$$\omega_1 = \omega\cos\frac{(C-A)}{2AC}Nt^2 \text{ and } \omega_2 = \omega\sin\frac{(C-A)}{2AC}Nt^2 \qquad ...(6)$$

Let (x,y,z) be a point on the instantaneous axis which pass through the origin and whose direction ratios are ω_1, ω_2 ω_3

$$\frac{x}{\omega_1} = \frac{y}{\omega_2} = \frac{z}{\omega_3} \text{ and } \frac{x^2}{\omega_1^2} = \frac{y^2}{\omega_2^2} = \frac{z^2}{\omega_3^2} \text{ or } \frac{x^2+y^2}{\omega^2} = \frac{z^2}{\omega_3^2}$$

$$\Rightarrow \omega_3^2 = [\omega^2/z^2 + (x^2+y^2)] \text{ and } (\omega_2/\omega_1) = (y/x).$$

$$\therefore (6)(\omega_2/\omega_1) = \tan[\{(C-A)/2AC\}N(C^2/N^2)\omega_3^2] \qquad (\because \omega_3 = Nt/C)$$

$$\Rightarrow (y/x) = \tan\,[\{(C-A)\,C/2AN\}\,(\omega^2 z^2/x^2+y^2)\}$$

$$\Rightarrow 2AN\,(x^2+y^2)\tan^{-1}(y/x) = C\,(C-A)\,\omega^2 z^2.$$

Example 16:

A thin ellipitic plate of semi axes a, b, (a>b) can turn freely about its centre which is fixed. It is set in motion with an angular velocity about an axis in its plane equally inclined to the axes of the ellipse. Show that the instantaneous axis will again be in the plane of the plate after a time,

$$2\int_0^{\lambda}(\lambda^4 - x^4)^{-1/2}\,dx \text{ where } \lambda^2 = \frac{n^2}{a^2}\frac{a^2-b^2}{a^2+b^2}$$

Solution:

Initially; we have

$\omega_1 = n \cos 45°; \omega_2 = n \sin 45°; \omega_3 = 0$

or $\omega_1 = \frac{n}{\sqrt{2}}, \omega_2 = \frac{n}{\sqrt{2}}, \omega_3 = 0$

Now Euler's equations of motion, under no forces, are

$$A\frac{d\omega_1}{dt} = (B - C)\omega_2\omega_3; B\frac{d\omega_2}{dt} = (C - A)\omega_3\omega_1$$

and $C\frac{d\omega_3}{dt} = (A - B)\omega_1\omega_2$

But for the elliptic plate, we have

$$A = \frac{Mb^2}{4}, B = \frac{Ma^2}{4}, \text{ and } C = A + B$$

$$\Rightarrow A\frac{d\omega_2}{dt} = -A\omega_2\omega_3, \text{ or } \dot{\omega}_1 = -\omega_2\omega_3, \quad ...(1)$$

$$\therefore B\frac{d\omega_3}{dt} = B\omega_3\omega_1, \text{ or } \dot{\omega}_2 = \omega_3\omega_1, \quad ...(2)$$

and $(A + B)\ \frac{d\omega_3}{dt} = (A - B)\omega_1\omega_2$ or $\dot{\omega}_3 = -\frac{2\lambda^2}{n^2}\omega_1\omega_2$ $\quad (\because C = A + B)$

$$\left[\because \lambda^2 = \frac{n^2}{2}\frac{a^2 - b^2}{a^2 + b^2} = \frac{n^2 B - A}{2\, B + A}\right]$$

Dividing (2) by (3), we get

$$\frac{\dot{\omega}_2}{\dot{\omega}_3} = \frac{n^2}{2\lambda^2}.\frac{\omega_3}{\omega_2} = 2\lambda^2\omega_2\dot{\omega}_2 = -n^2\omega_3\dot{\omega}_3$$

$\Rightarrow 2\lambda^2\omega\frac{2}{2} = -n^2\omega_3^2 + a$ where a is constant.

Initially. $\omega_2 = \frac{n}{\sqrt{2}}, \omega_3 = 0, \therefore a = \lambda^2 n^2$

$$\Rightarrow 2\lambda^2\omega_2^2 = \lambda^2 n^2 - n^2\omega_3^2 \Rightarrow 2\lambda^2\omega_2^2 = n^2(\lambda^2 - \omega_3^2) \quad ...(i)$$

Again, dividing (1) by (3), we have

$$\frac{\dot{\omega}_1}{\dot{\omega}_2} = \frac{n^2}{2\lambda^2}\frac{\omega_3}{\omega_1} \text{ or } 2\lambda^2\omega_1\dot{\omega}_1 = n^2\omega_3\dot{\omega}_3$$

$2\lambda^2\omega_1 = n^2\omega_3^2 + b$ integrating, where b is constant.

But initially $\omega_1 = \frac{n}{\sqrt{2}}, \omega_3 = 0, \quad \therefore b = \lambda^2 n^2$

$$\Rightarrow 2\lambda^2\omega_1^2 = n^2\omega_3^2 + \lambda^2 n^2 \text{ or } 2\lambda^2\omega_1^2 = n^2(\lambda^2 + \omega_3^2) \qquad ...(ii)$$

Now (i) × (ii) $\Rightarrow 4\lambda^4\omega_1^2\omega_2^2 = n^4(\lambda^2 - \omega_3^4) \Rightarrow \frac{2\lambda^2}{n^2}\omega_1\omega_2 = \pm(\lambda^4 - \omega_3^4)^{1/2}$

Using (3), we get $\omega_3 = (\lambda^4 - \omega_3^4)^{1/2}$ ...(iii)

But ω_3 will be maximum when $\dot{\omega}_3 = 0$, or when $\lambda^4 - \omega_3^4 = 0$ or $\omega_3 = \lambda$

Now integrating (iii,) we get

[choose plus sing, for ω_3 as ω_3 increases from 0 to λ]

and t_2 time from (ω_3 = l) to ω_3 (= 0) is $-\int_{\lambda}^{0} \frac{d\omega_3}{(\lambda^4 - \omega_3^{4 1/2}}$

$$= \int_0^{\lambda} \frac{d\omega_3}{(\lambda^4 - \omega_3^{4 1/2}}$$

[choose negative sign, for ω_3 as ω_3 decreases from λ to 0]

∴ Total time when the instantaneous axis will be again in the plane of the plate, is given by

$$t_1 + t_2 = 2\int_0^{\lambda} \frac{d\omega_3}{(\lambda^4 - \omega_3^4)^{1/2}} = 2\int_0^{\lambda} (\lambda^4 - z^4)^{-1/2} dx$$

Example 17:

A rigid body which is symmetric about an axis has one point fixed on this axis. Discuss the rotational motion of the body, assuming that there are no forces action other than the reaction forces at the fixed point.

Solution:

Let us assume the axis of symmetry coincident with one of the principal axes. Say the one having direction k. Then A = B and Euler's equations give:

$$A\dot{\omega}_1 + (C - A)\omega_3\omega_2 = 0 \qquad ...(1)$$

$$A\dot{\omega}_2 + (A - C)\omega_3\omega_1 = 0 \qquad ...(2)$$

and $$C\dot{\omega}_3 = 0 \qquad ...(3)$$

(3) ⇒ ω_3 = constant = K say, so that (1) and (2) become after dividing

by A, $\dot{\omega}_1 + \left\{\frac{C - A}{A}\right\} K\omega_2 = 0$ ...(4)

and $A, \dot{\omega}_2 + \left\{\frac{A-C}{A}\right\}K\dot{\omega}_1 = 0$ ...(5)

$$\therefore \ddot{\omega}_2 = -\left\{\frac{A-C}{A}\right\}K\dot{\omega}_1 = \left\{\frac{A-C}{A}\right\}K^2\left\{\frac{C-A}{A}\right\}\omega_2$$

$\Rightarrow \ddot{\omega}_2 + \kappa^2\omega_2 = 0$ where $\kappa = \left|\frac{A-C}{A}\right|K.$ ...(6)

Solving (6), we obtain $\omega_3 = \alpha \cos \kappa t + \beta \sin \kappa t.$ ...(7)

Now have the time scale in such a way that t = 0 Þ $\omega_2 = 0$.

$\therefore$ (7) gives $\omega_2 = \beta \sin \kappa t.$ $(\because \alpha = 0)$ ...(8)

Then from (5), we have $\dot{\omega}_2 = \left\{\frac{C-A}{A}\right\}K\omega_1 = \kappa\omega_1$

$\Rightarrow \beta\kappa \cos \kappa t = \kappa\omega_1$

$\Rightarrow \omega_1 = \beta \cos \kappa t$

$\therefore \vec{\omega} = \omega_1 i + \omega_2 j + \omega_3 k$

$= \beta \cos \kappa t\, i + \beta \sin \kappa t\, j + Kk$ ...(10)

and $\omega = |\vec{\omega}| = \sqrt{(\beta^2 + K^2)}$ = constant.

This shows that the angular velocity is constant in magnitude equal to $\omega = \sqrt{(\beta^2 + K^2)}$ and processes round the k axis with frequency

$f = \frac{\kappa}{2\pi} = \frac{|C-A|}{2\pi A}K$ as shown in the adjoining diagram,

Example 18:

A body under the action of no forces moves so that the resolved part of its angular velocity about one of the principal axes at the centre of gravity is constant. Show that the angular velocity of the body must be constant.

Solution:

If ω_1, ω_2, ω_3, are the angular velocities about the axes the C.G. and A,B, C are the principal moments at the C.G., then Euler's equations of motion under no forces are

$A\omega_1 - (B - C)\,\omega_2\omega_3 = 0$ ...(1)

$B\omega_2 - (C - A)\,\omega_3\omega_1 = 0$ ...(2)

$C\omega_3 - (A - B)\,\omega_1\omega_2 = 0$ ...(3)

Multiplying (1) by ω_1, (2) by ω_2 and (3) by ω_3 and adding we get

$A\omega_1\omega_2 + B\,\omega_2\omega_3 + C\,\omega_3\omega_3 = 0,$

which on integration gives

$$A\omega_1^2 + B\omega_2^2 + C\omega_3^2 = \text{constant.} \qquad ...(4)$$

Now as given in the question, let ω_1 = const, then two cases arise.

Case I. *When $B = C$*

Then from (1) we $\omega_2\omega_3 = 0$ get

$\because$ either $\omega_2 = 0$ or $\omega_3 = 0$.

Now, if $\omega_1 = 0$, then from (3), we get ω_3 = constant.

if $\omega_3 = 0$, then from (2), we get ω_2 = constant,

Thus under these conditions, we have

$$\omega_1^2 + \omega_2^2 + \omega_3^2 = \textit{const.}$$

Example 19:

A body moves about a point O under no forces, the principal moments of inertia at O being 3A, 5A and 6A. Initially the angular velocity has components $\omega_1 = n, \omega_2 = 0, \omega_3 = n$ *about the corresponding principal axes. Show that at any latter time, t* $\omega_2 = \frac{3n}{\sqrt{5}}\,\text{t han}\left(\frac{nt}{\sqrt{5}}\right)$ *and that the body ultimately rotates about is mean axis.*

Solution:

Euler's equations of motion under no forces are

$$A\dot{\omega}_1 - (B - C)\omega_2\omega_3 = 0;\ B\dot{\omega}_2 - (C - A)\omega_3\omega_1 = 0; \qquad ...(1)$$

and $C\dot{\omega}_3 - (A - B)\omega_1\omega 2 = 0$

Substituting in (1), the given values for A, B, C we get

$$3A\dot{\omega}_1 - (5A - 6A)\omega_2\omega_3 = 0 \Rightarrow 3\omega_1 = -\,\omega_2\omega_3 \qquad ...(2)$$

$$5A\dot{\omega}_1 - (6A - 3A)\omega_3\omega_1 = 0 \Rightarrow 5\omega_2 = 3\omega_3\omega_1 \qquad ...(3)$$

$$6A\dot{\omega}_3 - (3A - 5A)\omega_1\omega_2 = 0 \Rightarrow 3\omega_3 = \omega_2\omega_2 \qquad ...(4)$$

From (2) and (3), we get $9\omega_1\dot{\omega}_1 + 5\omega_1\dot{\omega}_2 = 0.$

which on integration yields $9\omega_1^2 + 5\omega_2^2 = a$ (says)

But initially $\omega_1 = n$ and $\omega_2 = 0$, $\therefore\ a = 9n^2.$

Thus we have $9_1^2 + 5\omega_2^2 = 9n^2$. ... (5)

From (2) (4) we get $\omega_1\dot{\omega}_3 - \omega_2\dot{\omega}_3 = 0\ 0$,

which on integration gives $\omega_1^2 - \omega_3^2 = b$ (say)

Initially $\omega_1 = n$ and $\omega_3 = n$, $\therefore$ $b = 0$.

which on integration gives $\omega_1^2 - \omega_3^2 = b$ (say)

Initially $\omega_1 = n$ and $\omega_3 = n$, $\therefore$ $b = 0$.

Hence $\omega_1^2 = \omega_3^2$

Now equation (3) is,

$$5\dot{\omega}_2 = 3\omega_3\omega_1 = 3\omega_1^2 \Rightarrow 5\frac{d\omega_2}{dt} = \frac{1}{3}(9n^2 = 5\omega_2^2) \text{ since } \omega_1 = \omega_2$$

$$\Rightarrow t = 15\int\frac{d\omega_2}{9n^2 - 5\omega_2^2} = \int\frac{d\omega_2}{\frac{9}{5}n^2 - \omega_2^2} = \frac{\sqrt{5}}{n}\tanh^{-1}\left(\frac{\sqrt{5}}{3n}\omega_2\right)$$

$$\Rightarrow \omega_2 = \frac{3n}{\sqrt{5}}\tanh\left(\frac{nt}{\sqrt{5}}\right)$$

Further when $t \to \infty$, $\tanh\left(\frac{nt}{\sqrt{5}}\right) \to 1 \therefore \omega_2 \to \frac{3n}{\sqrt{5}.}$

Putting $\omega_2 = \frac{3n}{\sqrt{5}}$ in (5), we get $\omega_1 = 0$

and then from (6), we have $\omega_3 = 0$

i.e. the rotation ultimately is about the mean axis.

Example 20:

If a top is made by running a thin pin through the centre of a circular disc, of radius 3 inches, so that length of the pin below the disc is 2 inches, prove that, for steady motion in which rim does not touch the ground, the number of revolutions per second about the axis must exceed $\frac{80}{3\pi(13)^{1/4}}$.

Solution:

In this case we shall discuss the steady motion when the rim of the disc just goes on touching the ground. Let axis OC of the top make an angle α with the vertical OZ, then it makes the angle $\frac{1}{2}\pi - \alpha$ with the horizontal OX. Again

$$\cos\left(\frac{\pi}{2} - \alpha\right) = \cos NOC = \frac{OC}{ON} \doteq \frac{2}{\sqrt{(13)}}$$

i.e. $\Rightarrow \sin\alpha = \dfrac{2}{\sqrt{(13)}}, \cos\alpha = \dfrac{3}{\sqrt{(13)}}$

The condition for steady motion is $C^2n^2 > 4Amgh\cos\alpha$ ...(1)

where A and C have their usual meanings.

Here A = moment of inertia about OA

= moment of inertia about a line through O $\perp$ to OC

$$= m\left\{\frac{1}{2}\left(\frac{1}{4}\right)^2 + \left(\frac{1}{6}\right)^2\right\} \qquad [\because \text{ radius of the disc}]$$

$$= 3'' = \frac{1}{4}\text{ft. and } CO = 2'' = \frac{1}{6}\text{ft.} = m\left(\frac{1}{64}+\frac{1}{36}\right) = \frac{25m}{576}$$

$$C = \text{moment of inertia about } OC = \frac{1}{2}m\left(\frac{1}{4}\right)^2 = \frac{m}{32}$$

and h = distance of C.G. of the top from the vertex

$= \text{distance of C.G. of the disc from } O = 2'' = \dfrac{1}{6}\text{ft.}$

Putting these values in (1) the required condition is

$$\left(\frac{m}{32}\right)^2 n^2 > 4\left(\frac{25m}{576}\right)mg\left(\frac{1}{6}\right)\frac{3}{\sqrt{(13)}} \text{ or } n^2 > \frac{4.25.1.3}{576.0}, \frac{32.32}{\sqrt{(13)}}$$

$$\text{or } n^2 > \frac{25.32.32}{2(13)^{1/2}} \quad n > \frac{160}{2(13)^{1/4}} \text{ or } \frac{n}{2\pi} > \frac{160}{3\pi(13)^{1/4}}$$

$$\text{or number of revolutions per second} > \frac{80}{3\pi(13)^{1/4}}$$

i.e. number of revolutions per second must exceed

$$= \frac{80}{3\pi(13)^{1/4}} = 4.5 \text{ approx.}$$

Example 21:

A body moves under no forces about a point O, the mean axis principal moments of inertia at O being 6A, 3A and A. Initially angular velocity of the body has the components $\omega_1 = n, \omega_2 = 0, \omega_3 = 3n$ *about the principal axes. Show that at any time t*

$$\omega_2 = -(\sqrt{5})\, n \tanh(\sqrt{5nt})$$

and ultimately body rotates about the mean axis.

Solution:

Euler's equations of motion under no forces are

$A\dot{\omega}_1 - (B - C)\omega_2\omega_3 = 0;$

$B\dot{\omega}_2 - (C - A)\omega_3\omega_1 = 0;$ and $C\dot{\omega}_3 - (A - B)\omega_1\omega_2 = 0$

Substituting in these equations, the given values of A,B and C, we get

$$6A\dot{\omega}_1 - (3A - A)\omega_2\omega_3 = 0 \Rightarrow 3\dot{\omega}_1 = \omega_2\omega_3 \qquad ...(1)$$

$$3A\dot{\omega}_2 - (A - 6A)\omega_3\omega_1 = 0 \Rightarrow 3\dot{\omega}_2 = -5\omega_3\omega_1 \qquad ...(2)$$

$$A\dot{\omega}_3 - (6A - 3A)\omega_1\omega_2 = 0 \Rightarrow \dot{\omega}_3 = 3\omega_1\omega_2 \qquad ...(3)$$

From (1) and (2), we get

$$5\omega_1\dot{\omega}_1 + \dot{\omega}_2\omega_2 0 \Rightarrow 5\omega_1^2 + \omega_2^2 = a \text{ (say)}$$

Initially $\omega_1 = n$ and $\omega_2 = 0$, $\quad \therefore a = 5n^2 \Rightarrow 5\omega_1^2 + \omega_2^2 = 5n^2$

Again from (1) and (3), we get

$$9\omega_1\dot{\omega}_1 - \omega_2\dot{\omega}_3 = 0 \Rightarrow 9\omega_1^2 - \omega_3^2 = b \text{ (say)}$$

Initially $\omega_1 = n$ and $\omega_3 = 3n$, $\therefore b = 0$.

Hence $9\omega_1^2 - \omega_3^2 = 0 \Rightarrow \omega_3 = 3\omega_1$

Using (5) in (2), we get

$$\dot{\omega}_2 = -5\omega_1^2 \Rightarrow \dot{\omega}_2 = -(5n^2 - \omega_2^2)$$

i.e. $\dfrac{d\omega_2}{dt} = -(5n^2 - \omega_2^2) \Rightarrow t = -\displaystyle\int \frac{d\omega_2}{5n^2 - \omega_2^2}$

$\therefore t = -\tanh^{-1}\dfrac{\omega_2}{n\sqrt{5}} + C$ But $C = 0$ since $\omega_2 = 0$ when $t = 0$.

Further as $t \rightarrow \propto . \tanh(n\sqrt{5}) \rightarrow 1, \because \omega_2 = -n\sqrt{5}$

Substituting this value of ω_2 in (4), we get $\omega_1 = 0$

And then from (5), we get $\omega_3 = 0$.

Thus we observe that the body ultimately rotates about its mean axis.

Example 22:

Obtain the direction cosines of a set of principal axes at O, one of the corners of the cube.

Solution:

Let O be a corner of the cube of edge 2a and G, the C.G. consider O as origin and three coterminous edges Ox, Oy, and Oz as the fixed axes. Hence co-ordinates of C.G. are (a.a,a). M.I. of the cube about any line through G is $\left(\frac{2}{3}\right)Ma^2(l^2+m^2+n^2)$ where l, m, n are the d.c.s and of the line, but $l^2+m^2+n^2$ = 1, hence the required m.i. becomes $\left(\frac{2}{3}\right)Ma^2$. Also the P.I. about any two perpendicular line through G is zero. Thus by parallel-axes theorem the P.I.'s about OG and any two perpendicular lines through O are zero. This implies that OG and any two perpendicular lines through O are the principal axes. Now consider a line OA ⊥ to the plane containing z-axis, OG, OD where OD is the diagonal of the base face of the cube. Then OA lies in the xy-plane and is ⊥ to OG and OD both. Therefore OA is another principal axis, the third principal axis OB is then evidently perpendicular to OA and OG. This suggests us that OA, OB and OG are principal axes at O. Let A,B,C be the M.I.'s of the cube about these principal axes, then we have

$$A = B = (2/3)Ma^2 + \sqrt{(3a^2)}\ M$$

$$= (11/3)\ Ma^2, C = (2/3)Ma^2, C = (2/3)Ma^2 \qquad ...(1)$$

Now, d.c.'s of the diagonal of the base face of the cube viz. of the line OD are $(1\sqrt{2}),(1\sqrt{2}),0;$dc'sof OZ are (0, 0, 1);

and dc's of OG are $(1\sqrt{3}),(1\sqrt{3}),(1\sqrt{3})$.

whence by solid geometry, the dc's of the line

OB(⊥ to OG and OA) are $(1\sqrt{2}),(1\sqrt{2}),0$

Again by solid geometry the dc's of the line

OB(⊥ to OG and OA) are $(1\sqrt{2}),(-1\sqrt{2}),0$

Thus the ds's of OA are $(1\sqrt{2}),(-1/\sqrt{2}),0$

the ds's of OB are $(1\sqrt{6}),(1\sqrt{6}),(-2/\sqrt{6})$;

and the ds's of OG are $(1\sqrt{3}),(1/\sqrt{6}),(1/\sqrt{6})$.

Example 23:

A solid cube is in motion about an angular point, which is fixed. If there are no external forces and are the angular velocities about the edges through the fixed point, prove that $\omega_1+\omega_2+\omega_3 = cons.$

and $\omega_1^2+\omega_2^2+\omega_3^2 = const.$

Solution:

If O is any fixed angular point of the cue and G its centre of gravity. Then the line OG and two perpendicular lines through perpendicular to OG form a set of principal axes. The direction cosines of principal axes through O are

$$\left(\frac{1}{\sqrt{3}},\frac{1}{\sqrt{3}},\frac{1}{\sqrt{3}},\right)\left(\frac{1}{\sqrt{2}},-\frac{1}{\sqrt{2}},0\right)\left(\frac{1}{\sqrt{6}},\frac{1}{\sqrt{6}},-\frac{2}{\sqrt{6}}\right).$$

Obviously the angular velocities about OG and the other two principal axes respectively are

$$\frac{\omega_1+\omega_2+\omega_3}{\sqrt{3}};\frac{\omega_1-\omega_2}{\sqrt{2}};\frac{\omega_1+\omega_2-2\omega_3}{\sqrt{6}}$$

Now, in this case $A=\frac{2Ma^2}{3},B\frac{11Ma^2}{3},C\frac{11Ma^2}{3}$

Euler's equations of motion under no forces are

$$A\frac{d}{dt}\left(\frac{\omega_1+\omega_2+\omega_3}{\sqrt{3}}\right)-(B-C)\left(\frac{\omega_1-\omega_2}{\sqrt{2}}\right)\left(\frac{\omega_1+\omega_2-2\omega_3}{\sqrt{6}}\right)=0 \quad ...(1)$$

$$B\frac{d}{dt}\left(\frac{\omega_1-\omega_2}{\sqrt{2}}\right)-(C-A)\left(\frac{\omega_1+\omega_2-2\omega_3}{\sqrt{6}}\right)\left(\frac{\omega_1+\omega_2+\omega_3}{\sqrt{3}}\right)=0 \quad ...(2)$$

$$B\frac{d}{dt}\left(\frac{\omega_1+\omega_2-2\omega_2}{\sqrt{6}}\right)-(A-B)\left(\frac{\omega_1+\omega_2+2\omega_3}{\sqrt{3}}\right)\left(\frac{\omega_1-\omega_2}{\sqrt{2}}\right)=0 \quad ...(3)$$

Since B = C, therefore from (1), we have

$$\frac{d}{dt}(\omega_1+\omega_2+\omega_3)=0\Rightarrow\omega_1+\omega_2+\omega_3=\text{const.}$$

Substituting the values of A,B,C in (2) and (3) respectively, we get

$$11(\dot{\omega}_1-\dot{\omega}_2)=3(\omega_1+\omega_2-2\omega_3)(\omega_1+\omega_2+-2\omega_3) \quad ...(2')$$

and $11(\dot{\omega}_1+\dot{\omega}_2-2\dot{\omega}_3)=-9(\omega_1+\omega_2+2\omega_3)(\omega_1+\omega_2)$ ...(3′)

Now dividing (2′) by (3¢) and simplifying, we get

$$(\omega_1+\omega_2-2\omega_3)(\dot{\omega}_1+\dot{\omega}_2-2\dot{\omega}_3)(\omega_1-\omega_2)(\dot{\omega}_1-\dot{\omega}_2)=0$$

Integrating it, we get

$$(\omega_1+\omega_2-2\omega_3)^2+3(\omega_1-\omega_2)^2=\text{ const.}$$

or $6(\omega_1^2+\omega_2^2+\omega_3^2)-2(\omega_1+\omega_2+\omega_3)^2=\text{const.}$

or $6(\omega_1^2+\omega_2^2+\omega_3^2)-2(\omega_1+\omega_2+\omega_3)^2=\text{const.}$

or $\omega_1^2 + \omega_2^2 + \omega_3^2 = \text{const.}$

Example 24:

A rigid body rotates freely about its centre of mass under no forces. The body is rotating with angular velocity about the axis of greatest moment C when a small angular momentum AP is applied about the axis of least moment A. Show that the instantaneous axis moves in the body with period.

$$\frac{2\pi}{T} \text{ where } T^2 = \frac{(C-B)(C-A)}{AB}\omega^2.$$

Solution:

Euler's equations motion under no forces are

$$A\dot{\omega}_1 - (B-C)\omega_2\omega_3 = 0 \quad \text{...(1)}$$

$$B\dot{\omega}_2 - (C-A)\omega_3\omega_1 = 0 \quad \text{...(2)}$$

and $C\dot{\omega}_3 - (A-B)\omega_1\omega_2 = 0$ ...(3)

Dividing (1) by (2), we get

$$A(C-A)\dot{\omega}_1\dot{\omega}_1 + B(C-B)\dot{\omega}_2\dot{\omega}_2 = 0$$

Integrating it, we have

$$A(C-A)\omega_1^2 + B(C-B)\omega_2^2 = a, \text{say} \quad \text{...(4)}$$

Initially $A\omega_1 = AP \Rightarrow \omega_1 = p, \omega_2 = 0$ and $\omega_3 = \omega$

$\therefore A(C+A)p^2 = a$

Substituting the value that of the constant a in (4), we get

$$A(C-A)\omega_1^2 + B(C-B)\omega_2^2 = A(C-A)p^2 \quad \text{...(5)}$$

Since it is given that C> B> A, hence both sides of (3) are positive, It also suggests that ω_1 and ω_2 will be very small corresponding to p. Thus $\omega_1\omega_2$ can be neglected.

Then $(3) \Rightarrow C\dot{\omega}_3 = 0 \Rightarrow \dot{\omega}_3 = \omega_3 = \text{const.} = w.$

Now consider the equation (1) *i.e.*

$$A\dot{\omega}_1 = - \Rightarrow (C-B)\omega_2\omega_3 = -(C-B)\omega_2\omega$$

$$\therefore A\ddot{\omega}_1 = -(C-BB)\omega\dot{\omega}_2[\omega = \text{const.}]$$

$$\therefore A\ddot{\omega}_1 = -(C-B)\frac{(C-A)}{B}\omega^2\omega_1 \quad \text{[using (2)]}$$

$$\Rightarrow \ddot{\omega}_1 = -\frac{(C-B)(C-A)\omega^2}{AB}\omega_1 = -T^2\omega_1$$

$$\text{where } T^2 = \frac{(C-B)(C-A)}{AB}\omega^2.$$

Example 25:

A circular disc, of radius a, has a thin rod pushed through its centre perpendicular to its plane, the length of the rod being equal to the radius of the disc. Show that the system can not spin with the rod vertical unless the angular velocity is greater then $\sqrt{\left(\frac{20g}{a}\right)}$

Solution:

With usual notations, in this case

$$A = \left(\frac{Ma^2}{a}\right) + Ma^2 = \left(\frac{5Ma^2}{4}\right), C = \left(\frac{Ma^2}{2}\right) \text{ and } h = a$$

For the top to execute steady motion with axis vertical, the necessary condition is that

$$C^2n^2 > 4AMgh \quad \sqrt{\left(\frac{4AMgh}{C^2}\right)}$$

$$\text{or } n > \sqrt{\left(\frac{20g}{a}\right)} \text{ putting for A, C and h.}$$

Example 26:

A rigid body rotates about a fixed point under the action of no forces. Investigate the following relations, the invariable line being taken as axis of z

$$\dot{\theta} = -G\sin\theta\cos\psi\sin\psi\left(\frac{1}{A} - \frac{1}{B}\right)$$

$$\dot{\phi} = G\left(\frac{\cos^2\phi}{A} + \frac{\sin^2\psi}{B}\right), \dot{\psi} + \cos\theta\dot{\phi} = \frac{G\cos\theta}{C}$$

where G denotes the angular momentum of the body and other symbols have their usual meanings.

Solution:

Let the fixed point about which the body rotate be O. Let OA, OB, OC be the principal axes at O. The components of angular momentum about OA, OB, OC respectively are given as $A\omega_1$, $B\omega_2$, $C\omega_3$ since G is the angular momentum about Oz, which makes an angle θ with OC, therefore we have

$C\omega_3 = G\cos\theta,\ A\omega_1 = -G\sin\theta\cos\psi,\ B\omega_2 = G\sin\theta\sin\psi$

Now Eulers geometrical equations are

$$\omega_1 = \dot{\theta}\sin\psi - \dot{\phi}\sin\theta\cos\phi \qquad ...(1)$$

$$\omega_2 = \dot{\theta}\cos\psi + \dot{\phi}\sin\theta\sin\phi \qquad ...(2)$$

$$\omega_3 = \dot{\psi} + \dot{\phi}\cos\theta \qquad ...(3)$$

Now multiplying (1) by sin y, (2) by cos y and adding, we get

$$\omega_1\sin\psi + \omega_2\cos\psi = \dot{\theta}$$

or $\dot{\phi}\sin\theta = \omega_2\sin\psi + \omega_1\cos\psi$

$$= \frac{G\sin\theta\sin^2\psi}{B} + \frac{G\sin\theta\cos^2\psi}{A}$$

$$= G\sin\theta\left(\frac{\sin^2\psi}{B} + \frac{\cos^2\psi}{A}\right) \therefore \dot{\phi} = G\left(\frac{\sin^2\psi}{B} + \frac{\cos^2\psi}{A}\right) \qquad ...(5)$$

From (3), we have

$$\dot{\psi} + \dot{\phi}\cos\theta = \omega_3 = \frac{G\cos\theta}{C} \therefore \dot{\psi} + \dot{\phi}\cos\theta = \frac{G\cos\theta}{C} \qquad ...(6)$$

Thus the required relations are obtained.

EXERCISES

1. A uniform circular disc free to turn about its centre is fixed, is set rotating with angular velocity ω about an axis which makes an angle 45° with the axis of the disc. Prove that in the subsequent motion the axis of the disc describes a right cone about an axis making an angle $\tan^{-1}\frac{1}{2}$ with the initial axis of rotation with constant velocity $\frac{1}{2}\omega^2\sqrt{(10)}$.

2. A symmetrical top is set in motion on a rough horizontal plane with an angular motion about its figure, the axis being inclined at angle i to the vertical. Show that between the great approach to end recess from the vertical, the centre of gravity describes an are

$$h\tan^{-1}\left(\frac{\sin i}{p - \cos i}\right),$$

where p and h have the usual meanings.

3. A top of vertex O, is set in motion so that the angular momentum Cn is equal to that about the vertical drawn upward through O. The inclination

of the axis to the vertical was initially i, and every point on the axis was moving horizontally. Determine if the axis will become vertical in a finite time, and if it does, show that the velocity of the centre of gyration P at that instant is equal to

$$2\sqrt{(gl)}\tan\frac{1}{4}i\left\{\sqrt{\left(p-\cos^2\frac{1}{4}i\right)}\right\} \text{ where } 4A^2 gp = C^2 n^2 l \text{ and } OP = l.$$

4. A uniform thin circular disc is set rotating with an angular velocity ω about an axis through the centre making an angle i with the normal. Prove that the semi-vertical angle θ of the come described by the axis of the disc is given by $\tan\ \theta = \left(\frac{1}{2}\right)\tan i.$

   If ω be the angular velocity, prove that the above cone is described in the period $2\pi/[\omega\sqrt{1+3\cos^2 i})].$

5. An elliptic lamina acted on by an external force F is set rotating about a line through its centre which makes equal with the major and minor axes and with the normal to its plane. Prove that, if e be the eccenricity of the ellipse and $\omega_1, \omega_2, \omega_3$ the angular velocities about the major and minor axes and the lamina to the planes then.

   $\omega_1 = F\cos\theta,\ \omega_2 = F\sin\theta,\ \omega_2 = \theta\ F\sqrt{\{(1-e^2\sin^2\theta)/(2-2^2)\}},$ where F is a constant.

6. Prove that a top can be set spinning steadily with its axis of the figure horizontal; by an impulsive couple about an axis making with the vertical and angle

$$\tan^{-1}\left(\frac{C^2n^2}{Amgh}\right)$$

7. When the axis of a symmetrical top is stationary and the spin is large and equal to u, a blow J is applied perpendicular to the axis at a distance c from the fixed point. Prove that the maximum angular deviation of the axis is approximately

   $2\tan^{-1}(Jc/Cn)$

   C being the moment of inertia of the top about its axis of symmetry.

8. A top is a right cone of height 1 and vertical [angle, 90° and move about its vertex under gravity. Initially it started rotating with angular velocity $5\sqrt{(3g/l)},$ the axis being instantaneously at rest and inclined as 60° to

the upward vertical. Show that in the subsequent motion the angle between the axis and vertical oscillations between 60°. and 120°.

9. A top is processing steadily with one end of its axis fixed and its axis horizontal. If a small weight W be suddenly attached to the other end of its axis without impulse, show that an oscillatory motion is set up in which the maximum dip of the axis below the horizontal is approximately.

   $2W\,a/(A^2\,\Omega + C^2n^2)$,

   Where Ω is the initial angular velocity of precession and a the length of the axis.

10. State and prove Euler's Dynamical equations under the action of external forces.

11. A ring of wire of radius c rests on the top of smooth fixed sphere of radius a and is set rotating about the vertical diameter of the sphere. If the ring is slightly displaced, prove that the motion is unstable if the angular velocity is less than

$$\left[2g\frac{2a^2-c^2}{c^4}\sqrt{\{(a^2-c^2)\}}\right]^{1/2}$$

12. A solid of revolution whose principal moments at the centre of inertia are A, A, C, (C>A) is set spinning with angular velocity ω about an axis passing through the centre of inertia and making and angle i with the figure. Prove that the instantaneous axis of rotation describes in body a right cone of semi-vertical angle i in period $\dfrac{2\pi A\sec i}{(C-A)\omega}$, and that the axis of the figure describes a right cone in space of semi-vertical angle θ where tan θ (A/C) tan i, in the period $2\pi/\Omega$:

    where Ω sin ω sin i or $\Omega = (\omega/A)\ \sqrt{(A^2\sin^2 i + C^2\cos^2 i)}$.

5

Lagrange's Equations and Mechanical System

INTRODUCTION

Holonomic System and Non Holonomic System

Let q_1 , q_2 ;.....,q_n denote the generalised co-ordinates describing a system and let t denote the time. If all the constraints of the system can be expressed as equations having the form $(q_1,q_1,....q_n ;t) = 0$ or their equivalent, then the system is said to be Holonomic otherwise it is be Non- Holonomic system.

Conservation and Non-conservative System

If the forces acting on the system are derivable from a potential function [or potential energy] V, then the system is called conservative, otherwise it is non-conservative.

Scleronomic System

The mechanical system in which t, the time, does not enter explicity in equations (1) or (2) is called a scleronomic system.

Rheonomic System

The mechanical system in which the moving constraints are involved and the time t does enter explicity is called a Rheonomic system.

KINETIC ENERGY AND GENERALISED VELOCITIES

The K.E of the system is $T = \frac{1}{2}\Sigma_{v=1}^{n} mv\, \dot{r}_v$ ...(1)

The K.E of the system can be written as a quadratic form in the generalised co-ordinates q_α If the system is independent of time explicitly *i.e.* scleronomic then the quadratic form has only terms of the type $q_{\alpha\beta}\, q_\alpha\, q_\beta$ In case the system is Rhenomic, linear terms in q_α are also present.

GENERALISED FORCES

If W is the total work done on a system of particles by forces Fv acting on the v-th particle, then $dW = \Sigma_{\alpha=1}^{n} \phi_\alpha dq_\alpha$ where $\phi_\alpha = \Sigma_{v=1}^{N} F_v \bullet \frac{\partial r_v}{\partial q_\alpha}$ is called the *generalised force* associated with generalised co-ordinator q_α

Suppose that a system undergoes increments $dq_1, dq_2, ...dq_n$; then the v-th particle undergoes a displacement.

$$dr_v = \Sigma_{\alpha=1}^{n} \frac{\partial r_v}{\partial q_\alpha} dq_\alpha \qquad ...(1)$$

$\therefore$ Total work done is given by

$$dW = \Sigma_{v=1}^{N} F \bullet dr_v = \Sigma_{v=1}^{N} \left\{ \Sigma_{\alpha=1}^{n} F_v \bullet \frac{\partial r_v}{\partial q_\alpha} \right\} dq_\alpha \qquad ...(2)$$

Now, let $\phi_\alpha = \Sigma_{v=1}^{N} \left(\Sigma_{v=1}^{N} F_v \bullet \frac{\partial r_v}{\partial q_\alpha} \right)$

Now, let $\phi_\alpha = \Sigma_{v=1}^{N} F_v \bullet \frac{\partial r_v}{\partial q_\alpha}$

then (1) $dW = \Sigma_{\alpha=1}^{N} \left(\Sigma_{v=1}^{N} F_v \bullet \frac{\partial r_v}{\partial q_\alpha} \right) d_{a\alpha} = \Sigma_{\alpha=1}^{n} \phi_\alpha d_{a\alpha}$ ...(3)

We have $dW = \Sigma_{\alpha=1}^{N} \frac{\partial W}{\partial q_\alpha}, \therefore \frac{\partial W}{\partial q_\alpha} = \phi_\alpha$ ...(4)

PRINCIPLE OF ENERGY FROM THE LAGRANGE'S EQUATIONS

Lagrange's equations are:

$$\frac{d}{dt}\left(\frac{\partial T}{\partial \dot{q}_\alpha} \right) - \frac{\partial T}{\partial q_\alpha} = -\frac{\partial V}{\partial q_\alpha}; \quad (\alpha = 1, 2, ...n) \qquad ...(1)$$

Also by equation we know that

$$T = \frac{1}{2}(a_{11}\dot{q}_1{}^2 + a_{22}\dot{q}^2 + ... + a_m\dot{q}_n{}^2 + 2a_{12}\dot{q}_1\dot{q}_2 + ...)$$

that is, T can be expressed as a quadratic expression in generalised velocities. Hence applying Euler's theorem. We get

$$\Sigma_{\alpha=1}^{n} \dot{q}_\alpha \frac{\partial T}{\partial \dot{q}_\alpha} = 2T \qquad ...(2)$$

Also $\frac{dT}{dt} = \Sigma_{\alpha=1}^{n} \frac{\partial T}{\partial q_\alpha} \dot{q}_\alpha + \Sigma_{\alpha=1}^{n} \frac{\partial T}{\partial \dot{q}_\alpha} \ddot{q}_\alpha$ ...(3)

Now multiplying the n equations of (1) by $\dot{q}_1, \dot{q}_2, ... \dot{q}_n$, respectively and then adding. we get

$$\left\{\dot{q}_1 \frac{d}{dt}\left[\frac{\partial T}{\partial \dot{q}_1}\right] + ... + \dot{q}_n \frac{d}{dt}\left[\frac{\partial T}{\partial \dot{q}_1}\right]\right\} - \left\{\dot{q}_1 \frac{\partial T}{\partial q_1} + ... + \dot{q}_n \frac{\partial T}{\partial \dot{q}_n}\right\}$$

$$= -\left\{\dot{q}_1 \frac{\partial V}{\partial q_1} + ... + \dot{q}_n \frac{\partial V}{\partial q_n}\right\}$$

$$\Rightarrow \frac{d}{dt}\left\{\Sigma_{\alpha=1}^{n} \dot{q}_\alpha \frac{\partial T}{\partial \dot{q}_\alpha}\right\} - \left(\Sigma_{\alpha=1}^{n} \ddot{q}_\alpha \frac{\partial T}{\partial \dot{q}_\alpha}\right) - \left\{\Sigma_{\alpha=1}^{n} \dot{q}_n \frac{\partial T}{\partial q_\alpha}\right\}$$

$$= -\left(\Sigma_{\alpha=1}^{n} \dot{q}_\alpha \frac{\partial V}{\partial q_n}\right)$$

$$\Rightarrow \frac{d}{dt}(2T) - \frac{dT}{dt} = -\frac{dV}{dt} \Rightarrow \frac{dT}{dt} + \frac{dV}{dt} = 0 \Rightarrow \frac{d}{dt}(T+V) = 0 \Rightarrow T+V = \text{constant}.$$

SMALL OSCILLATIONS

To explain how Lagrange's equations are used in case of small oscillations. (when there exist three generalised co-ordinates),

In order to investigate the theory of small oscillations by the use of Lagrange's equation about the position of equilibrium, the generalised co-ordinates q_1, q_2, q_3 must be chosen in such a way that they vanish in the position of equilibrium.

But the system makes small oscillations about the position of equilibrium, so q_1, q_2, q_3 and their differentials $\dot{q}_2$, $\dot{q}_2$, $\dot{q}_3$ will remain small during the whole motion. Hence we shall reject all the powers of small quantities except the lowest one.

Now, let T be kinetic energy, and W the work-function of the system; then we have

$$T = A_{11}\dot{q}_1^{\,2} + A_{22}\dot{q}_2^{\,2} + A_{33}\dot{q}_3^{\,2} + 2A_{12}\dot{q}_2\dot{q}_1 + 2A_{23}\dot{q}_2\dot{q}_3 + 2A_{13}\dot{q}_1\dot{q}_3 \quad ...(1)$$

$$W = C + B_1q_1 + B_2\,q_2 + B_3q_3 + B_{11}q_1^{\,2} + B22\,q_2^{\,2} + q_3^{\,2} \quad ..(2)$$

Now assume that q_1, q_2, q_3 can be expressed in terms of X, Y, Z by the equations of the form

$$q_1 = \lambda_1X + \lambda_2Y + \lambda_3Z; q_2 = \mu_1X + \mu_2Y + \mu_3Z;$$

$$q_3 = v_1X + v_2Y + v_3Z;$$

and further assume $\lambda_1, \lambda_2, \lambda 3, \mu_1, \mu_2, \mu_3, v_1, v_2, v_3$ in such a way that if we put value of q_1, q_2, q_3 in (1) and (2), then T does not involve $\dot{Y}\dot{Z}, \dot{Z}\dot{X}, \dot{X}\dot{Y}$ and W does not involve YZ, ZX, XY. Then X,T, Z, are called the *normal or Principal Co-ordinates.*

Thus equation (1) and (2) $\Rightarrow$

$$T = A'_{11}\dot{X}^2 + A'_{22}\dot{Y}_{22}\dot{Y}^2 + A'_{33}\dot{Z}^2$$

$$W = C' + B_1 X + B'_2 Y + B'_3 Z + B'_{11} X^2 + B_{22} Y^2 + B'_{33} Z^2$$

Now, using Lagrange's equations,

$$\text{viz.} \frac{d}{dt}\left(\frac{dT}{d\dot{X}}\right) - \left(\frac{dT}{dX}\right) = \left(\frac{dW}{dX}\right) \text{etc., we obtain}$$

$$2A'_{11}\ddot{X} = B'_1 + 2B'_{11}X; \;\; 2A'_{22}\ddot{Y} = B'_2 + 2B'_{22}Y;$$

$$2A'_{33}\ddot{Z} = B'_3 + 2B'_{33}Z$$

which can be put in the forms

$$\ddot{X} = -\omega^2{}_1 X, \; \ddot{Y} = -\omega^2{}_3 Y, \; \ddot{Z} = -\omega^2{}_3 Z \text{ etc.}$$

which represents simple Harmonic Motions, Or in other words they give small oscillations about the position equilibrium.

LAGRANGE'S EQUATIONS WITH IMPULSIVE FORCES

When the forces are finite, we have by **7.07**

$$\frac{d}{dt}\left(\frac{\partial T}{\partial \dot{q}_\alpha}\right) - \frac{\partial T}{\partial q_\alpha} = \phi_\alpha \text{ where } \phi_\alpha = \sum_v V_v \cdot \frac{\partial r_v}{\partial q_\alpha}$$

Integrating both sides of (1) w.r.t. "t" form t = 0 to t = τ, we get

$$\left[\left(\frac{\partial T}{\partial \dot{q}_\alpha}\right)\right]_{t=0}^{t=\tau} - \int_0^\tau \frac{\partial T}{\partial q_\alpha} dt = \int_{t=0}^{\tau} \phi_\alpha dt = \sum_v \left\{\left(\int_{t=0}^{\tau} F_v dt\right) \cdot \frac{\partial r_v}{\partial q_\alpha}\right\} \quad ...(1)$$

Taking the limit as $\tau \to 0$, we get

$$\lim_{\tau \to 0}\left\{\left(\frac{\partial T}{\partial \dot{q}_\alpha}\right)_{t=\tau} - \left(\frac{\partial T}{\partial \dot{q}_\alpha}\right)_{t=0}\right\} - \lim_{\tau \to 0}\int_0^\tau \frac{\partial T}{\partial q_\alpha} dt$$

$$= \sum_v \left\{\left(\lim_{\tau \to 0}\int_0^\tau F_v dt\right) \cdot \frac{\partial r_v}{\partial q_\alpha}\right\}$$

$$\Rightarrow \left(\frac{\partial T}{\partial \dot{q}_\alpha}\right)_1 - \left(\frac{\partial T}{\partial \dot{q}_\alpha}\right)_0 = \Sigma^* I_v \cdot \frac{\partial r_v}{\partial q_\alpha} = {**}P_\alpha\,(\text{say}) \qquad ...(2)$$

where subscripts 0 and 1 denote respectively quantities before and after the application of the impulsive force.

These equations are known as Lagrange's equations under impulsive forces.

LAGRANGE'S EQUATION

Let F be the net external force acting on the Vth particle of a system, then bt Newton's second law applied to Vth particle, we have

$$m_v \ddot{r}_v = F_v$$

$$\Rightarrow m_v \ddot{r}_v \bullet \frac{\partial r_v}{\partial q_\alpha} = F_v \bullet \frac{\partial r_v}{\partial q_\alpha} \qquad ...(1)$$

$$\Rightarrow \Sigma_{v=1}^{N} m_v \ddot{r}_v \bullet \frac{\partial r_v}{\partial q\alpha} = \Sigma_v^N F_v \bullet \frac{\partial r_v}{\partial q\alpha}$$

$$\Rightarrow \frac{d}{dt}\left(\Sigma_{v=1}^{N} m_v \dot{r}_v \bullet \frac{\partial r_v}{\partial q_\alpha}\right) - \Sigma_{V=1}^{N} m_v \dot{r}_v \bullet \frac{d}{dt}\left(\frac{\partial r_v}{\partial q\alpha}\right) = \Sigma_{v=1}^{N} F_v \bullet \frac{\partial r_v}{\partial q_\alpha}$$

But $r_v = r_v(q_1, q_2, ..., q_n; t)$ ...(2)

$$\therefore \dot{r}_v = \frac{\partial r_v}{\partial q_1}\dot{q}_1 + \frac{\partial r_v}{\partial q_2}\dot{q}_2 + ... + \frac{\partial r_v}{\partial q_n}\dot{q}_n + \frac{\partial r_v}{\partial q\alpha} \qquad ...(3)$$

$$\Rightarrow \frac{\partial \dot{r}_v}{\partial \dot{q}\alpha} = \frac{\partial r_v}{\partial q\alpha} \qquad \text{[Cancellation law of the dots]} \qquad ...(4)$$

Also, $\frac{\partial \dot{r}_v}{\partial q\alpha}(\dot{r}_v) = \frac{\partial}{\partial q\alpha}\left(\frac{\partial r_v}{\partial q_1}\dot{q}_1 + \frac{\partial r_v}{\partial q_2}\dot{q}_2 + ... + \frac{\partial r_v}{\partial q_n}q_{\dot{n}} + \frac{\partial r_v}{\partial t}\right)$

$$= \frac{\partial^2 r_v}{\partial q_\alpha \partial q_1}\dot{q}_1 + \frac{\partial^2 r_v}{\partial q_\alpha \partial q_1}\dot{q}_2 + ... + \frac{\partial^2 r_v}{\partial q_\alpha \partial q_n}\dot{q}_n + \frac{\partial}{\partial q_\alpha}\left(\frac{\partial r_v}{\partial t}\right)$$

$$= \frac{\partial}{\partial q_1}\left(\frac{\partial r_v}{\partial q_\alpha}\right)\dot{q}_1 + \frac{\partial}{\partial q_2}\left(\frac{\partial r_v}{\partial q_\alpha}\right)\dot{q}_2 + ... + \frac{\partial}{\partial q_\alpha}\left(\frac{\partial r_v}{\partial q_n}\right)\dot{q}_n + \frac{\partial}{\partial t}\left(\frac{\partial r_v}{\partial q_\alpha}\right)$$

or $= \frac{\partial}{\partial q_\alpha}\left(\frac{\partial r_v}{\partial t}\right) = \frac{d}{dt}\left(\frac{\partial r_v}{\partial q_\alpha}\right) \Rightarrow \frac{d}{dt}\left(\frac{\partial}{\partial q_\alpha}\right) \equiv \frac{\partial}{\partial q\alpha}\left(\frac{d}{dt}\right)$ ...(5)

[interchange law of the order of operators]

$$\text{Now } \frac{d}{dt}\left\{\Sigma_{v=1}^{N} m_v \dot{r}_v \bullet \frac{\partial r_v}{\partial q_\alpha}\right\} - \Sigma_{v=1}^{N} m_v \dot{r}_v \bullet \frac{\partial \dot{r}_v}{\partial q_\alpha} = \Sigma_{v=1}^{N} F_v \bullet \frac{\partial r_v}{\partial q_\alpha} \qquad ...(6)$$

$$\text{and } T = \frac{1}{2}\sum_v m_v \dot{r}_v^2 = \frac{1}{2}\sum_v m_v (\dot{v}_v \cdot \dot{r}_v) \qquad ...(7)$$

$$\Rightarrow \frac{\partial T}{\partial q_\alpha} = \sum_v m_v \dot{r}_v \cdot \frac{\partial \dot{r}_v}{\partial q_\alpha} \qquad ...(8)$$

$$\text{and } \Rightarrow \frac{\partial T}{\partial \dot{q}_\alpha} = \sum_v m_v \dot{r}_v \cdot \frac{\partial \dot{r}_v}{\partial \dot{q}_\alpha} = \Sigma m_v \dot{r}_v \cdot \frac{\partial v_v}{\partial q_\alpha} \text{ [using (12)]} \qquad ...(9)$$

$$\therefore (14) \Rightarrow \frac{d}{dt}\left(\frac{\partial T}{\partial \dot{q}_\alpha}\right) - \frac{\partial T}{\partial q_\alpha} = \phi_\alpha = \frac{\partial W}{\partial q_\alpha} \text{ using (7)} \qquad ...(10)$$

Note. The quantity $p_\alpha = \frac{\partial T}{\partial \dot{q}_\alpha}$ is called the generalised momentum associated with the general co-ordinates q_α.

DEGREES OF FREEDOM

The number of independent co-ordinates required to specify the position of a system of one or more particles is called the number of degrees of freedom of the system.

Example 1:

A particle moving freely in space require 3 co-ordinates, e.g. (x, y, z), to specify its position. Thus the number of degrees of freedom is 3.

Example 2:

A system containing of N-particles moving freely in space require 3N co-ordinates to specify the position. The number of degrees of freedom is 3N.

A rigid body which can move freely in space has 6 degrees of freedom i.e. 6. co-ordinates are required to specify the position.

Let 3 non-collinear points of a rigid body be fixed in space, then the rigid body also fixed in space. Let these points have co-ordinates (x_1,y_1,z_1); (x_2,y_2,z_2); (x_3,y_3,z_3) respectively, a total of 9. Since the body is rigid, we must have

$$(x_1 - x_2)^2 + (y_1 - y_2)^2 + (z_1 - z_2)^2 = \text{constant.}$$

$$(x_2 - x_3)^2 + (y_2 - y_3)^2 + (z_2 - z_3)^2 = \text{constant.}$$

$$(x_3 - x_1)^2 + (y_3 - y_1)^2 + (z_3 - z_1)^2 = \text{constant.}$$

Hence 3 co-ordinates can be expressed in terms of the remaining six. Thus six independent co- ordinates are needed to describe the motion *i.e.* there exit six degrees of freedom.

TRANSFORMATION EQUATION

Let $rv = x_v i + y_v j + z_v k$ be the position vector of v-th particle with respect to xyz co-ordinates system. The relationships of th generalised co-ordinates $q_1, q_2, \ldots, q_n$ the position co-ordinates are given by the transformation equations.

$$\left.\begin{aligned} x_v &= x_v(q_1, q_2, \ldots, q_n; t) \\ y_v &= y_v(q_1, q_2, \ldots, q_n; t) \\ z_v &= z_v(q_1, q_2, \ldots, q_n; t) \end{aligned}\right\} \qquad \ldots(1)$$

where t denotes the time. In vector (1) can be written as

$$r_v = r_v(q_1, q_2, \ldots, q_n; t) \ \ldots(2)$$

where the functions in (1) or (2) are continuous and have continuous derivatives.

LAGRANGIAN FUNCTION

If the forces are derivable from a potential function V, then

$$\phi_\alpha = \frac{\partial W}{\partial q_\alpha} = -\frac{\partial V}{\partial q_\alpha},$$

since the potential, or potential energy is a function of q's only (and possibly the name t) then, we have

$$\frac{d}{dt}\left(\frac{\partial T}{\partial \dot{q}_\alpha}\right) - \frac{\partial T}{\partial q_\alpha} = -\frac{\partial V}{\partial q_\alpha} \Rightarrow \frac{d}{dt}\left[\frac{\partial}{\partial \dot{q}_\alpha}(T - V)\right] - \left(\frac{\partial T}{\partial q_\alpha} - \frac{\partial V}{\partial q_\alpha}\right) = 0$$

$$\Rightarrow \frac{d}{dt}\left(\frac{\partial L}{\partial \dot{q}_\alpha}\right) - \frac{\partial L}{\partial q_\alpha} = 0, \text{ where } L = T - V \qquad \ldots(19)$$

The function L defined by L = T – V is said to be Lagrangian function.

GENERALISED MOMENTUM

We defined $p_\alpha = \dfrac{\partial T}{\partial \dot{q}_\alpha}$ *to be the generalised momentum associated with generalised co-ordinates* q_α *or the conjugate momentum.*

In case the system is conservative, we have

$$T = L + V \Rightarrow (\partial T / \partial \dot{q}_\alpha) = (\partial L / \partial \dot{q}_\alpha) + (\partial V / \partial \dot{q}_\alpha) = (\partial L / \partial \dot{q}_\alpha)$$

because V, the P.E. of the system does not depend upon $\dot{q}_\alpha$

$$\therefore P_\alpha = (\partial L / \partial q_\alpha).$$

KINETIC ENERGY AS A QUADRATIC FUNCTION OF VELOCITIES

If at time t, the position of the vth particle (mass m), of a holonomic system is defined by r_v then K.E. is given by

$$T = \frac{1}{2}\Sigma_{v=1}^{N} m_v \dot{r}_v^2, \text{ where } r_v = r_v(q_1, ...q_n; t) \qquad ...(1)$$

$$\text{So that } \dot{r}_v = \dot{q}_1 \frac{\partial r_v}{\partial q_1} + \dot{q}_2 \frac{\partial r_v}{\partial q_2} + ... + \dot{q}_n \frac{\partial r_v}{\partial q_n} + \frac{\partial r_v}{dt}$$

$$\Rightarrow T = \frac{1}{2}\Sigma_{v=1}^{N} m_v \left(\dot{q}_1 \frac{\partial r_v}{\partial q_1} + \dot{q}_2 \frac{\partial r_v}{\partial q_2} + ... + \dot{q}_n \frac{\partial r_v}{\partial q_n} + \frac{\partial r_v}{\partial t} \right)^2$$

$$= \frac{1}{2}[(a_{11}\dot{q}_1^2 + a_{22}\dot{q}_2^2 + ... + a_{nn}\dot{q}_n^2 + 2a_{12}\dot{q}_1\dot{q}_2 + 2a_{1n}\dot{q}_1\dot{q}_n + ...)$$

$$+ 2(a_1\dot{q}_1 + a_2\dot{q}_2 + . + a_n\dot{q}_n) + a] \qquad ...(2)$$

$$\left.\begin{array}{l} \text{where } 5a_{rs} = \Sigma_{v=1}^{N} m_v (\partial r_v / \partial q_r).(\partial r_v / \partial q_s)(s \geq r) \\ a_{rr} = \Sigma_{v=1}^{N} m_v (\partial r_v / \partial q_r)^2, a = \Sigma_{v=1}^{N} m_v (\partial r_v / \partial t)^2, = \Sigma_{v=1}^{N} m_v \left(\frac{\partial r_v}{\partial q_r}\right) \cdot \left(\frac{\partial r_1}{\partial t}\right) \end{array}\right\}$$

From (2), we see that T is a quadratic function of the of the generalised velocities.

The case t is not explicity involved, is of considerable importance. Hence we have $\frac{\partial r_v}{\partial t} = 0$ and therefore (2) implies that

$$T = \frac{1}{2}(a_{11}\dot{q}_1^2 + a_{22}\dot{q}_2^2 + ... + a_{nn}\dot{q}_n^2 + 2a_{12}\dot{q}_1\dot{q}_2 + ...) \qquad ...(3)$$

$$= \frac{1}{2}\Sigma_{s=1}^{n}\Sigma_{r=1}^{n} a_{rs}\dot{q}_r\dot{q}_s \text{ where } a_{rs} = a_{sr}. \qquad ...(4)$$

Now using Euler's theorem for homogeneous functions, we get

$$\dot{q}_1 \frac{\partial T}{\partial \dot{q}_1} + \dot{q}_2 \frac{\partial T}{\partial \dot{q}_2} + ... + \dot{q}_n \frac{\partial T}{\partial \dot{q}_n} = 2T$$

$$\Rightarrow 2T = \Sigma_{\alpha=1}^{n} \dot{q}_\alpha \frac{\partial T}{\partial \dot{q}_\alpha} p_\alpha \dot{q}_\alpha$$

i.e. $2T = p_1\dot{q}_1 + p_2\dot{q}_2 + ... + p_n\dot{q}_n$

Generalised Co-ordinates

Suppose that a particle or a system of N-particles moves subject to possible constraints, as for example a particle moving along a circular wire or a rigid body moving along an inclined plane, then there will be necessarily a minimum number of independent co-ordinates then needed to specify the motion. These co-ordinates may be distances, angles or quantities relating to them.

MISCELLANEOUS EXAMPLES

Example 1:

Use Lagrange's equations to find the differential equation for a compound pendulum which oscillates in a vertical plane about a fixed horizontal axis.

Solution:

Let the plane of oscillation be represented by xy-plane, where B is its intersection with the axis of rotation and G is the centre of gravity. Let the mass of the pendulum be M and let its moment of inertia about the axis of rotation be MK^2. Then potential energy relative to the horizontal plane through N is V = – Mgh cos.

$$T = \frac{1}{2}Mk^2\dot{\theta}^2 \therefore L = T - V = \frac{1}{2}Mk^2\dot{\theta}^2 + Mgh\cos\theta \qquad ...(1)$$

$$\Rightarrow \frac{\partial L}{\partial \dot{\theta}} = Mk^2\theta \text{ and } \frac{\partial L}{\partial \theta} = -Mgh\sin\theta \qquad ...(2)$$

Now Lagrange's θ equation gives

$$\frac{d}{dt}\left(\frac{\partial L}{\partial \theta}\right) - \frac{\partial L}{\partial \theta} = 0 \Rightarrow \frac{d}{dt}(Mk^2\dot{\theta}) + Mgh\sin\theta = 0$$

$$i.e.\ Mk^2\ddot{\theta} + Mgh\sin\theta = 0 \Rightarrow \ddot{\theta} = -\frac{gh}{k^2}\sin\theta$$

When θ is small, we have $D^2\theta = -\frac{gh}{k^2}\theta \qquad (\because \sin\theta = \theta)$

$$\text{or } \left(D^2 + \frac{gh}{k^2}\right)\theta = 0.$$

This is the differential equation of the pendulum.

Example 2:

A particle of mass m moves in a conservative force field. Find a the Lagrangian function, (b) the equations of motion in cylindrical co-ordinates (ρ, ϕ, z).

Solution:

We have $OP = OP_0 + P_0P = OA + Ap_0 + P_0P = \vec{\rho}$ (say)

$\therefore \vec{\rho} = \rho\sin\phi j + \rho\cos\phi i + zk$ where i, j, k

are the unit vector along OX , OY and OZ respectively.

Hence the unit vector along the direction of ρ increasing is

given by $\vec{\rho_1} = \frac{\partial\vec{\rho}}{\partial\rho} / \left|\frac{\partial\vec{\rho}}{\partial\rho}\right| = \sin\phi j + \cos\phi j$

Similarly $\vec{\phi_1} = \frac{\partial\vec{\rho}}{\partial\phi} / \left|\frac{\partial\vec{\rho}}{\partial\phi}\right|$

$$= \frac{\rho\cos\phi j - \rho\sin\phi i}{\rho} = -\sin\phi i + \cos\phi j$$

Now $v = \frac{d\vec{\rho}}{dt} = \frac{d}{dt}(\rho\sin\phi j + \rho\cos\phi i + zk)$

$$= \rho\cos\phi\dot{\phi}j + \dot{\rho}\sin\phi j - \rho\sin\phi\dot{\phi}i + \dot{\rho}\cos\phi i + k\dot{z}$$

$$= \dot{\rho}\cos\phi i + \dot{\rho}\sin\phi j + \rho\dot{\phi}(\cos\phi j - \sin\phi i) + z\dot{k}$$

$$= \rho\vec{\rho_1} + \rho\dot{\phi}\vec{\phi} + \dot{z}k$$

$\therefore T = \frac{1}{2}m\{\dot{\rho}^2 + \rho^2\dot{\phi}^2 + \dot{z}^2\}$ and $V = V(\rho, \phi, z)$

(a) Hence the Lagrangian function is

$$L = T - V = \frac{1}{2}m\left[\dot{\rho}^2 + \rho^2\dot{\phi}^2 + \dot{z}^2\right] - V(\rho, \phi, z)$$

(b) Lagrange's equations are

$\frac{d}{dt}\left(\frac{\partial L}{\partial\dot{\rho}}\right) - \frac{\partial L}{\partial\rho} = 0$ *i.e.* $\frac{d}{dt}(m\dot{\rho}) - \left(m\rho\dot{\phi}^2 \frac{\partial V}{\partial\rho}\right) = 0$

i.e. $m(\ddot{\rho} - \rho\dot{\phi}^2) = -\frac{\partial V}{\partial\rho}$ ...(1)

$\frac{d}{dt}\left(\frac{\partial L}{\partial\dot{\phi}}\right) - \frac{\partial L}{\partial\phi} = 0$ *i.e.* $\frac{d}{dt}(m\rho^2\dot{\phi}) + \frac{\partial V}{\partial\phi} = 0,$

or $\frac{d}{dt}(\rho^2\dot{\phi}) = -\frac{\partial V}{\partial\phi}$ ...(2)

and $$\frac{d}{dt}\left(\frac{\partial L}{\partial \dot{z}}\right)-\frac{\partial L}{\partial z}=0\ \frac{d}{dt}(m\dot{z})+\frac{\partial V}{\partial z}=0 \text{ or } m\ddot{z}=-\frac{\partial V}{\partial z} \quad ...(3)$$

Example 3:

(i) *Set up the Lagrangian for a simple pendulum, and*

(ii) *Obtain an equation describing its motion.*

Solution:

(i) Choose as generalised coordinates, the angle θ made by the string OB of the pendulum and the vertical OA. Let l be the length of OA, then K.E. is given by

$$T=\frac{1}{2}mv^2=\frac{1}{2}m(l\dot{\theta})^2=\frac{1}{2}ml^2\dot{\theta}^2$$

where m is the mass of the bob.

The potential energy of mass m is given by

$V = mg\,(OA - OC) = mg\,(l - l\cos\theta) = mgl\,(1 - \cos\theta)$

$$\therefore L=T-V=\frac{1}{2}ml^2\dot{\theta}^2-mgl(1-\cos\theta)$$

(ii) Hence Lagrange's θ equation gives

$$\frac{d}{dt}\left(\frac{\partial L}{\partial \dot{\theta}}\right)-\frac{\partial L}{\partial \theta}=0 \Rightarrow \frac{d}{dt}(ml^2\dot{\theta})-(-mgl\sin\theta)=0$$

$$\frac{d}{dt}\left(\frac{\partial L}{\partial \dot{\theta}}\right)-\frac{\partial L}{\partial \theta}=0 \Rightarrow l\dot{\theta}=-g\sin\theta \Rightarrow \ddot{\theta}=-\frac{g}{l}\sin\theta,$$

which is the required equation of motion.

Example 4:

A particle Q moves on a smooth horizontal circular wire of radius, a which is free to rotate about a vertical axis through a point 0, distance c from the centre C. If the $\angle QCA=\theta$, show that

$$a\ddot{\theta}+\dot{\omega}(a-c\cos\theta)=c\omega^2\sin\theta.$$

where ω is the angular velocity of the wire.

Solution:

Let $0Q = r$, and $\angle AOQ=\alpha$

$$\Rightarrow r^2=a^2+c^2-2ac\cos\theta \quad ...(1)$$

$r\cos(\alpha-\theta)=a-c\cos\theta.$ ...(2)

The particle Q moves on circle of radius a, *so its velocity along the tangent QT will be $a\theta$ but Q revolves about O with angular velocity ω, which causes a velocity aw at the right angles to 0Q.*

$\Rightarrow v\frac{1}{2}=$ (velocity)^2 of the particle at Q

$=a^2\dot{\theta}^2+r^2\omega^2+2ar\dot{\theta}\cos(\alpha-\theta)\omega$

$\angle AOQ=\alpha, \angle CQO=\alpha-\theta$

Now $T=\frac{1}{2}mV_q^{\ 2}=\frac{1}{2}[a^2\dot{\theta}^2+r^2\omega^2$ $\left[\begin{array}{l}\angle NQT=\alpha-\theta, OQ=r\\ HQ=a-c\cos\theta\end{array}\right.$

$+2a\,r\,\omega\theta\cos(\alpha-\theta)]$

Note: *If r is the position vector of the particle at any time t, then $\partial r/\partial r$ is the vector tangent to the curve q = constant *i.e.* a vector in the direction of r (increasing r). A unit vector in this direction is thus given by

$$r_1=\frac{\partial r}{\partial r}/\left|\frac{\partial r}{\partial r}\right|$$

Similarly, $\partial r/\partial\theta$ is the vector tangent to the curve r = constant, A unit vector in this direction is given by $\vec{\theta}_1=\frac{\partial r}{\partial\theta}/\left|\frac{\partial r}{\partial\theta}\right|.$

$=\frac{1}{2}m[a^2\dot{\theta}^2+(a^2+c^2$ $=r\cos(\alpha-\theta)]$

$-2ac\cos\theta)\omega^2+2a\omega\dot{\theta}(a-c\cos\theta)$ and work function = 0

($\because$ weight does no work)

$\therefore$ Lagrange's θ equation $\Rightarrow\frac{d}{dt}\left(\frac{\partial T}{\partial\dot{\theta}}\right)-\frac{\partial T}{\partial\theta}=0.$

$\Rightarrow\frac{d}{dt}[a^2\dot{\theta}+a\omega(a-c\cos\theta)]-ac\,\omega^2\sin\theta-ac\,\omega\dot{\theta}\sin\theta=0$

$\Rightarrow a^2\ddot{\theta}+a\dot{\omega}(a-c\cos\theta)]+a\omega c\dot{\theta}\sin\theta-ac\,\omega^2\sin\theta-ac\,\omega\dot{\theta}\sin\theta=0$

$\Rightarrow a\ddot{\theta}+\dot{\omega}(a-c\cos\theta)=c\,\omega^2\sin\theta$

Example 5:

A uniform rod, of mass 3m and length 2l, has its middle point fixed and a mass m attached at one extremity. The rod when in horizontal position is set rotating about a vertical axis through its centre with on angular velocity equal to $\sqrt{\left(\frac{2ng}{l}\right)}$ show that the heavy end of the rod will fall till the inclination of the rod to the vertical is $\cos^{-1}[\sqrt{(n^2+1)}-n$ and will the rise again.

Solution:

The mass is attached at L. On the rod ML. Take a point p such that

$OP=\xi$, the element $PQ=d\xi$. Further at any time t, let the plane through it and the vertical have turned through an angle ϕ from its initial position and let the rod be inclined at an angle θ to the vertical. Taking O, the mid point of the rod, as the origin and OX, OY (a line perpendicular to the plane of the paper) and OZ as axes of reference then co-ordinates of the point P on the rod are:

$$x=\xi\sin\theta\cos\phi, y=\zeta\sin\theta\sin\phi, z=\xi\cos\theta$$

$$\therefore \dot{x}=\xi\cos\theta\cos\dot{\theta}, -\sin\phi\dot{\phi}\theta.$$

$$\dot{y}=\xi\cos\theta\sin\phi\dot{\theta}, +\xi\sin\theta\dot{\theta}. \text{ Thus,}$$

$$v_\rho^2=(\text{velocity})^2 \text{ of } P=\dot{x}^2+\dot{y}^2+\dot{z}^2=\xi^2(\dot{\theta}^2+\dot{\phi}^2\sin^2\theta).$$

$$\therefore v_L^2=l^2(\dot{\theta}^2+\dot{\phi}^2\sin^2\theta)=(\text{velocity})^2 \text{ of mass m,}$$

Now mass of the element $PQ=\frac{3m}{2l}d\xi=dm, \text{say}.$

$\therefore$ Its kinetic energy

$$=\frac{1}{2}dm.v_\rho^2=\frac{1}{2}\cdot\frac{3m}{2l}d\xi(\dot{\theta}^2+\dot{\phi}^2\sin^2\theta)\xi^2$$

$$=\frac{3m}{4l}(\dot{\theta}^2+\dot{\phi}^2\sin^2\theta)\xi^2 d\xi$$

and K.E. of the rod $=\frac{3m}{4l}(\dot{\theta}^2+\dot{\phi}^2\sin^2\theta)\int_{-l}^{l}\xi^2 d\xi$

$$=\frac{1}{2}m(\dot{\theta}^2+\dot{\phi}^2\sin^2\theta)l^2.$$

Again, (velocity)² of the particle $m=l^2(\dot{\theta}^2+\dot{\phi}^2\sin^2\theta)$.

$\therefore$ Kinetic energy of the particle of mass $m=\frac{1}{2}ml^2(\dot{\theta}^2+\dot{\phi}^2\sin^2\theta)$.

$\therefore$ Total K.E. = T = K.E. of the rod + K.E. of the particle

$$=\frac{1}{2}ml^2(\dot{\theta}^2+\dot{\phi}^2\sin^2\theta)+\frac{1}{2}ml^2(\dot{\theta}^2+\dot{\phi}^2\sin^2\theta)$$

i.e. $T = ml^2(\dot{\theta}^2 + \dot{\phi}^2 \sin^2\theta)$

Also the work function is given by W = mgl cos θ + C.

Lagrange's φ-equation is $\frac{d}{dt}\left(\frac{\partial T}{\partial \dot{\phi}}\right) - \frac{\partial T}{\partial \phi} = \frac{\partial W}{\partial \phi}$

which gives $\frac{d}{dt}(2ml^2\dot{\phi}\sin^2\theta) = 0.$

Integrating it, we get $\dot{\phi}\sin^2\theta = K$ (constant).

Initially, $\theta = \frac{\pi}{2}$ and $\dot{\phi} = \sqrt{\left(\frac{2ng}{l}\right)}$. $\therefore K = \sqrt{\left(\frac{2ng}{l}\right)}$.

$\therefore \dot{\phi}\sin^2\theta = \sqrt{\left(\frac{2ng}{l}\right)}$. ...(1)

and Lagrange's θ-equation is $\frac{d}{dt}\left(\frac{\partial T}{\partial \dot{\theta}}\right) - \frac{\partial T}{\partial \theta} = \frac{\partial W}{\partial \theta}$

i.e. $\frac{d}{dt}(2ml^2\dot{\theta}) - 2ml^2\dot{\phi}^2\sin\theta\cos\theta = mgl\sin\theta$

or $2l\ddot{\theta} - 2l\dot{\phi}^2\sin\theta\cos\theta = -g\sin\theta.$ ...(2)

Substituting value of $\dot{\phi}$ from (1) in (2), we have

$2l\ddot{\theta} - 4ng\cot\theta\,\text{cosec}^2\theta = -g\sin\theta.$ ...(2)

Integration provides us $2l\dot{\theta}^2 + 4ng\cot^2\theta = 2g\cos\theta + k.$

Initially $\theta = \frac{\pi}{2}, \dot{\theta} = 0, \therefore k = 0.$

$\therefore 2l\dot{\theta}^2 + 4ng\cot^2\theta = 2g\cos\theta.$ ...(3)

The rod will fall till $\dot{\theta} = 0$.

i.e. $4ng\cot^2\theta = 2g\cos\theta$ or $2n\cos^2\theta - \cos\theta\sin^2\theta = 0.$

$\therefore$ either $\cos\theta = 0 \Rightarrow \theta = \frac{\pi}{2}$ which gives initial position.

as $2n\cos\theta - \sin^2\theta = 0 \Rightarrow \cos^2\theta + 2n\cos\theta - 1 = 0.$

Solving it, $\cos\theta = \frac{-2n \pm \sqrt{(4n^2+4)}}{2} = \left\{-n + \sqrt{(n^2+1)}\right\}$

[the other value being inadmissible because θ can not be obtuse] or $\theta = \cos^{-1}[-n + \sqrt{(n^2 + 1)}]$. This proves the required result. If we substitute this value of θ in equation (2′), then we find that θ comes out to be positive. Hence at that time the rod begins to rise.

Example 6:

A bear, of mass M, slides on a smooth fixed wire, whose inclination to the vertical is α, and has hinged to it a rod, of mass m and length 2l, which can move freely in the vertical plane through the wire. If the system starts from rest with the rod hanged vertically, show that

$$\{4M + m(1 + 3\cos^2\theta\}l\dot{\theta}^2 = 6(M + m)\, g \sin\alpha(\sin\theta - \sin\alpha)$$

where θ is the angle between the rod and the lower part of the wire.

Solution:

Let OL be the fixed wire. At any time t. Let the bead of mass M be at A where OA = x, also let θ be the angle which the rod AB makes with the lower part of the fixed wire.

Take O as origin and the fixed wire OL as x axis and a line though O and prep to OL as y axis; the co-ordinates of G, the C.G. of the rod AB, are

$$\{x + l\cos\theta, l\sin\theta\}$$

i.e. $x_G = x + l\cos\theta)$ and $y_G = l\sin\theta$.

$$\ddot{x}_G = (\dot{x} - l\sin\theta\dot{\theta});\ \dot{y}_G = l\cos\theta\dot{\theta}$$

$$\therefore \text{(velocity)}^2 \text{ of } G = v_G^2 = \dot{y}_G^2 + \dot{y}_G^2 = (\dot{x} - l\sin\theta\dot{\theta})^2 + (l\cos\theta\dot{\theta})^2.$$

Now let T be the kinetic energy and W the work function of the work function of the system Then we easily get

Total energy = T = K.E. of the bead + K.E, of the rod

$$= \frac{1}{2}M\dot{x}^2 + \frac{1}{2}m\left[\frac{l^2}{3}\dot{\theta}^2 + (\dot{x} - l\sin\theta\dot{\theta})^2 + (l\cos\theta\dot{\theta}^2\right]$$

$$= \frac{1}{2}(M + m)\dot{x}^2 - ml\dot{x}\dot{\theta}\sin\theta + \frac{2}{3}ml^2\dot{\theta}^2.$$

Also, the work function is given by

$$W = M\,g\,x\cos\alpha + mg\ \{x\cos\alpha + l\cos(\theta - \alpha)\}$$

$$= (M + m)\,gx\cos\alpha + mgl\cos\cos(\theta - \alpha)$$

$\therefore$ Lagrange's x-equation gives, $\dfrac{d}{dt}\left(\dfrac{\partial T}{\partial \dot{x}}\right) - \dfrac{\partial T}{\partial x} = \dfrac{\partial W}{\partial x}$

i.e $\frac{d}{dt}[M+m)\dot{x} - ml\dot{\theta}\sin\theta] = (M+m)g\cos\alpha$

or $(M+m)\ddot{x} - ml\ddot{\theta}\sin\theta - ml\dot{\theta}^2\cos\theta = (M+m)go\cos\alpha.$ ...(i)

Again Lagrange's θ – equation gives

$$\frac{d}{dt}\left(\frac{\partial T}{\partial \dot{\theta}}\right) + \frac{\partial T}{\partial \theta} = \frac{\partial W}{\partial \theta}$$

i.e. $\frac{d}{dt}[-ml\dot{x}\sin\theta + \frac{4}{3}ml^2\dot{\theta}] + ml\dot{x}\dot{\theta}\cos\theta = -mgl\sin(\theta - \alpha),$

or $-\ddot{x}ml\sin\theta - \dot{x}\dot{\theta}ml\cos\theta + \frac{4}{3}ml^2\ddot{\theta} + \dot{x}\dot{\theta}ml\cos\theta = -mgl\sin(\theta - \alpha).$

or $-\ddot{x}\sin\theta + \frac{4}{3}l\ddot{\theta} = -g\sin(\theta - \alpha)$ ...(ii)

Eliminating $\ddot{x}$ between (i) and (ii), we get

$\ddot{\theta}[-ml\sin^2\theta + \frac{4}{3}(M+m)l] - ml\dot{\theta}^2\sin\theta\cos\theta$

$= (M+m)g[\cos a\sin\theta - \sin(\theta - a)]$

or $l\ddot{\theta}[3M + m + 3m\cos^2\theta] - 3ml\dot{\theta}^2\sin\theta\cos\theta$

$= 3(M+m)g\cos\theta\sin\alpha.$

Whence on integrating, we get

$l\dot{\theta}^2[4M + m + 3m\cos^2\theta] = 6(M+m)g\sin a\sin\theta + C$ ...(iii)

When $\theta = a, \dot{\theta} = 0,$

$\therefore C = -6(M+m)g\sin^2\alpha.$

Putting the value of C in (iii), we get

$l\dot{\theta}^2(4M + m + 3m\cos^2\theta) = 6(M+m)g\sin\alpha(\sin\theta - \sin\alpha).$

Example 7:

A uniform rod, of length 2a, which has one end attached to a fixed point by a light inextensible string, of length $\frac{5}{12}$a, performing small oscillations in a vertical plane about its position of equilibrium. Find the position at any time, and show that the period of its principal oscillations are $2\pi\sqrt{\left(\frac{5a}{3g}\right)}$ and $\pi\sqrt{\left(\frac{a}{3g}\right)}$.

Solution:

Figure is self explanatory at any time t, let the string and the rod be inclined at θ and ϕ to the vertical OY.

Co-ordinates of G are given by

$$x_G = \frac{5}{12} a \sin\theta + a \sin\phi,$$

$$y_G = \frac{5}{12} a \cos\theta + a \cos\phi,$$

$$\dot{x}_G = \frac{5a}{12} a \cos\theta\dot{\theta} + a \cos\phi\dot{\phi},$$

$$\dot{y}_G = -\left(\frac{5a}{12} \sin\theta\dot{\theta} + a \sin\phi\dot{\phi}\right)$$

$$\therefore \dot{x}_G{}^2 + \dot{y}_G{}^2 = (\text{velocity})^2 \text{of G}$$

$$= \frac{25a^2}{144}\dot{\theta}^2 + a^2\dot{\phi}^2 + \frac{5a^2}{6}\dot{\theta}\dot{\phi}\cos(\theta+\phi) = \frac{25a^2}{144}\dot{\theta}^2 + a^2\dot{\phi}^2 + \frac{5}{6}a^2\dot{\theta}\dot{\phi}$$

[$\because$ θ and ϕ are small so $\cos(\theta + \phi) = 1$]

Again let T, be the kinetic energy and W, the work function of the system, then we easily get

$$T = \frac{1}{2} m \left[\frac{a^2}{3}\dot{\phi}^2 + \left(\frac{25}{144}a^2\dot{\theta}^2 + a^2\dot{\phi}^2 + \frac{5}{6}a^2\dot{\theta}\dot{\phi}\right)\right]$$

$$= \frac{ma^2}{288}[25\dot{\theta}^2 + 192\dot{\phi}^2 + 120\dot{\theta}\dot{\phi}]$$

$$\text{and } W = mg\left[\frac{5}{12} a \cos\theta + a \cos\phi\right]$$

$\therefore$ Lagrange's θ-equation gives

$$\frac{d}{dt}\left(\frac{\partial T}{\partial \dot{\theta}}\right) - \frac{\partial T}{\partial \theta} = \frac{\partial W}{\partial \theta} \Rightarrow \frac{d}{dt}\left\{\frac{ma^2}{144}(25\dot{\theta} + 60\dot{\phi})\right\} = -\frac{5mga}{12}\theta$$

{$\sin\theta = \theta$ is small}

$$\Rightarrow 5\dot{\theta} + 12\ddot{\phi} = -\frac{12g}{a}\theta. \qquad \text{...(1)}$$

and Lagrange's ϕ- equation gives

$$\frac{d}{dt}\left\{\frac{ma^2}{144}(192\dot{\phi} + 60\dot{\theta}\right\} = -mga\phi \Rightarrow 5\ddot{\theta} + 16\ddot{\phi} = -12\frac{g}{a} \qquad \text{...(2)}$$

Equations (1) and (2) $\Rightarrow$ $(5D^2 + 12c)\,\theta + 12D^2\theta = 0$ ...(3)

and $5D^2\theta + (6D^2 + 12c)\,\phi = 0$ where $(g/a) = c$. ...(4)

Now eliminating f between these two equations, we get

$[(5D^2 + 12c)(16D^2 + 12c) - 60D^4]\,\theta = 0.$ ...(5)

Let $\theta = A\cos(pt + B)$ $\therefore D\theta = -pA\sin(pt + B),$

$D^2\theta = -p^2A\cos(pt + B) = -p^2\theta$ and $D^4\theta = p^4\theta.$

Substituting these values in (5), we get

$(5p^4 - 63c^2 - 36c^2)\,\theta = 0 \Rightarrow (5p^4 - 63cp^2 + 36c^2) = 0$ $(\because \theta \neq 0)$

$$\Rightarrow (5p^2 - 3c)(p^2 - 12c) = 0 \Rightarrow \left(5p^2 - \frac{3g}{a}\right)\left(p^2 - \frac{12g}{a}\right) = 0$$

$$\therefore p_1^2 = \frac{3g}{5a} \text{ and } p_2^2 = \frac{12g}{a}$$

The periods of oscillations are $\frac{2\pi}{p_1}$ and $\frac{2\pi}{p_2}$

i.e. $2\pi\sqrt{\left(\frac{5a}{3g}\right)}$ and $2\pi\sqrt{\left(\frac{a}{12g}\right)}$ *i.e.* $2\pi\sqrt{\left(\frac{5a}{3g}\right)}$ and $\pi\sqrt{\left(\frac{a}{3g}\right)}$

Example 8:

A uniform rod, of mass 5m and length 2a, turns freely about one end which is fixed, to its other extremity is attached one end of a light string, of length 2a, which comes at its other end a particle of mass m, show that the periods of the small oscillations in a vertical plane are the same as those of simple pendulums of length $\frac{2a}{3}$ *and* $\frac{20a}{7}$

Solution:

Let the string BC and the rod AB make angles ϕ and θ with the vertical at any time t. The particle of mass m is tied to the end C of the string.

Now $x_c = 2a\sin\theta + 2a\sin\phi,$

$$\dot{x}_c = 2a(\cos\theta\dot{\theta} + \cos\phi\dot{\phi})$$

$$y_c = 2a\cos\theta + 2a\cos\phi.$$

$$\dot{y}_c = -2a(\sin\theta\dot{\theta} + \sin\phi\dot{\phi})$$

$$\therefore (\text{velocity})^2 \text{ of } m = \dot{x}_c^{\,2} + \dot{y}_c^{\,2}$$

$$= 4a^2(\dot{\theta} + \dot{\phi}^2 + 2\dot{\theta}\dot{\phi}).$$

Again co ordinates of G are $(a\sin\theta, a\cos\theta)$

$\therefore (\text{velocity})^2 \text{ of} G = a^2\dot{\theta}^2$

Now let T be the kinetic energy, and W the work function of the system, then we have

Total K.E. = K.E. of rod + K.E. of particle of mass m.

$$T = \frac{1}{2}5m\left(\frac{a^2}{3}\dot{\theta}^2 + a^2\dot{\theta}^2\right) + \frac{1}{2}m.4a^2(\dot{\theta}^2 + \phi^2 + 2\dot{\theta}\phi)$$

$$= ma^2\left(\frac{16}{3}\dot{\theta}^2 + 2\dot{\phi}^2 + 4\dot{\theta}\dot{\phi}\right)$$

and W = 5m ga cos θ + mg.2a (cosθ + cosϕ)

$$= 7mga\cos\theta + 2mga\cos\phi = 7mag\left(1 - \frac{\theta^2}{2}\right) + 2mag\left(1 - \frac{\phi^2}{2}\right)$$

∴ Lagrange's θ equation is given by

$$\frac{d}{dt}\left(\frac{\partial T}{\partial\dot{\theta}}\right) - \frac{\partial T}{\partial\theta} = \frac{\partial W}{\partial\theta}$$

$$\Rightarrow \frac{d}{dt}\left(\frac{32}{3}\dot{\theta} + 4\dot{\phi}\right) = -\frac{7g}{a}\theta \Rightarrow 32\ddot{\theta} + 12\ddot{\phi} = -21\frac{g}{a}\theta \quad \text{...(1)}$$

Lagrange's ϕ equation is given by

$$\frac{d}{dt}(4\dot{\phi} + 4\dot{\theta}) = -\frac{2g}{a}\phi \Rightarrow i.e.\ 2\ddot{\theta} + 2\ddot{\phi} = -\frac{g}{a}\phi \quad \text{...(2)}$$

$$\therefore (1) \text{ and } (2) \Rightarrow (32D^2 + 21c)\theta\ 12D^2\phi = 0 \quad \text{...(3)}$$

$$\text{and } 2D^2\theta + (2D^2 + c)\phi = 0 \text{ where } \frac{f}{a} = c$$

Now eliminating 'ϕ' between (3) and (4), we get

$$[(32D^2 + 21c)(2D^2 + c) - 24D^2]\theta = 0$$

$$\text{or } [40D^4 + 74cD^2 + 21c^2]\theta = 0 \quad \text{...(5)}$$

Now let $\theta = A\cos(pt + B) \Rightarrow D\theta = -pA\sin(pt + B)$.

$D^2\theta = -p^2A\cos(pt + B) = -p^2\theta$ and $D^4\theta = p^4\theta$.

substituting these in (5) we get

$$(40p^4 - 74cp^2 + 21c^2)\theta = 0\ i.e.\ 40p^4 - 74cp^2 + 21c^2 = 0 \quad \text{as } \theta \neq 0$$

$$\text{or } (2p^2 - 2c)(20p^2 - 7c) = 0$$

$$\text{i.e, } \left(2p^2 - \frac{3g}{a}\right)\left(20p^2 - 7\frac{g}{a}\right) = 0. \Rightarrow p_1^{\ 2} = \frac{3g}{2a} \text{ and } \Rightarrow p_2^{\ 2} = \frac{7g}{20a}.$$

*Hence length of equivalent pendulums are

$$\frac{g}{p_1^2} \text{ and } \frac{g}{p_2^2} \text{ i.e. } \frac{2a}{3} \text{ and } \frac{20}{7}a.$$

Example 9:

A uniform rod, of length 2a, can turn freely about one end, which is fixed. Initially it is inclined at an angle α, to the downward drawn vertical and it is set rotating about a vertical axis through its fixed end with angular velocity w. Show that, during the motion, the rod is always inclined to the vertical at an angle which is > or< α, according as $\omega^2 >$ or $< \frac{3}{4a\cos\alpha}$ *and that in each case its motion is inclined between the inclination α and*

$$\cos^{-1}[-n+\sqrt{(2n\cos\alpha+n^2)}] \text{ when } n=\frac{a\omega^2\sin^2\alpha}{3g}$$

If it be slightly disturbed when revolving steadily at a constant angle α show that the time of a small oscillation is

$$2\pi\sqrt{\left[\frac{4a\cos\alpha}{3g(1+3\cos^2\alpha}\right]}$$

Solution:

The rod OA is turning about the end O. Take a point P on the rod such that $OP=\xi$, and the element $PQ=\xi$,

$$\therefore \text{ mass of element } PQ=\frac{m}{2a}d\xi,$$

where m is the mass of the rod Further at any time t, let the rod be inclined at an angle θ to the vertical and let the plane through the rod and the vertical have turned though angle ϕ from its initial position OX, then co-ordinates of the point P are

$$x_p=\xi\sin\theta\cos\phi,\ y_p=\xi\sin\theta\sin\phi,\ z=\xi\cos\theta$$

$$\therefore v_p^{\ 2}=(\text{velocity})^2\text{of } p=\dot{x}_p^{\ 2}+\dot{y}_p^{\ 2}+\dot{z}_p^{\ 2}=\xi^2(\dot{\theta}^2+\dot{\phi}^2\sin^2\theta)$$

$$\text{and kinetic energy of the element } PQ=\frac{1}{2}\frac{m}{2a}v^2_p$$

$$=\frac{1}{2}\frac{m}{2a}d\xi(\dot{\theta}^2+\dot{\phi}^2\sin^2\theta)\xi^2$$

Now, let T, be the K.E. of the rod OA, the b we have

$$T = \frac{1}{2}\frac{m}{2a}(\dot{\theta}^2 + \dot{\phi}\sin^2\theta)\int_0^{2a}\xi^2 d\xi = \frac{2ma^2}{3}(\dot{\theta} + \dot{\phi}^2\sin^2\theta)$$

or $T = \frac{2ma^2}{3}(\dot{\theta}^2 + \dot{\phi}^2\sin^2\theta)$

Also the work function, $W = m\,ga\cos\theta + C.$

Lagrange's ϕ-equation gives

$$\frac{d}{dt}\left(\frac{4ma^2}{3}\dot{\phi}\sin^2\theta\right) = 0 \textit{ i.e. } \frac{d}{dt}[\dot{\phi}\sin^2\theta] = 0 \quad ...(1)$$

$$\Rightarrow \dot{\phi}\sin^2\theta = k\,(\text{constant}). \quad ...(2)$$

Initially $\theta = \alpha, \dot{\phi} = \theta, \quad \therefore k = \omega\sin^2\alpha.$

Thus (2) gives $\dot{\phi}\sin^2\theta = \omega\sin^2\alpha,$

and Lagrange's θ-equation is $\frac{d}{dt}\left(\frac{\partial T}{\partial \theta}\right) - \frac{\partial T}{\partial \theta} = 0.$

* when $\theta = A(pt + B)$, the period of motion is given $T = \frac{2\pi}{p}$. If l is the length of the simple equivalent pendulum, we have

$$T = 2\pi\sqrt{(l/g)} \Rightarrow l = \frac{g}{p^2}$$

$$\Rightarrow \frac{d}{dt}\left(\frac{4ma^2}{3}\dot{\theta}\right) - \frac{2ma^2}{3}\dot{\phi}^2.2\sin\theta\cos\theta = -\,m\,ga\sin\theta.$$

$$\Rightarrow \ddot{\theta} - \dot{\phi}^2\sin\theta\cos\theta = -\frac{g}{4a}\sin\theta. \quad ...(4)$$

Eliminating $\dot{\phi}$ between (4) and (3), we have

$$\ddot{\theta} - \frac{\omega^2\sin^4\alpha}{\sin^3\theta}\cos\dot{\theta} = -\frac{3g}{4a}\sin\theta. \quad ...(5)$$

$$\Rightarrow \dot{\theta} + \frac{\omega^2\sin^4\alpha}{\sin^3\theta} = \frac{3g}{4a}\cos\theta + A. \quad ...(6)$$

Initially $\theta = \alpha, \theta = 0, \therefore A = \omega^2\sin^2\alpha - \frac{3g}{2a}\cos\alpha$

Substituting this value of A in (6) , we get

$$\dot{\theta}^2 + \frac{\omega^2\sin^4\alpha}{\sin^2\theta} = \frac{3g}{2a}\cos\theta + \omega^2\sin^2\alpha - \frac{3g}{2a}\cos\alpha,$$

$$\text{or } \dot{\theta}^2 = \omega^2 \sin^2\alpha\left(1 - \frac{\sin^2\alpha}{\sin^2\theta}\right) + \frac{3g}{2a}(\cos\theta - \cos\alpha)$$

$$= \frac{3ng}{a}\left(1 - \frac{\sin^2\alpha}{\sin^2\theta}\right) + \frac{3g}{2a}(\cos\theta - \cos\alpha)\left[\because n = \frac{a\omega^2\sin^2\alpha}{3g}\right]$$

$$= \frac{3g}{2a}.\frac{\cos\alpha - \cos\theta}{\sin^2\theta}[2n(\cos\alpha + \cos\theta) - \sin^2\theta]$$

$$\text{i.e. } \dot{\theta}^2 = \frac{3g}{2a}.\frac{\cos\alpha - \cos\theta}{\sin^2\theta}[(\cos^2\theta + 2n\cos\theta + 2n\cos\alpha - 1)] \qquad ...(7)$$

From (7), we see that $\dot{\theta} = 0$, when

$$(\cos\alpha - \cos\theta)[(\cos^2\theta + 2n\cos\theta + 2n\cos\alpha - 1)] = 0$$

i.e. if either r $\cos\alpha - \cos\theta$ *i.e.* $\theta = \alpha$ (the initial position)

or $\cos^2 + 2n\cos\theta + 2n\cos\alpha - 1 = 0$

$$\text{i.e. } \cos\theta = \frac{-2n \pm \sqrt{[4n^2 + 4(1 - 2\cos\alpha)]}}{2}$$

$$\text{or } \cos\theta = -n + \sqrt{(1 - 3n\cos\alpha + n^2)} \qquad ...(8)$$

(the other value being inadmissible because that gives value of cos θ numerically greater that unity.

Hence the motion is included between $\theta = \alpha$ and $\theta = \theta_1$ where

$$\cos\theta_1 = \{\sqrt{(1 - 2n\cos\alpha + n^2)} - n\}$$

The rod will move above or below its initial position, if $\theta_1 >$ or $< \alpha$

or if $\cos\theta_1 <$ or $> \cos\alpha$

i.e. if $1 - 2n\cos\alpha + n^2 <$ or $> (n + \cos\alpha)^2$

i.e. if $\dfrac{3ng}{a\omega^2} <$ or $> 4n\cos\alpha$ *i.e.* if $\omega^2 >$ or $< \dfrac{3g}{4a\cos\alpha}$.

Small Oscillations about the Steady Motion

The motion will be steady if the rod goes round, inclined at the same angle α with the vertical or mathematically if θ =α(throughout the motion), then $\ddot{\theta} = 0$. Making these substitutions in (5), we get,

$$-\frac{\omega^2\sin^4\alpha}{\sin^3\theta}\cos\theta = -\frac{3g}{4a}\sin\theta \text{ i.e. } \omega^2 = \frac{3g}{4a\cos\alpha}$$

when ω^2 has this value and there are small oscillations about the position $\theta = \alpha$, then putting $\theta = \alpha + \psi$ in equation (5) we get

$$\ddot{\psi} = \frac{3g}{4a\cos\alpha}\frac{\sin^4\alpha}{\sin^3(\alpha+\psi)}\cos(\alpha+\psi) - \frac{3g}{4a}\sin(\alpha+\psi)$$

$$= \frac{3g}{4g}\left[\frac{\sin^4\alpha(\cos\alpha\cos\psi - \sin\alpha\sin\psi}{\cos\alpha(\sin\alpha\cos\psi + \cos\alpha\sin\psi)^3} - (\sin\alpha\cos\psi + \cos\alpha\sin\psi\right]$$

$$= \frac{3g}{4g}\left[\frac{\sin^4\alpha(\cos\alpha - \psi\sin\alpha\sin\psi}{\cos\alpha(\sin\alpha\cos\psi\cos\alpha)^3} - (\sin\alpha + \psi\cos\alpha\right], \text{ approximately}$$

$$= -\frac{3g\sin\alpha}{4a}[(1-\psi\tan\alpha)(1+\psi\cot\alpha)^{-3} - (1+\psi\cot\alpha], \text{ approx.}$$

$$= -\frac{3g\sin\alpha}{4}(4\cot\alpha + \tan\alpha)\psi,$$

$$\text{app.} = \frac{-3g(1+3\cos^2\alpha)}{4a\cos\alpha}\psi = -\mu\psi \text{ say}$$

$$\therefore \text{ time of small oscillation } = \frac{2\pi}{\sqrt{\pi}} = 2\pi\sqrt{\left(\frac{4a\cos\alpha}{3g(1+3\cos^2\alpha)}\right)}.$$

Example 10:

A uniform bar of length 2 a is hung from a fixed point by a string of length b fastened to one end of the bar. Show that when the system makes small normal oscillations in a vertical plane, the length 1 of the equivalent simple pendulum is a root of the quadratic,

$$l^2 - \left(\frac{4}{3}a + b\right)l + \frac{ab}{3} = 0.$$

Solution:

Figure is self explanatory.

At any time t, let the string OA and the rod AB make angles θ and ϕ with the vertical.

$$x_G = b\sin\theta + a\sin\phi.$$

$$y_G = b\cos\theta + a\sin\phi.$$

$$\therefore \dot{x}_G^2 + \dot{y}_G^2 = (\text{velocity})^2 \text{ of G}$$

$$= b^2\dot{\theta}^2 + a^2\dot{\phi}^2 + 2ab\dot{\theta}\dot{\phi}\cos(\theta - \phi)$$

$$= b^2\dot{\theta}^2 + a^2\dot{\phi}^2 + 2ab\dot{\theta}\dot{\phi} \qquad [\because \theta \text{ and } \phi \text{ are small}]$$

Now let T be the kinetic energy and W the work function of the system, then we easily obtain

$$W = mg\,[b\cos\theta + a\cos\phi]$$

$$\text{and } T = \frac{1}{2}m\left[\frac{a^2}{3}\dot{\phi}^2 + b^2\dot{\theta}^2 + a^2\dot{\phi}^2 + 2ab\dot{\theta}\dot{\phi}\right]$$

$$= \frac{1}{2}m\left[\frac{4a^2}{3}\dot{\phi}^2 + b^2\dot{\theta}^2 + 2ab\dot{\theta}\dot{\phi}\right]$$

$$\therefore \text{ Lagrange's } \theta\text{-equation is } \frac{d}{dt}\left(\frac{\cdot\partial T}{\partial\dot{\theta}}\right) - \frac{\partial T}{\partial\theta} = \frac{\partial W}{\partial\theta}$$

$$\Rightarrow \frac{d}{dt}\{m(b^2\dot{\theta} + ab\dot{\phi})\} = -\,mgb\theta. \qquad \{\because \sin\theta = 0\}$$

$$\Rightarrow b\ddot{\theta} + a\dot{\phi} = -\,g\theta \qquad \text{...(1)}$$

Lagrange's ϕ - equation is given by

$$\frac{d}{dt}\left\{\left(m\frac{4}{3}a^2\dot{\phi} + ab\dot{\theta}\right)\right\} = -\,mag\phi$$

$$\Rightarrow 4a\ddot{\phi} + 3b\ddot{\theta} = -\,3g\phi. \qquad \text{...(2)}$$

Equations (1) and (2) again can be written as

$$(bD^2 + g)\,\theta + aD^2\,\phi = 0 \qquad \text{...(3)}$$

$$\text{and } 3bD^2\;\theta + (4\,aD^2\;3g)\,\phi = 0. \qquad \text{...(4)}$$

Eliminating θ between these equations, we obtain

$$[(bD^2 + g)(4aD^2 + 3g) - 3abD^4]\theta = 0$$

$$\textit{i.e. } [(abD^4 + (4a + 3b)gD^2 + 3g^2]\theta = 0. \qquad \text{...(5)}$$

$$\text{Now let } \theta = A\cos\left[\sqrt{\left(\frac{g}{l}\right)}t + B\right]$$

where l is the length of the simple equivalent pendulum.

$$\text{Then } D\theta = -\sqrt{\left(\frac{g}{l}\right)}A\sin\left[\sqrt{\left(\frac{g}{l}\right)}t + B\right]$$

$$D^2\theta = -\frac{g}{l}A\cos\left[\sqrt{\left(\frac{g}{l}\right)}t + B\right] = \frac{g}{l}\theta \text{ and } D^2\theta = \frac{g^2}{l^2}\theta,$$

$$\therefore (5) \Rightarrow \left[ab\frac{g^2}{l^2} - (4a + 3b)\frac{g^2}{l} + 3g^2\right]\theta = 0$$

$$\Rightarrow 3l^2 - (4a + 3b)\,l + ab = 0 \qquad \{\because \theta \neq 0\}$$

$$\Rightarrow l^2 - \left(\frac{4}{3}a + b\right)l + \frac{ab}{3} = 0.$$

Example 11:

A uniform straight rod of length 2a, is freely movable about its centre and a particle of mass one-third that of the rod is attached by a light inextensible string, of length a, to one end of the rod; show that one period of principle oscillation is $\left(\sqrt{5+1}\right)\pi\sqrt{\left\{\frac{a}{g}\right\}}$.

Solution:

Figure is self explanatory.

At time t, let θ and ϕ be the inclinations of the rod and the string to the vertical. Co - ordinates of C are

$x_C = a \sin \theta + a \sin \phi$ and

$y_C = a \cos \theta + a \cos \phi$.

$$\therefore \dot{x}_C = a\cos\theta\dot{\theta} + a\cos\phi\dot{\phi} \text{ and } \dot{y}_C = -a\sin\theta\theta - a\sin\phi\dot{\phi}$$

$$\Rightarrow \dot{x}_C^2 + \dot{y}_C^2 = a^2\dot{\theta}^2 + a^2\dot{\phi}^2 + 2a^2\cos(\phi - \phi)\dot{\theta}\dot{\phi}$$

$$= a^2\,\dot{\theta}^2 + a^2\,\dot{\phi}^2 + 2a^2\dot{\theta}\dot{\phi}$$

[neglecting higher powers of small quantities]

$\therefore$ (velocity)^2 of the particle C = $v_C^2 = a^2\dot{\theta}^2 + a^2\,\dot{\phi}^2 + 2a^2\dot{\theta}\dot{\phi}$.

And velocity of the C.G. of the rod *i.e.* of O, is zero.

Now let T, be the kinetic energy and W, the work function of the system, then we easily get $W = \frac{mg}{3}(a\cos\theta + a\cos\phi) + C$

$$\text{and } T = \frac{1}{2}m\frac{a^2}{3}\dot{\theta}^2 + \frac{1}{2}\left(\frac{m}{3}\right)[a^2\dot{\theta}^2 + a^2\dot{\phi}^2 + 2a^2\dot{\theta}\dot{\phi}]$$

$$= \frac{ma^2}{6}[2\dot{\theta}^2 + \dot{\phi}^2 + 2\dot{\theta}\dot{\phi}]$$

$\therefore$ Lagrange's θ-equation is given by

$$\frac{d}{dt}\left(\frac{2ma^2}{3}\dot{\theta} + \frac{ma^2}{3}\dot{\phi}\right) = -\frac{mga}{3}\theta \Rightarrow 2\ddot{\theta} + \ddot{\phi} = -\frac{g}{a}\theta \qquad ...(1)$$

While Lagrange's ϕ-equation gives $\frac{d}{dt}\left[\frac{ma^2}{3}\dot{\phi}+\frac{ma^2}{3}\dot{\theta}\right]=-\frac{mga}{3}\theta$

i.e $\quad \ddot{\theta}+\ddot{\phi}=-\frac{g}{a}\phi.$

Equations (1) and (2) again give

$(2D^2 + c)\,\theta + D^2\,\phi = 0$ ...(3) and $D^2\,q + (D^2 + c)\,\phi = 0$ ...(4)

where $c=\frac{g}{a}$.

Eliminating ϕ in between (3) and (4), we get

$\{d^2 + c)\,(2D^2 + c) - D^4\}\,\theta = 0$ *i.e.* $[D^4 + 3cD^2 + c^2]\,\theta = 0.$ ...(5)

To solve (5), let $\theta = A\cos(pt + B)\ \therefore\ D\,\theta = pA\sin(pt + B),$

$D^2\theta = -p^2A\cos(pt + B) = -p^2\theta$ and $D^4\theta = p^4\theta.$

With these substitutions, (5) gives

$(p^4 - 3cp^2 + c^2)\,\theta = 0 \Rightarrow p^4 - 3cp^2 + c^2 = 0 \qquad (\because \theta \neq 0)$

$$\therefore p^2=\frac{3c\pm\sqrt{(9c^2-4c^2)}}{2}=\left(\frac{3\pm\sqrt{5}}{2}\right)c=\frac{(3\pm\sqrt{5}}{2}\cdot\frac{g}{a}.$$

$\therefore$ one period of principal oscillations corresponding to p_1, is given by

$$\frac{2\pi}{p_1}=2\pi\sqrt{\left(\frac{2}{3-\sqrt{5}}\cdot\frac{a}{g}\right)}=2\pi\sqrt{\left(\frac{a}{g}\right)}\sqrt{\left\{\frac{2(3+\sqrt{5}}{9-5}\right\}}$$

$$=2\pi\sqrt{\left(\frac{a}{g}\right)}\sqrt{\left(\frac{6+2\sqrt{5}}{4}\right)}=2\pi\sqrt{\left\{\frac{a}{g}\right\}}\sqrt{\left\{\frac{(\sqrt{5}+1)^2}{4}\right\}}$$

$$=(\sqrt{5}+1)\pi\sqrt{\left(\frac{a}{g}\right)}.$$

Example 12:

A mass m hangs from a fixed point by a light string of length l and a mass m′ hangs from m by a second string of length l′. For oscillations in a vertical plane, show that the periods of the principal oscillations are the values of $\frac{2\pi}{n}$ where n is given by the equation

$$n^4-gn^2\frac{m+m'}{m}\left(\frac{1}{l}+\frac{1}{l'}\right)+g^2\frac{m=m'}{ml'l}=0.$$

Solution:

At any time t, let the strings be inclined at angles θ and φ to the vertical. Co-ordinates of m are (l sin θ, l cos θ).

∴ (velocity)² of $m = l^2\dot{\theta}^2$ while co-ordinates of m′ are

$x_B = l\sin\theta + l'\sin\phi, \Rightarrow$

$\dot{x}_B = l\cos\theta\dot{\theta} + l'\cos\phi\dot{\phi}$

$y_B = -\,l\cos\theta + l'\cos\phi \Rightarrow$

$\dot{y}_B = -\,(l\sin\theta\dot{\theta} + l'\sin\phi\dot{\phi})$

∴ (velocity)² of $m' = \dot{x}_B^2 + \dot{y}_B^2 = l^2\dot{\theta}^2 + l^2\dot{\phi}^2 + 2ll\theta\dot{\phi}$ [∵ θ find φ are small]

Now let T, be the kinetic energy and , W the work function, of the system, then we have

W = mgl cos θ + m′g (l cos θ + l ′ cos φ)

= gl (m + m′) cos θ + m′ gl′ cos φ

and $T = \frac{1}{2}ml^2\dot{\theta}^2 + \frac{1}{2}m'[l^2\dot{\theta}^2 + l'\theta^2 + 2ll\dot{\theta}\dot{\phi}]$

$\frac{1}{2}[(m+m')l^2\dot{\theta}^2 + m'l'^2\dot{\theta}^2 + 2m'\,ll'\dot{\theta}\dot{\phi}$

Lagrange's θ-equation is given by $\frac{d}{dt}\left(\frac{\partial T}{\partial \dot{\theta}}\right) - \frac{\partial T}{\partial \theta} = \frac{\partial W}{\partial \theta}$

$= \frac{d}{dt}[(m+m')\,l^2\dot{\theta} + m'll'\dot{\phi}] = -\,gl(m+m')\,\theta.$

$\Rightarrow \{m + m'\}\,l\ddot{\theta} + m'l'\ddot{\phi} = -\,g\,(m+m')\theta$ ...(1)

While Lagrange's φ - equation gives

$\frac{d}{dt}[m'l'^2\dot{\phi} + m'll'\dot{\theta}] = -\,m'gl'\phi$

$\Rightarrow l'\ddot{\phi} + l\ddot{\theta} = -\,g\phi$ ...(2)

Equation (1) and (2) again give

$(m+m')(lD^2 + g)\theta + m'l'D^2\phi = 0$ ...(3)

$lD^2\theta + (l'D^2 + g)\phi = 0$ ...(4)

Eliminating φ, we get $[(m+m')(l'D^2 + g) - m'll'D^4]\theta = 0$

i.e. $[mll'D^4 + (m+m')(l+l')gD^2 + (m+m')g^2]\theta = 0$ ...(5)

Now let θ = A cos (nt + B); ∴ Dθ = – nA sin (nt + B)

$D^2\theta = -n^2A \text{ cps } (nt + B) = -m^2\theta$ and $D^4\theta = n^4\theta$ ...(6)

$\therefore$ (5) and (6) give $mall' n^4 - (m + m')(l + l') gn^2 + (m + m') g^2 = 0$

$$\text{or } n^4 - \frac{m+m'}{m}\left(\frac{1}{l}+\frac{1}{l'}\right)gn^2 + \frac{(m+m')g^2}{mall'} = 0. \quad ...(7)$$

Example 13(a):

A mass M hangs from a fixed point at the end of a very long string whose length l is a, to M is suspended a mass m by a string whose length l is small compound with a; prove that the time of a small oscillation of m is

$$2\pi\sqrt{\left(\frac{M}{M+m}\cdot\frac{1}{g}\right)}$$

Solution:

Here, we have $m = M$, $m' = m$, $l = a$, $l' = l$.

$$\therefore n^4 - \frac{M+m}{M}\left(\frac{1}{a}+1\right)\frac{g}{l}n^2 + \frac{(M+m)g^2}{Ml^2} = 0$$

$$i.e.\ \therefore n^4 - \frac{M+m}{M}\left(\frac{1}{a}+1\right)\frac{g}{l}n^2 + \frac{(M+m)g^2}{Ml^2}\cdot\frac{1}{a} = 0 \quad ...(8)$$

But a is large compared to $l \therefore \frac{1}{a} \to 0$

Hence the equation (8), gives

$$n^4 - \frac{M+m}{M}\cdot\frac{g}{l}\cdot n^2 = 0 \ \ i.e.\ \ n^2 = \frac{M+m}{M}\cdot\frac{g}{l}$$

$$\therefore \text{Time of a small oscillation} = \frac{2\pi}{n} = 2\pi\sqrt{\left\{\frac{M}{M+m}\cdot\frac{1}{g}\right\}}.$$

Example 13(b):

At the lowest point of a smooth circular tube, of mass M and radius a, is placed a particle of mass M', the tube hangs in a vertical plane from its highest point, which is fixed, and can turn freely in its own plane about this point. If the system be slightly displaced, show that the periods of the two independent oscillations of the system are

$$2\pi\sqrt{\left(\frac{2a}{g}\right)} \text{ and } 2\pi\sqrt{\left(\frac{Mag^{-1}}{M+M'}\right)}.$$

And that for one principal mode of oscillations, the particle remains at res relative to the tube and for the other, the centre of gravity of the particle and the tube and for the other, the centre of gravity of the particle and the tube remain at rest.

Solution:

Let C be the centre of the tube and A the position of the particle M′ at time t when OC and CA make angle 3 θ and φ with the vertical

$\therefore\ x_A = a\sin\theta + a\sin\phi,$

$y_A = a\cos\theta + a\cos\phi.$

and (velocity)² of A = $\dot{x}_A^2 + \dot{y}_A^2$

$= (a\cos\theta\theta + a\cos\phi\dot{\phi})^2 + (-a\sin\theta\dot{\theta} - a\sin\phi\dot{\phi})^2$

$= a^2\dot{\theta}^2 + a^2\ \dot{\phi}h2 + 2a^2\dot{\theta}\dot{\phi}\cos(\theta - \phi) = a^2\dot{\theta}^2 + 2a^2\dot{\theta}\dot{\phi}$

[neglecting small quantities of the higher order].

Also $C \equiv (a\sin\theta, a\cos\theta)$

(velocity)² of $C = (a\cos\theta\dot{\theta})^2 + (-a\sin\theta\dot{\theta})^2 = a^2\dot{\theta}^2$

Now let T, be the kinetic energy and W the work function of the system then we readily obtain $W = Mga\cos\theta + M'g\ = (M + M')\,ga\cos\theta + M'ga\cos\phi + K$

T = K.E. of circular tube + K.E. of particle

$= \frac{1}{2}M(a^2\dot{\theta}^2 + a^2\,\dot{\theta}^2) + \frac{1}{2}M(a^2\dot{\theta}^2 + a^2\dot{\phi}^2 + 2a^2\dot{\theta}\dot{\phi})$

$= \frac{2M + M'}{2}a^2\dot{\theta}^2 + \frac{1}{2}M'a^2\dot{\phi}^2 + M'a^2\dot{\theta}\dot{\phi}.$ ...(2)

$\therefore$ Lagrange's θ-equation gives

$\frac{d}{dt}[(2M + M')\,a^2\dot{\theta} + M'a^2\dot{\phi}] = -\,(M + M')ga\theta.$

$\Rightarrow (2M + M')\ddot{\theta} + M'\,\ddot{\phi} + = -\,(M + M')\frac{g}{a}\theta.$ ...(3)

Also Lagrange's φ - equation gives

$\frac{d}{dt}[M'\,a^2\dot{\phi} + M'\,a^2\dot{\theta}] = -\,M'ga\,\ddot{\phi} \Rightarrow \phi + \ddot{\theta} = -\frac{g}{a}\phi$ ...(4)

Equations (3) and (4) can be re-written as

$[(2M + M')\,D^2 + (M + M')\,c]\,q + M'\,D^2\,\phi = 0$ ...(5)

and $D^2\theta + (D^2 + c)\,\phi = 0$ where $c = \frac{g}{a}.$

Eliminating ϕ between these two equations, we get

$$[\{(2M + M')D^2 + (M + M')c\}(D^2 + c) - M'D^4]\theta = 0$$

i.e. $[2MD^4 + c(3M + 2M')D^2 + c^2(M + M')]\theta = 0.$ ...(7)

To solve (7),

let $\theta = A\cos(pt + B);\ D\theta = -pA\sin(pt + B)$

$D^2\theta = -p^2A\cos(pt + B) = -p^2\theta$ and $D^4\theta = \theta^4\theta$ ...(8)

$\therefore$ (7) and (8) give $[2Mp^4 - c(3M + 2M')p^2 + c^2(M + M')\theta = 0$

i.e. $2Mp^4 - c(3M + 2M')p + c^2(M + M') = 0.$ $[\because \theta \neq 0]$

which again gives $(2p^2 - c)[Mp^2 - c(M + M')] = 0$

$\therefore p_1^2 = \dfrac{c}{2}$ and $p_2^2 = \dfrac{c(M + M')}{M}\dfrac{g}{a}.$ $\left(\because c = \dfrac{g}{a}\right)$

Hence periods of oscillations are given by

$\dfrac{2\pi}{p_1}$ and $\dfrac{2\pi}{p_2}$ i.e. by $2\pi\sqrt{(2a/g)}$ and $2\pi\sqrt{\left(\dfrac{M}{(M + M'}\dfrac{a}{g}\right)}.$

Multiplying (6) by λ and adding to (5), we have

$D^2\{(2M + M' + l)\theta + (M' + \lambda)\phi\} = -\ \{(M + M')\theta + \lambda\phi\}$ ...(9)

Now chose λ such that

$$\frac{2M + M' + \lambda}{M' + \lambda} = \frac{M + M'}{\lambda} \Rightarrow \lambda = M' \text{ and } \lambda = -(M + M').$$

taking $\lambda = M'$, equation (9) reduces to

$$D^2\{(M + M')\theta + M'\phi\} = -\frac{1}{2}c\{M + M')\theta + M'\phi\}$$

and when $\lambda = -(M + M')$, equation (9) reduces to

$$D^2\{\theta - \phi) = -\frac{M + M'}{M}c(\theta - \phi)$$

$\therefore$ Principal co-ordinates are $q - f$ and $\{(M + M')\theta + M\text{–}\phi\}$. For the first mode, $\theta - \phi = 0$, *i.e.* $\theta = \phi$. This shows that the particle is at rest relative to the tube. For the second mode we have $(M + M')\theta + M'\phi = 0$. Further, the x-coordinates of C.G. of the particle and the tube

$$= \frac{Ma\sin\theta + M'(a\sin\theta + a\sin\phi)}{M + M'}$$

$$= \frac{a}{M + M'}\{M\theta + M'(\theta + \phi)\}$$ (since θ and ϕ are small,

$\therefore$ sin θ = θ and sin ϕ = ϕ)

$$= \frac{a}{M+M'}\{(M+M')\theta + M'\phi\} = 0 \text{ [using above results]}$$

⇒ The common C.G. of the particle and the tube remains at rest.

Example 14:

A perfectly rough sphere lying inside a hollow cylinder, which rests on a perfectly rough plane, is slightly displaced from its position of equilibrium. Show that the time of a small oscillation is

$$2\pi\sqrt{\left(\frac{a-b}{g}.\frac{4M}{10M+7m}\right)}$$

where a is the radius of the cylinder, b that of the sphere, and M, m are the masses of the cylinder and the sphere.

Solution:

Let CA be a line fixed in the cylinder and C'N, a line fixed in the sphere, which were initially vertical. Let these lines, after a time t, make angles ψ and ϕ with the vertical. Further let θ be the angle which the line joining centre makes with the vertical.

Initially O was the point of contact with the horizontal plane, which is regarded as origin. There is no slipping; between the cylinder, horizontal plan e and between the cylinder, sphere; therefore, we get OB = are AB = aψ and are AP = arc NP

i.e. a (θ + ψ) = b (θ + ϕ), *i.e.* $b\dot{\phi} = (a-b)\ \dot{\theta} + a\dot{\psi}$.

Referred to the horizontal and vertical through O as co-ordinate axes, we have

$xc' = a\psi + a(a-b)\sin\theta,\ yc' = a-(a-b)\cos\theta.$ (co-ordinates of c'

$\therefore$ (velocity)^2 of c'

$$\dot{x}c'^2 + \dot{y}c^2 = \{a\dot{\psi} + (a-b)\cos\theta\dot{\theta}\}^2 + \text{man}\{a-b)\sin\theta\dot{\theta}\}^2$$

$$= a^2\dot{\psi}^2 + (a-b)^2\dot{\theta}^2 + 2a(a-b)\dot{\psi}\dot{\theta}\cos\theta$$

$$= a^2\dot{\psi}^2 + (a-b)^2\dot{\theta}^2 + 2a(a-b)\dot{\psi}\dot{\theta}. \quad ...(1)$$

Also co-ordinates of C are (a ψ, a).

$\therefore$(velocity)^2 $(a\dot{\psi})^2 = a^2\dot{\psi}^2.$ ...(2)

Now let T be the kinetic energy and $\dot{W}$,the work function. Then we have

$$W = (a-b)mg\cos\theta + C \qquad ...(3)$$

and $T = \frac{1}{2}M\left(v^2C + a^2\dot{\psi}^2\right) + \frac{1}{2}m\left(v^2c' + \frac{2b^2}{5}\dot{\phi}^2\right)$

$$\Rightarrow T = \frac{1}{2}M(a^2\dot{\psi}^2 + a^2\dot{\psi}^2)$$

$$+\frac{1}{2}m\left[\frac{2b^2}{5}\dot{\phi} + a^2\dot{\psi}^2 + (a-b)^2\dot{\theta}^2 + 2a(a-b)\dot{\psi}\dot{\theta}\right]$$

$$= Ma^2\dot{\psi}^2 + \frac{1}{2}m[\frac{2}{5}\{(a-b\theta + a\dot{\psi}\} + a^2\dot{\psi}^2 + (a-b)^2\dot{\theta}^2 + 2a(a-b)\dot{\psi}\dot{\theta}]$$

$$= \frac{10M+7m}{10}a^2\dot{\psi}^2 + \frac{7m}{10}[(a-b)^2\theta^2 + 2a(a-b)\dot{\psi}\dot{\theta}] \qquad ...(4)$$

$\therefore$ Lagrange's θ-equation gives

$$\frac{d}{dt}\left[\frac{7m}{5}\{(a-b)^2\dot{\theta} + a(a-b)\dot{\psi}\}\right] = -(a-b)mg\sin\theta$$

$$\Rightarrow 7(a-b)\ddot{\theta} + 5a\ddot{\psi} = -5g\theta \text{ (as } \theta \text{ is small).} \qquad ...(5)$$

Also Lagrange's ψ-equation $\frac{d}{dt}\left[\frac{10M+7m}{5}a^2\dot{\psi} + \frac{7m}{5}a(a-b)\theta\right] = 0$

$$\Rightarrow (10M+7m)a\ddot{\psi} + 7m(a-b)\ddot{\theta} = 0$$

$$\Rightarrow (7m(a-b)\ddot{\theta} + (10M+7m)a\ddot{\psi} = 0 \qquad ...(6)$$

Eliminating $\ddot{\psi}$ between(5) and (6) ,we get

$$[7(10M+7m) - 49m](a-b)\ddot{\theta} = -5(10M+7m)g\theta$$

or $70M(a-b)\ddot{\theta} = -5(10M+7m)g\theta$

or $\ddot{\theta} = -\frac{(10M+7m)}{14M}\cdot\frac{g}{a-b)}\theta = -\mu\theta$ say,

$\therefore$ time of a small oscillation $= \frac{2\pi}{\sqrt{\mu}} = 2\pi\sqrt{\left(\frac{a-b}{g}\cdot\frac{14M}{10M+7m}\right)}$.

Example 15:

A perfectly rough sphere, of mass m and radius b, rest at the lowest point of a fixed spherical cavity of radius a. To the highest point of the movable sphere is attached a particle of mass m and the system is disturbed. Show that the oscillations are the same as those of a simple pendulum of length

$$a - b)\frac{4m' + \frac{7}{5}m}{m + m'\left(2 - \frac{a}{b}\right)}$$

Solution:

O is the centre of the fixed cavity, C the centre of the sphere and at B, a particle of m′ is attached. Initially NB was vertical, and N coincided with A.

After the time t, let BN or CN, a line fixed in the sphere make an angle ϕ with the vertical. But there is no slipping between the cylinder and the sphere.

$\therefore$ arc AP = arc NP,

i.e. $a\theta = b(\theta + \phi)$

i.e. $b\phi = (a - b)\theta \Rightarrow b\dot{\phi} = (a - b)\dot{\theta}$

Now co-ordinates of iB (i.e. of m ′) are

$x_B = (a - b)\sin\theta + b\sin\phi, y_B = (a - b)\cos\theta - b\cos\phi,$

and co-ordinates of the centre C are

$\{(a - b)\sin\theta, (a - b)\cos\theta\}$

$\therefore (\text{velocity})^2 \text{ of } B = \dot{x}_B{}^2 + \dot{y}_B{}^2$

$= \{(a - b)\cos\theta\dot{\theta} + b\cos\theta\dot{\theta}\}^2 + \{-(a - b)\sin\theta\dot{\theta} + b\sin\theta\dot{\theta}\}^2$

$= (a - b)^2\dot{\theta}^2 + b^2\dot{\phi}^2 + 2b(a - b)\dot{\theta}\dot{\phi}$

[neglecting small quantities of second and higher orders]

and (velocity)^2 of C = $\{(a - b)\cos\theta\dot{\theta}\}^2 + \{-(a - b)\sin\theta\dot{\theta}\}^2$

$= (a - b)^2\dot{\theta}^2$

Now let T, be the kinetic energy and W the work function of the system, then we readily obtain

$W = mg(a - b)\cos\theta + m'g\{(a - b)\cos\theta - b\cos\phi\} + D$

$= (a - b)(m + m')g\cos\theta - m'gb\cos\phi + D$

$= (a - b)(m + m')g\cos\left(\frac{b}{a - b}\phi\right) - m'\,gb\cos\phi + D \qquad \{\because (a - b)\theta = b.\phi\}$

and $T = \frac{1}{2}m\left[\frac{2b^2\dot{\phi}^2}{5} + (a - b)^2\dot{\theta}^2\right] + \frac{1}{2}m'\{(a - b)^2\dot{\theta}^2 + b^2\dot{\phi}^2 + 2b(a - b)\dot{\theta}\dot{\phi}$

$$=\frac{1}{2}m\left[\frac{2b^2\dot{\phi}^2}{5}+b^2\dot{\phi}^2\right]+\frac{1}{2}m'\{b^2\dot{\phi}^2+b^2\dot{\phi}^2+2b^2\dot{\phi}^2]\qquad [\because (a-b)\dot{\theta}=b\dot{\phi}]$$

$$=\frac{b^2}{10}(7m+20m')\dot{\phi}^2.$$

Further Lagrange's ϕ-equation gives $\frac{d}{dt}\left[\frac{b^2}{5}(7m+20m')\dot{\phi}\right]$

$$=-(a-b)(m+m')g\frac{b}{a-b}\sin\left(\frac{b}{a-b}\phi\right)+m'gb\sin\phi$$

$$=-(m+m')bg\sin\frac{b\phi}{a-b}+m'gb\sin\phi$$

$$\Rightarrow\frac{b^2}{5}(7m+20m')\ddot{\phi}=-(m+m')bg.\frac{b}{a-b}\phi+m'gb\phi \text{ app.}$$

$$\Rightarrow\left(4m'+\frac{7m}{5}\right)\ddot{\phi}=-\frac{g}{a-b}\left[m+m'\left(2-\frac{a}{b}\right)\right]\phi,\text{ app.}$$

$$\Rightarrow\ddot{\phi}=-\frac{g}{a-b}\cdot\frac{m+m'\left(2-\frac{a}{b}\right)}{4m'+\frac{7}{5}m}\phi=-\mu\phi\quad[\text{form }\ \ddot{x}=-\mu x\,]$$

$\therefore$ Length of the simple equivalent pendulum $=\frac{g}{\mu}$

$$=(a-b)\frac{4m'+\frac{7}{5}m}{m+m'\left(2-\frac{a}{b}\right)}.$$

Example 16:

A perfectly tough sphere rests at the lowest point of a fixed spherical cavity of double its own radius. To the highest point of the movable sphere is attached a particle of mass 7/20 times that of the sphere and the system is disturbed. Show that the oscillations are the same as those of a simple pendulum of length 14/5 times the radius of the sphere.

Solution:

Do yourself.

Example 17:

A plank, of mass M, radius of gyration k and length 2b, can swing like a sea-saw across a perfectly rough fixed cylinder of radius a. At its ends hang two particles each of mass m, by strings of length l. Show that, as the system swings, the lengths of its simple equivalent pendulum are

$$l \text{ and } \frac{Mk^2 + 2mb^2}{(m + 2m)a.}$$

Solution:

Let the strings be AP and BQ such that their free extremities are having particles of mass m. Initially G was at D and AP, BQ were vertical.

After a time t, when the plank ψ with the horizontal. Let the strings be inclined at θ and ϕ to the vertical.

But there is no slipping between the plank and the cylinder, so we have

$$\text{Now } xp = a\sin\psi + (b - a\psi)\cos\psi + l\sin\theta = b + l\theta - \frac{1}{2}b\psi^2 \quad ...(1)$$

$$yp = a\cos\psi - (b - a\psi)\sin\psi - l\cos$$

$$= a(1 - \frac{1}{2}\psi^2) - (b - a\psi)\psi - l(1 - \frac{1}{2}\theta^2) \quad ...(2)$$

$$\therefore \dot{x}_p = l\dot{\theta} - b\psi\dot{\psi} \text{ and } \therefore \dot{y}_p = -b\dot{\psi} + a\psi\dot{\psi} + l\theta\dot{\theta}$$

$$\therefore (\text{Velocity})^2 \text{ of } = \dot{x}_p{}^2 + \dot{y}_p{}^2 = l^2\dot{\theta}^2 + b^2\dot{\psi}^2 \, p$$

$$\text{Thus kinetic energy of m at } P = \frac{1}{2}m(l^2\dot{\phi}^2 + b^2\dot{\psi}^2)$$

$$\text{*Similarly kinetic energy of m at } Q = \frac{1}{2}m(l^2\dot{\phi}^2 + b^2\dot{\psi}^2)$$

$$\text{Also y-coordinate of Q } = a - l + b\psi + \frac{1}{2}a\psi^2 + \frac{1}{2}l\phi^2 = yq \text{ say}$$

[put–ψ for ψ and ϕ for θ in (2)]

the co-ordinates of G are {a sin ψ –aψ cosψ, acosψ + aψ sinψ} (neglecting higher powers of ψ as it is small)

$$i.e. \left\{\frac{a\psi^2}{2}, a + \frac{a\psi^2}{2}\right\}.$$

$$\therefore (\text{velocity})^2 \text{of } G = a^2\psi^2\dot{\psi}^2 [\because \dot{x} = a\dot{\psi}\psi, \dot{y} = a\psi\dot{\psi}]$$

(neglecting smaller, quantities). Thus kinetic energy of the plank

$$= \frac{1}{2}M(k^2\dot{\psi}^2 + a^2\psi^2\dot{\psi}^2) = \frac{1}{2}Mk^2\dot{\psi}^2, \text{ approximately}$$

* To obtain the corresponding quantities for the point Q. We shall write– ψ for ψ and ϕ for θ.

Now let T be the kinetic energy and W, the work function of the system, then we obtain

$$W = -mg\left(a - l - b\psi + \frac{a\psi^2}{2} + l\frac{\theta^2}{2}\right)$$

$$-mg\left(a - l + b\psi + \frac{a\psi^2}{3} + l\frac{\theta^2}{2}\right) - mg\left(a + \frac{a\psi^2 2}{}\right)$$

$$= K - \frac{1}{2}a(M + 2m)g\psi^2 - \frac{1}{2}m\,gl\left(\theta^2 - \phi^2\right)$$

and $T = \frac{1}{2}m(l^2\dot{\theta}^2 + b^2\dot{\psi}^2) + \frac{1}{2}m(l^2\dot{\phi}^2 + b^2\dot{\psi}^2) + \frac{1}{2}MK^2\dot{\psi}^2$

Lagrange's θ-equation is $\frac{d}{dt}\left(\frac{\partial T}{\partial \dot{\theta}}\right) - \frac{\partial T}{\partial \theta} = \frac{\partial W}{\partial \theta}$

$\Rightarrow \frac{d}{dt}(ml^2\dot{\theta}^2) = -mgl\theta \Rightarrow \ddot{\theta} = -\frac{g}{l}\theta,$

$\therefore$ length of the simple equivalent pendulum = l

Also Lagrange's $\psi - aq^n$ gives

$$\frac{d}{dt}\{(Mk^2 + 2mb^2)\dot{\psi}\} = -a(M + 2m)g\psi \Rightarrow \ddot{\psi} = -ag\frac{M + 2m}{Mk^2 + 2mb^2}\psi$$

Example 18:

A smooth circular wire, of mass 8 m and radius a, swings in a vertical plane, being suspended by an inextensible string of length a attached to one point of it, a particle of mass m can slide on the wire. Prove that the periods of small oscillations are

$$2\pi\sqrt{\left(\frac{8a}{3g}\right)}, 2\pi\sqrt{\left(\frac{a}{3g}\right)}, 2\pi\sqrt{\left(\frac{8a}{9g}\right)},$$

Solution:

At any time t, let the string OA, and the radius AC be inclined at angles θ and ϕ with he vertical and further let the radius to the particle (m) be inclined at an angle ψ with the vertical.

Now co-ordinates of C are

$$(a\sin\theta + a\sin\phi, a\cos\theta + a\cos\phi).$$

$\therefore (\text{velocity})^2$ of $C = a^2\dot{\theta}^2 + a^2\dot{\phi}\cos(\theta - \phi)$

$$= a^2\dot{\theta}^2 + a^2\dot{\phi}^2 + 2a^2\dot{\theta}\dot{\phi} \text{ approximately} \begin{bmatrix} \because \dot{x}_C = a\cos\theta\dot{\theta} + a\cos\phi\dot{\phi} \\ \dot{y}_C = -a(\sin\theta\dot{\theta} + \sin\phi\dot{\phi} \end{bmatrix}$$

Also co-ordinates of the particle m (*i.e.* of the pt. P) are

$x_P = a(\sin\theta + \sin\phi + \sin\psi)$,

$y_P = a(\cos\phi + \cos\phi + \cos\psi)$.

$\because (\text{celocity})^2 \text{ of } m = a^2(\dot{\theta}^2 + \dot{\phi}^2 + \dot{\psi}^2 + 2\dot{\theta}\dot{\phi} + 2\dot{\theta}\dot{\phi} + 2\dot{\phi}\dot{\psi} + 2\dot{\psi}\dot{\theta})$ app,

Let T, be the kinetic energy and W, the work function of the system, then we readily get

$$W = 8mg(a\cos\theta + a\cos\phi) + mg(a\cos\theta + a\cos\phi + a\cos\psi)$$

$$= mga[9\cos\theta + 9\cos\phi + \cos\psi)$$

$$\text{and } T = \frac{1}{2}.8m[a^2\dot{\phi}^2 + (a^2\dot{\theta}^2 + a^2\dot{\phi}^2 + a^2\dot{\phi}^2 + 2a^2\dot{\theta}\dot{\phi})]$$

$$+\frac{1}{2}ma^2[\dot{\theta}^2 + \dot{\phi}^2 + \dot{\psi}^2 + 2\dot{\theta}\dot{\phi} + 2\dot{\phi}\dot{\psi} + 2\dot{\psi}\dot{\theta}]$$

$$i.e.\ T = \frac{1}{2}m[9\dot{\theta}^2 + 17\dot{\phi}^2 + \dot{\psi}^2 + 18\dot{\theta}\dot{\phi} + 2\dot{\phi}\dot{\psi} + 2\dot{\psi}\dot{\theta}] \quad ...(2)$$

$\therefore$ Lagrange's q and y equations give

$$9\ddot{\theta} + 9\ddot{\phi} + \ddot{\psi} = -9\frac{g}{a}\theta \quad ...(3)$$

$$9\ddot{\theta} + 17\ddot{\phi} + \ddot{\psi} = -9\frac{g}{a}\phi \quad ...(4)$$

$$\text{and } \ddot{\theta} + \ddot{\phi} + \ddot{\psi} = \frac{g}{a}\psi,$$

which can be rewritten as

$$(9D^2 + 9c)\theta + 9D^2\phi + D^2\psi = 0 \quad ...(6)$$

$$9D^2\theta + (17D^2 + 9c)\phi + D^2\psi = 0 \quad ...(7)$$

$$\text{and } D^2\theta + D^2\phi + (D^2 + c)\psi = 0 \quad ...(8)$$

Eliminating ϕ and ψ in ...(6), (7) and (8), we get

$$\begin{vmatrix} 9D^2 + 9c & 9D^2 & D^2 \\ 9D^2 & 17D^2 + 9c & D^2 \\ D^2 & D^2 & D^2 + c \end{vmatrix}\theta = 0$$

$$i.e.\ (8D^2 + 9c)[9c(2D^2 + c) + D^2(8D^2 + 9c)]\theta = 0$$

$$i.e.\ (8D^2 + 9c)[8D^4 + 27cD^2) + 9c^2]\theta = 0$$

i.e. $(8D^2 + 9c)(8D^4 + 3c)D^2 + 3c]\theta = 0$ ...(9)

Now let $\theta = A\cos(pt + B)$, then

$D\theta = -pA\sin(pt + B),\ D^2\theta = -p^2A\cos(pt + B) = -p^2\theta$ ...(10)

$\therefore$ (9) gives $(8p^2 - 9c)(8p^2 - 3c)(p^2 - 3c) = 0$ $[\because \theta \neq \theta]$

$$\Rightarrow \left(8p^2 - \frac{9g}{a}\right)\left(8p^2 - \frac{3g}{a}\right)\left(p^2 - \frac{3g}{a}\right) = 0.$$

i.e. $p_1^2 = \dfrac{9g}{8a}, p_2^2 = \dfrac{3g}{8g}, p_3^2 = \dfrac{3g}{a}$

Thus periods of small oscillations are $\dfrac{2\pi}{p_1}, \dfrac{2\pi}{p_2}, \dfrac{2\pi}{p_3}$

i.e. $2\pi\sqrt{\left(\dfrac{8a}{9g}\right)}, 2\pi\sqrt{\left(\dfrac{8a}{3g}\right)}, 2\pi\sqrt{\left(\dfrac{a}{9g}\right)}$

Example 19:

A plank, 2a feet long, is placed symmetrically across a light cylinder, of radius a which rests and is free to roll on a perfectly rough horizontal plane. A heavy particle whose mass is an times that of plank is embedded in the cylinder at its lowest point. If the system is slightly displaced, show that its periods of oscillation are the values of

$$\frac{2\pi}{p}\sqrt{\left(\frac{a}{g}\right)}$$

given by the equation $4p^4 - (n + 12)p^2 + 3(n - 1) = 0.$

Solution:

C is the centre of the cylinder whose radius is a.

Let the mass of the plank be m, therefore mass of the particle embedded at the lowest point of the cylinder is nm.

Suppose θ and ϕ are the angles through which the cylinder and the plank have turned in time t, from their initial positions.

Initially G was coinciding with K which was at the highest point of the cylinder.

There is no slipping between the plank and the cylinder, hence Further there is no sliding between the cylinder and the horizontal plane,

$PG = \text{arc } KP = a(\phi = \theta)$

hence OC_1 = arc $C_1S = a\theta$, where O was the initial point of contact of the cylinder and the horizontal plane.

Taking O as origin and the horizontal and vertical through O as co-ordinates axes, we obtain $x_G = aq + a \sin f - a (f - q) \cos f$

$= a\theta + a\phi - a(\phi - \theta) = 2a\theta$

and $y_G = a + a \cos \phi + a (\phi - \theta) \sin \phi = 2a - a\theta\phi + a \dfrac{\phi^2}{2}$

Further co-ordinates of the particle nm (*i.e.* of the point S)

arc $\equiv \{a\theta - a\sin\theta, a - \cos\theta\}$

$\therefore$ (velocity)^2 of (nm) = $(a\dot{\theta} - a\cos\theta\dot{\theta})^2 + (a\sin\theta\dot{\theta})^2$,

$$= (a^2 + a^2 - 2a^2\cos\theta)\dot{\theta}^2.$$

Now let T, be the kinetic energy and W, the work function of the system, then we readily obtain

$$W = -n\,mg\,(a - a\cos\theta) - mg\left(2a - a\theta\phi + a\frac{\phi^2}{2}\right)$$

$$= -\ nmga\frac{\theta^2}{2} - mga\left(\frac{\phi^2}{2} - \theta\phi\right) - 2mga$$

and $T = \dfrac{1}{2}m\left[\dfrac{a^2}{3}\dot{\phi}^2 + 4a^2\dot{\theta}^2\right] + \dfrac{1}{2}nm[2a^2 - 2a^2\cos\theta]\dot{\theta}^2$

$$= \frac{1}{2}m\left[\frac{a^2}{3}\dot{\phi}^2 + 4a^2\dot{\theta}^2\right] + \frac{1}{2}nm.2a^2\dot{\theta}^2\frac{\theta^2}{2}$$

$= \dfrac{1}{2}m\left[\dfrac{a^2}{3}\dot{\phi}^2 + 4a^2\dot{\theta}^2\right]$, approximately, (neglecting the second part)

$\therefore$ Lagrange's q - equation is given by $\dfrac{d}{dt}\left(\dfrac{\partial T}{\partial\dot{\theta}}\right) - \dfrac{\partial T}{\partial\dot{\theta}} = \dfrac{\partial W}{\partial\theta}$

$$\Rightarrow \frac{d}{dt}(4ma^2\dot{\theta}) = -\,nmga\theta + mga\phi \Rightarrow 4\ddot{\theta} = -\frac{g}{a}(n\theta - \phi) \qquad ...(1)$$

Further Lagrange's f-equation gives

$$\frac{d}{dt}\left(\frac{ma^2}{3}\dot{\phi}\right) = -\,mg(\phi - \theta) \Rightarrow \ddot{\phi} = -\frac{3g}{a}(\phi - \theta) \qquad ...(2)$$

Equations (1) and (2) can be re-written at $(4D^2 + nc)\,\theta - c\phi = 0$ ...(3)

and $3cq - (D^2 + 3c)\,\phi = 0$ where $c = (g/a)$ ...(4)

Eliminating f in between (3) and (4), we get

$[(4D^2 + nc)(D^2 + 3c) - 3c^2]\theta = 0$

$\Rightarrow [4D^2 + c(n + 12)D^2 + 3(n - 1)c^2]\theta = 0$ ...(5)

Now if the periods of oscillation are the values of

$\frac{2\pi}{p}\sqrt{\left(\frac{a}{g}\right)}$ *i.e.* values of $\frac{2\pi}{\sqrt{\left(\frac{g}{a}p^2\right)}}$,

then the solution of the above equation must be of the form

$$\theta = A\cos\left[\sqrt{\left(\frac{g}{a}\right)}pt + B\right]$$

$$\therefore D\theta = -\sqrt{\left(\frac{g}{a}\right)}pA\sin\left[\sqrt{\left(\frac{g}{a}\right)}pt + B\right]$$

$$D^2\theta = -\left(\frac{g}{a}\right)pA\cos\left[\sqrt{\left(\frac{g}{a}\right)}pt + B\right]$$

$$= -\left(\frac{g}{a}\right)p^2\theta = -cp^2\theta \text{ and } D^4\theta = c^2p^4\theta.$$

Substituting these values in (5) we get

$[4c^2p^4 - c^2(n + 12)p^2 + 3(n - 1)c^2]\theta = 0$

$\Rightarrow 4p^4 - (n + 12)p^2 + 3(n - 1) = 0$ $(\because \theta \neq 0)$

Example 20:

To a point of a solid homogeneous sphere, of mass M, is freely hinged one end of a homogeneous rod, of mass nM; and the other end is freely hinged to a fixed point. If the system makes small oscillations under gravity about the position of equilibrium, the centre of a sphere and the rod being always in a vertical plane passing through the fix point, show that the periods of the principal oscillations are the values of $\frac{2\pi}{p}$ *given by the equation*

$2ab(6 + 7n)p^4 - p^2g\{10a(3 + n) + 21b(2 + n)\} + 15g^2(2 + n) = 0,$

where a is the length of the rod and b is the radius if the sphere.

Solution:

At any time t, let the rod and the sphere have turned through angel θ and ϕ to vertical, where OA = a, AC = b.

Now co - ordinates of C are

$x_C = a \sin\theta + b \sin\phi,\ y_C = a\cos\theta + b\cos\phi$

$\therefore$ (velocity)^2 of C $= \dot{x}_C^2 + \dot{y}_c^2 = a^2\dot{\theta}^2 + b^2\dot{\phi}^2 + 2ab\dot{\theta}\dot{\phi}\cos(\theta - \phi)$

$= a^2\dot{\theta}^2 + b^2\dot{\phi}^2 + 2ab\dot{\theta}\dot{\phi}$ [$\because$ θ and ϕ are small]

Also the co-ordinates of G are $\left(\frac{a}{2}\sin\theta, \frac{a}{2}\cos\theta\right)$

$\therefore$ (velocity)^2 of G $= \frac{a^2}{4}\dot{\theta}^2$

Now let T be, the kinetic energy and W, the work function of the system, then we readily obtain

$$W = n\ Mg\frac{a}{2}\cos\theta + Mg\,(a\cos\theta + b\cos\phi)$$

$$= \frac{1}{2}a(n+2)\,Mg\cos\theta + bMg\cos\phi$$

and $T = \frac{1}{2}M\left[\frac{2b^2}{5}\dot{\phi}^2 + (a^2\dot{\theta}^2 + b^2\dot{\theta}^2 + 2ab\dot{\theta}\dot{\phi}\right] + \frac{1}{2}n.M\left[\frac{a^2}{12}\dot{\theta}^2 + \frac{a^2}{4}\dot{\theta}^2\right]$

$$= \frac{1}{6}a^2(n+3)\,M\dot{\theta}^2 + \frac{7b^2}{10}M\dot{\phi}^2 + abM\dot{\theta}\dot{\phi}$$

$\therefore$ Lagrange's q-equation is given by $\frac{d}{dt}\left(\frac{\partial T}{\partial\dot{\theta}}\right) - \frac{\partial T}{\partial\theta} = \frac{\partial W}{\partial\theta}$

$\Rightarrow \frac{d}{dt}\left\{\frac{1}{3}a^2(n+3)\,M\dot{\theta} + abM\dot{\phi}.\right\} = -\frac{1}{2}a(n+2)Mg\,\theta$ app.

$\Rightarrow 2a(n+3)\dot{\theta} + 6b\ddot{\phi} = -3(n+2)g\theta$ ...(1)

While Lagrange's ϕ-equation gives

$\frac{d}{dt}\left\{\frac{7b^2}{5}M\dot{\phi} + abM\dot{\theta}\right\} = -3(n+2)g\theta$

$\Rightarrow 7b\ddot{\phi} + 5a\ddot{\theta} = -5g\phi$ ...(2)

These can be re-written as

$\{2a\,(n+3)D^2 + 3\,(n+2)\,g\}\,\theta + 6b\,D^2\,\phi = 0$ ...(3)

and $5aD^2\theta + (7bD^2 + 5g)\,\phi = 0$ ...(4)

Eliminating ϕ between (3) and (4), we get

$[2ab\,(7n+6)\,D^4 + \{10a\,(n+3) + 21b\,(n+2)\}\,gD^2$
$+ 15\,(n+2)\,g^2]\,\theta = 0$ ...(5)

To solve (5), let us put

$\theta = A\cos(pt + B), \therefore D\theta = -pA\sin(pt + B).$

$D^2\theta = -p^2 A\cos(p\dot{t} + \dot{B}) = -p^2\theta$ and $D^2\theta = p^4\theta.$

Substituting these values in (5), we have

$2ab(7n + 6)p^4 - \{10(n + 3)a + 2lb(n + 2)\}p^2 g + 15(n + 2)g^2 = 0.$

Example 21:

A hollow cylindrical garden roller is fitted with a counterpoise which can turn on the axis of the cylinder; the system is placed on the rough horizontal plane and oscillates under gravity, if $\dfrac{2\pi}{p}$ be the time of a small oscillation, show that p is given by the equation $p^2[2M + M')k^2 - M'h^2] = (2M + M')gh$ where M and M′ are the masses of the roller and counterpoise, k is the radius of gyration of M′ about the axis of the cylinder and h is the distance of its centre of mass from the axis.

Solution:

Suppose O′ is the centre of the cylindrical roller and G is the centre of gravity of the counterpoise, whose mass is M′. Obviously the line O′ K is a line fixed in the roller and the line O′ G is fixed in the counterpoise which were initially vertical. Suppose they make angles θ and ϕ with the vertical at the time t. Initially K was at O.

There is no slipping between the cylinder and the horizontal plane, hence OB = arc KB = aθ. Now assuming O as origin and the horizontal and vertical through O as co-ordinates axes, we have co-ordinates of G:-

$x_G = a\theta + h\sin\phi,\ y_G = a - h\cos\phi,$

$\therefore$ (velocity)^2 of $G = \dot{x}_G^2 + \dot{y}_G^2 = (a\dot{\theta} + h\cos\phi\dot{\phi})^2 + (h\sin\phi\dot{\phi})^2$

$= a^2\dot{\theta}^2 + h^2\dot{\phi}^2 + 2ah\dot{\theta}\dot{\phi}$ [neglecting higher powers of ϕ]

But radius of gyration of the counterpoise about O′ is k, where O′G = h. Hence the square of radius of gyration about G is $(k^2 - h^2)$.

Also (velocity)^2 of O′ = $(a\dot{\theta})^2 = a^2\dot{\theta}^2$. [$\therefore$ co-ordinates of O′ are (aθ, a)]

Now let T, be the kinetic energy and W, the work function of the system then we get

$$T = \frac{1}{2}M(a^2\dot{\theta}^2 + a^2\dot{\theta}^2) + \frac{1}{2}M'[(k^2 - h^2)\dot{\phi}^2 + a^2\dot{\theta}^2 + h^2\dot{\phi}^2 + 2ah\dot{\theta}h]$$

$$= \frac{M + M'}{2}a^2\dot{\theta}^2 + \frac{1}{2}M'k^2\dot{\phi}^2 + M'ah\dot{\theta}\dot{\phi}$$

$$M = M'gh\cos\phi + D.$$

$\therefore$ Lagrange's θ-equation gives $\frac{d}{dt}[(2M+M')a^2\dot{\theta}+M'ah\dot{\phi}]=0$

$\Rightarrow (2M+M')a\ddot{\theta}+M'h\ddot{\phi}=0$ ...(1)

while Lagrange's ϕ-equation gives

$\frac{d}{a}[M'k^2\dot{\phi}+M'ah\dot{\theta}]=-M'gh\phi \Rightarrow k^2\ddot{\phi}+ah\ddot{\theta}=-gh\phi$

$\Rightarrow ah\ddot{\theta}+k^2\ddot{\phi}=-gh\phi,$ ...(2)

$O'O=a,\quad O'G=h$

Now eliminating $\ddot{\theta}$ between (1) and (2), we get

$[Mh^2-(2M+M')k^2]\ddot{\phi}=(2M+M')gh\phi.$

or $[(2M+M')k^2-M'h^2]\ \ddot{\phi}=-(2M+M')gh\phi$ ...(3)

But, if $\frac{2\pi}{p}$ is the time of small oscillation, then we assume $\phi = A\cos(pt+b)$

$\therefore \dot{\phi}=-pA\sin(pt+B)$ and $\ddot{\phi}=-p^2A\cos(pt+B)=-p^2\theta$

$\therefore$ Equation (3) gives

$-p^2[(2M+M')k^2-M'h^2]=-(2M+M')gh$ $[\because \phi\neq 0]$

Example 22:

Two equal rods AB and BC; each of length l, smoothly jointed at B, are suspended from A and oscillate in a vertical plane through A. Show that the periods of normal oscillations are

$$\frac{2\pi}{n} \text{ where } n^2=\left(3\pm\frac{6}{\sqrt{7}}\right)\frac{g}{l}.$$

Solution:

At any time t, let the rod AB and BC be inclined at angles θ and ϕ to the vertical.

$\therefore$ Co-ordinates of G_1 are $\left(\frac{1}{2}\sin,\frac{1}{2}\cos\theta\right)$

$$\therefore v_{G_1}^2=\left(\frac{1}{2}\cos\theta\dot{\theta}\right)^2+\left(-\frac{1}{2}\sin\theta\dot{\theta}\right)^2=\frac{l^2}{4}\dot{\theta}^2$$

$$x_{G_s}=l\sin\theta+\frac{1}{2}\sin\phi, y_{G_s}=l\cos\theta+\frac{1}{2}\phi$$

$$\therefore x_{G_s}^2=l^2\dot{\theta}^2+\frac{1}{2}l^2\dot{\phi}^2+l^2\dot{\theta}\dot{\phi}\cos(\theta-\phi)=l^2\theta^2+\frac{1}{4}l^2\phi^2+l^2\dot{\theta}\dot{\phi}$$

[$\because \theta$ and ϕ are small]

now, let T, be the kinetic energy and W, the work function, of the system, then

we have $W = mgl\cos\theta + mg\left(l\cos\theta + \frac{1}{2}\cos\phi\right)$

$$= \frac{1}{2}mgl(3\cos\theta + \cos\phi)$$

and $T = \frac{1}{2}m\left[\frac{1}{12}l^2\dot{\theta}^2 + \frac{1}{2}l^2\dot{\theta}^2\right] + \frac{1}{2}m\left\{\frac{1}{12}l^2\dot{\phi}^2 + l^2(\dot{\theta}^2 + \frac{1}{4}\dot{\phi}^2 + \dot{\theta}\dot{\phi})\right\}$

$$= \frac{1}{2}ml^2\left(\frac{4}{3}\dot{\theta}^2 + \frac{1}{3}\dot{\phi}^2 + \dot{\theta}\dot{\phi}\right)$$

$\therefore$ Lagrange's θ-equations is

$$\frac{d}{dt}\left(\frac{\partial T}{\partial \dot{\theta}}\right) - \frac{\partial T}{\partial \theta} = \frac{\partial W}{\partial \theta} \Rightarrow \frac{d}{dt}\left(\frac{4}{d}l\dot{\theta} + \frac{1}{2}l\dot{\phi}\right) = -\frac{3}{2}g\theta$$

$$\Rightarrow 8\ddot{\theta} + 3\ddot{\phi} = -\frac{9g}{l}\theta$$

While Lagrange's ϕ equation gives

$$\frac{d}{dt}\left(\frac{1}{3}l\dot{\phi} + \frac{1}{2}l\dot{\theta}\right) = -\frac{1}{2}g\phi \Rightarrow 3\ddot{\theta} + 2\ddot{\phi} = -\frac{3g}{l}\phi$$

Equations (1) and (2) can be written as

$(8D^2 + 9c)\ \theta + 3D^2\phi = 0$

and $3D^2\theta + (2D^2 + 3c)\ \phi = 0$ where $c = \left(\frac{g}{l}\right)$

Eliminating ϕ between (3) and (4), we get

$(8D^2 + 9c)\ (2D^2 + 3c)\ - 9D^4]\theta = 0$

i.e. $(7D^2 + 42cD^2 + 27c^2)\ \theta = 0$

Let us assume, $\theta = A\cos(nt + B) = -\ n^2\theta$ and $D^4\theta = n^4\theta$

Substituting these values in (5), we get

$(7D^4 + 42cD^2 + 27c^2)\ \theta = 0$ *i.e.* $7n^2\ -\ 42cn^2\ - 27c^2) = 0$ [$\because \theta \neq 0$]

i.e. $n^2 = \dfrac{42c \pm \sqrt{[\{(42c)^2 - 4(7)(27c^2)\}]}}{14}$

$$= \left[3 \pm \frac{6}{14}\sqrt{(28)}\right]c = \left(3 \pm \frac{6}{\sqrt{7}}\right)c = \left\{3 \pm \frac{6}{\sqrt{7}}\right\}\frac{g}{l}.$$

Example 23:

A solid uniform sphere has a light rod rigidly attached to it which passes through its centre. This rod is so jointed to a fixed vertical axis that the angle q between the rod and the axis may alter but the rod must turn with the axis. If the vertical axis be forced to revolve constantly with uniform angular velocity, show that the equation of motion is of the form

$$\dot{\theta}^2 = n^2(\cos\theta\ 0\ \cos\beta)(\cos\alpha - \cos\theta)$$

Show also that the total energy imparted to the sphere as q increase from θ_1 to θ_2 *varies as* $\cos^2\theta_1 - \cos^2\theta_2$.

Solution:

First of all, we reduce the revolving axis OZ to rest by introducing an additional force $m\omega^2$ NO′, along NO′ *i.e.* a force $m\omega^2 l \sin\theta$ along NO′.

After the interval of time t. Let the rod make an angle q with the vertical ONZ (taken as Z-axis. (Further let ϕ be the angle through which the sphere turn in the horizontal plane, then the co-ordinates of C.G. of the sphere, are given by

$$x_{0'} = l\sin\theta\cos\phi.\ y_0' = l\sin\theta\sin\phi,\ z_{0'} = l\cos\theta$$

$$\therefore \dot{x}_0' = l\dot{\theta}\cos\theta\cos\phi - l\dot{\phi}\sin\theta\sin\phi,$$

$$\dot{y}_0 = l\dot{\theta}\cos\theta\cos\phi + l\dot{\phi}\sin\theta\cos\phi,\ \dot{z}_0' = -\ l\dot{\theta}\sin\phi$$

$$\Rightarrow \dot{x}_0'^2 + \dot{y}_0'^2 + \dot{z}_0'^2 = l^2\dot{\theta}^2 + l^2\sin^2\theta\dot{\phi}^2$$

$$\therefore T = \frac{1}{2}m\left[\frac{2a^2}{5}\dot{\theta}^2 + \dot{x}_{0'}^2 + \dot{y}_{0'}^2 + \dot{z}_{0'}^2\right]$$

$$= \frac{1}{2}m\left[\frac{2a^2}{5}\dot{\theta}^2 + l^2\dot{\theta}^2 + l^2\dot{\phi}\sin^2\theta\right]$$

$$= \frac{1}{2}m\left[\left(\frac{2a^2}{5}l^2\right)\dot{\theta}^2 + l^2\omega^2\sin^2\theta\right] \qquad (\because \phi \neq \omega \text{ given})$$

and $W^* = mgl\cos\theta + m\omega^2 l\sin\theta = mgl\cos\theta + m\omega^2 . 2\sin\theta\cos\theta$

$= -\ mgl\sin\theta + 2\ m\omega^2 l^2\sin\theta\cos\theta$

$$\Rightarrow \left(\frac{2a^2}{5} + l^2\right)\ddot{\theta} = -\ mgl\sin\theta + 3m\omega^2 l^2\sin\theta\cos\theta$$

Multiplying both sides by $2\dot{\theta}$ and integrating, we get

$$\left(\frac{2}{5}a^2 + l^2\right)\dot{\theta}^2 = 2gl\cos\theta - 3l^2\omega^2\cos^2\theta + D$$

$$\Rightarrow \dot{\theta}^2 = n^2(\cos\theta - \cos\beta)(\cos\alpha - \cos\theta)$$

where n is constant and α, β are the values of θ for which θ vanish.

SECOND PART

**K.E. + Potential Energy = T + K, when P.E. is K and is given by

$$K = T - W + C = \frac{1}{2}m\left[\left(\frac{2a^2}{5} + l^2\right)\dot{\theta}^2 + l^2\omega^2\sin^2\theta\right]$$

$$- mgl\cos\theta - ml^2\omega^2\sin^2\theta + C$$

$$= \frac{1}{2}m[\{2gl\cos\theta - 3l^2\omega^2\cos^2\theta + D\} + l^2\omega^2\sin^2\theta]$$

$$- mgl\cos\theta - ml^2\omega^2\sin^2\theta + C$$

$$= -\frac{1}{2}m\omega^2l^2\cos^2\theta - \frac{1}{2}m\omega^2 l + A \text{ where A is constant.}$$

$$= -\frac{1}{2}m\omega^2l^2\cos^2\theta + B \text{ where B is constant}$$

$\therefore$ Total energy imparted

$$= \left[-\frac{1}{2}m\omega^2l^2\cos^2\theta + B\right]_{\theta_1}^{\theta_2} = \frac{1}{2}m\omega^2l^2(\cos^2\theta_1 - \cos^2\theta_2)$$

which varies as $(\cos^2\theta_1 - \cos^2\theta_2)$.

Example 24:

Four uniform rods; each of length 2a, are hinged at their ends so as to form a rhombus ABCD. The angles B and D are connected by an elastic string and the lowest end A rests on a horizontal plane while the end C slides on a smooth vertical wire passing through A; in the position of equilibrium the string is stretched to twice its natural length and the angle BAD is 2α. Show that the time of a small oscillation about this position is

$$2\pi\left\{\frac{2a(1 + 3\sin^2\alpha)}{3g\cos 2\alpha}\cos\alpha\right\}^{1/2}$$

Solution:

In the position of equilibrium, rods are making angles α with the vertical. When the system is slightly displaced from the position of equilibrium, let the rods make angle $(\alpha + \theta)$ with the vertical θ being a small displacement. Now assuming the fixed and A as origin and the horizontal and vertical lines through it as co-ordinate axes, the co-ordinates of G_2 are $\{a\sin(\alpha + \theta), 3a\cos(\alpha + \theta)\}$

$\therefore$ (velocity)^2 of $G_2 = \{a\cos(\alpha+\theta)\dot{\theta}\}^2 + (-3a\sin(\alpha+\theta)\dot{\theta}\}^2$

$= a^2[(1+8\sin^2(\alpha+\theta]\dot{\theta}^2$

Co-ordinates of G_1 are $\{a\sin(\alpha+\theta), a\cos(\alpha+\theta)\}$

$\therefore$ (velocity)^2 of $G_1 = a^2\dot{\theta}^2$

$\therefore$ Kinetic energy of the four taken together is

$$T = 2.\frac{1}{2}m.\left[\frac{a^2}{3}\dot{\theta}^2 + a^2\dot{\theta}^2\right]$$

$$+ 2.\frac{1}{2}m\left[\frac{a^2}{3}\dot{\theta}^2 + a^2\{1+8\sin^2(\alpha+\theta)\}\dot{\theta}\right]$$

$$= \frac{8ma^2}{3}\left[1+\sin^2(\alpha+\theta)\right]\dot{\theta}^2 \quad (\because v_{G_1} = v_{G_4} \text{ and } v_{G_2} = v_{G_3})$$

The work function W is given by $W = 2\{-mga\cos(\alpha+\theta)\}$

$$+2\{-mg3a\cos(\alpha+\theta)\} - 2\int_0^{2a\sin(\theta+\alpha)} \lambda\left(\frac{y-c}{c}\right)dy$$

$$= -8mga\cos(\alpha+\theta) - \frac{\lambda}{c}\{2a\sin(\alpha+\theta) - \}^2$$

Lagrange's θ equation gives

$$\frac{d}{dt}\left|\frac{16ma^2}{3}\{1+3\sin^2(\alpha+\theta)\}\dot{\theta} - 16ma^2\sin(\alpha+\theta)\cos(\alpha+\theta)\dot{\theta}^2\right.$$

$$= 8mga\sin(\alpha+\theta) - \frac{4\lambda}{c}a\cos(\alpha+\theta)\{2a\sin(\alpha+\theta) - c\}$$

$$\Rightarrow \frac{16ma^2}{3}(1+3\sin^2(\alpha+\theta)\}\ddot{\theta}$$

$$= 8mga\sin(\alpha+\theta) - \frac{4\lambda a}{c}\cos(\alpha+\theta)\{2a\sin(\alpha+\theta) - c\} \qquad \text{...(1)}$$

Initially when $\theta = 0, \dot{\theta} = 0, \ddot{\theta} = 0, c = a\sin\alpha$, hence (1) gives

$$\lambda = \frac{2mgc}{a\cos\alpha}.$$

Putting this value of λ in equation (1), we get

$$\frac{16ma^2}{3}\{1+3\sin^2(\alpha+\theta)\}\ddot{\theta}$$

$$= mga8\sin(\alpha+\theta) - \frac{8mg\cos(\alpha+\theta)}{\cos\alpha}\{2a\sin(\alpha+\theta) - c\}$$

*The force $m\omega^2 l\sin\theta$ also contributes to W. The distance of the point of application at O′ of this force from the vertical OZ is equal to l sin θ, hence the contribution $m\omega^2 l^2 \sin^2\theta$ to W is as given in (2).

** If W is the work function of the system , then P.E. = C –W.

$$i.e.\ \frac{16ma^2}{3}\{1+3\sin^2\alpha\}\theta = 8mga(\sin\alpha+\theta\cos\alpha)$$

$$-\frac{8mg}{\cos\alpha}(\cos\alpha - 8\sin\alpha)\{2a(\sin\alpha+\theta\cos\alpha)-a\sin\alpha\}.$$

$$= -\frac{8mg\cos 2\alpha}{\cos\alpha}\theta \text{ app.} \Rightarrow \ddot{\theta} = -\frac{3g\cos 2\alpha}{2a\cos\alpha(1+3\sin^2\alpha)}\theta \text{ app.}$$

∴ Time of a small oscillation about the position of equilibrium is given by

$$2\pi\sqrt{\left\{\frac{2a\cos\alpha(1+3\sin^2\alpha}{3g\cos^2\alpha}\right\}}$$

Example 25:

A uniform rod AB, of length 2a can turn freely about a point distance c from its centre, and is at rest at an angle α to the horizon when a particle is hung by a light string of length l from one end. If the particle be displaced slightly in the vertical plane of the rod show that it will oscillate in the same time as a simple pendulum of length

$$l\frac{a^2+3ac\cos^2\alpha+3c^2\sin^2\alpha}{2^2+3ac}$$

Solution:

Let M be the mass and G the centre of gravity of the rod. Further let O be the point about which it nac turn where OG = c. Let BP be a string of length l and m the mass of the particle tied to the end P of the string BP.

Initially rod was making an angle α with the horizontal and the string was vertical.

Let B_0 be the initial position of B so that $\angle B_0OX = \alpha$, After a time t let the rod make an angle θ with OB_0 *i.e.* an angle)α + θ) with the horizontal and let the string be inclined at an angle θ with the vertical.

Initially the system was at rest, hence taking moments about O, (in this position), we get Mc = m(a – c).

Assuming O as origin and the horizontal and the vertical through O as co-ordinates axes, we obtain the following results

$x_p = (a-c)\cos(\alpha+\theta) + l\sin\phi.\ y_p = (a-c)\sin(\alpha+\theta) + l\cos\phi.$

$\therefore$ (velocity)^2 of $P = \{-(a-c)\sin(\alpha+\theta)\dot\theta + l\sin\theta\dot\theta\}^2$

$+\{(a-c)\cos(\alpha+\theta)\dot\theta - l\sin\phi\dot\phi\}^2$

$= (a-c)^2\dot\theta^2 + l^2\dot\phi^2 - 2l(a-c)\dot\theta\dot\phi\sin(\alpha+\theta+\phi)$

$= (a-c)^2\dot\theta^2 + l^2\dot\phi^2 + 2l(a-c)\dot\theta\dot\phi\sin\alpha$

(neglecting small quantities of the higher order)

Also co-ordinates of G (the C.G. of the rod) are

$\{-c\cos(\alpha+\theta), -c\sin(\alpha+\theta)\}$

$\therefore$ (velocity)^2 of G $= \{c\sin(\alpha+\theta)\dot\theta\}^2 + \{-c\cos(a+\theta)\dot\theta\}^2 = c^2\dot\theta^2.$

Now let T, be the kinetic energy and W, the work function of the system, then we easily obtain

$W = mg\ \{(a-c)\sin(\alpha+\theta) + l\cos\phi] - Mgc\sin(\alpha+\theta) + D$

$= mgl\cos\phi + D$

[the first and the last terms cancel, because m (a – c) = Mc]

and $T = \frac{1}{2}M\left[\frac{a^2}{3}\dot\theta^2 + c^2\dot\theta^2\right] + \frac{1}{2}m[(a-c)^2\dot\theta^2$

$+l^2\dot\phi^2 - 2l(a-c)\dot\theta\dot\phi\sin\alpha]$

$\therefore$ Lagrange's θ-equation gives

$$\frac{d}{dt}\left[\left\{M\left(\frac{a^2}{3}+c^2\right)+m(a-c)^2\right\}\dot\theta - ml(a-c)\dot\phi\sin\alpha\right] = 0$$

$$\Rightarrow \left\{M\left(\frac{a^2}{3}+c^2\right)+m\,(a-c)^2\right\}\dot\theta - ml(a-c)\ddot\phi\sin\alpha = 0.$$

$$\Rightarrow \left\{M\left(\frac{a^2}{3}+c^2\right)+mc\,(a-c)\right\}\ddot\theta - mcl\ddot\phi\sin\alpha = 0 \qquad [\because m(a-c) = Mc]$$

$\Rightarrow (a^2+3ac)\ddot\theta - 3lc\sin\alpha\ddot\phi = 0.$ While Lagrange's ϕ equation gives

$$\frac{d}{dt}\left\{ml^2\dot\phi - ml(a-c)\dot\theta\sin\alpha\right\} = -\,mgl\phi$$

$$\Rightarrow l\ddot\phi - (a-c)\ddot\theta\sin\alpha = -\,g\phi \qquad ...(2)$$

Eliminating θ in (1) and (2), we have

$$l\ddot\phi - (a-c)\sin\alpha\frac{3lc\sin\alpha}{a^2+3ac}\ddot\phi = -\,g\phi$$

i.e. $\Rightarrow \dfrac{a^2+3ac-3c(a-c)\sin^2\alpha}{a^2+3ac} l\ddot{\phi} = -g\phi$

$$\Rightarrow \ddot{\phi} = -\frac{g}{l(a^2+3ac\cos^2\alpha+3c^2\sin^2\alpha)}\phi = -\mu\phi \text{ (say)}$$

$\therefore$ Length of the simple equivalent pendulum $= \dfrac{g}{\mu}$

$$= \frac{a^2+3ac\cos^2\alpha+3c^2\sin^2\alpha}{a^2+3ac} l.$$

Example 26:

A uniform rod AB of length 8a is suspended from a fixed point O by means of light inextensible string, of length 13a, attached to B. If the system is slightly displaced in a vertical plane, show that $(\theta+3\phi)$ *and* $(12\theta-13\phi)$ *are principal co-ordinates where* θ and ϕ *are the angles which the rod and string respectively make with the vertical. Also show that periods of small oscillations are*

$$2\pi\sqrt{\left(\frac{a}{g}\right)} \text{ and } 2\pi\sqrt{\left(\frac{52}{3g}\right)}.$$

Solution:

We have

$x_G = 13a \sin\phi + 4a \sin\theta$

and $y_G = 13a \cos\phi + 4a \cos\phi$

$\therefore \dot{x}_G^2 + \dot{y}_G^2 = 169a^2\dot{\phi}^2 + 16a^2\dot{\theta}^2$

$+ 104a^2\dot{\theta}\dot{\phi}\cos(\theta-\phi)$

Thus $T = \dfrac{1}{2}m[k^2\dot{\theta}^2 + (\dot{x}_G^2+\dot{y}_G^2)] = \dfrac{1}{2}ma^2\left[\dfrac{64}{3}\dot{\theta}^2 + 169\dot{\phi}^2 + 104\dot{\theta}\dot{\phi}\right]$

and the work function

$W = mg\,(13a\cos\phi + 4a\cos\theta)$

$\therefore$ Lagrange's θ-equation gives

$$\frac{d}{dt}\left(\frac{\partial T}{\partial \dot{\theta}}\right) - \frac{\partial T}{\partial \theta} \Rightarrow 61\ddot{\theta} + 39\ddot{\phi} = -\frac{3g}{a}\theta \quad ...(1)$$

While Lagrange's θ-equation gives $4\ddot{\theta} + 13\ddot{\phi} = -\dfrac{g}{a}\phi$ ...(2)

Equations (1) and (2) $\Rightarrow D^2(\theta+3\phi) = -\dfrac{3g}{52a}(\theta+3\phi)$

and $D^2(12\theta - 13\phi) = -\frac{g}{a}(12\theta + 13\phi)$

Now putting $q + 3\phi = X$ and $12\theta - 13\theta = Y$ in these equations, we get D^2X

$$= -\frac{3g}{52a}X \text{ and } D^2Y = -\frac{g}{a}\cdot Y;$$

which obviously represent two independent simple harmonic motions. Hence X and Y are principal co-ordinates, that is

$(\theta + 3\phi)$ and $(12\theta - 13\phi)$ are principal co-ordinates.

Also, periods of small oscillations are given by

$$2\pi\sqrt{\left(\frac{52a}{3g}\right)} \text{ and } 2\pi\sqrt{\left(\frac{a}{g}\right)}$$

Example 27:

A ring slides on a smooth circular hoop of equal mass and of radius a which can turn in a vertical plane about a fix point O in its circumference. If θ *and* ϕ *be the inclinations to the vertical of the radius through O and of the radius through the ring, prove that principal co-ordinates are* $(2\theta + \phi)$ *and* $\phi - \theta)$*, and the periods of small oscillations are*

$$2\pi\sqrt{\left(\frac{a}{2g}\right)} \text{ and } 2\pi\sqrt{\left(\frac{2a}{g}\right)}.$$

Solution:

Let the mass of the hoop and the ring be m. In a displaced position, let O′ be the centre of the hoop and P the position of the ring which slides on the hoop.

$\therefore$ Co-ordinates of the point P are given by

$x_p = a(\sin\theta + \sin\phi)$

$y_p = a(\cos\theta + \cos\phi)$

$\Rightarrow \dot{x}_P^2 + \dot{y}_P^2 = a^2(\dot{\theta}^2 + \dot{\phi}^2 + 2\dot{\theta}\dot{\phi})$

when θ and ϕ are small.

also co-ordinates of $O' = a^2\dot{\theta}^2$

Thus T = Energy of circular hoop + Energy of ring

$= \frac{1}{2}m(a^2\dot{\theta}^2 + a^2\dot{\theta}^2) + \frac{1}{2}ma^2(\dot{\theta}^2 + \dot{\phi}^2 + 2\dot{\theta}\dot{\phi})$

$= \frac{1}{2}ma^2(3\dot{\theta}^2 + \dot{\phi}^2 + 2\dot{\theta}\dot{\phi})$

Also the work function

W = mga cos θ + mg (a cos θ + a cos ϕ) = mga (2 cos θ + cos ϕ).

∴ Lagrange's θ-equation gives $3\ddot{\theta}+\ddot{\phi}=-\frac{2g}{a}\theta$ ...(1)

Lagrange's f-equation gives $\ddot{\theta}+\ddot{\phi}=-\frac{g}{a}\phi$ ...(2)

Now adding (1) and (2) and multiplying (2) by 2 and subtracting (1) from that, we get $D^2(2\theta+\phi)=-\frac{g}{2a}(2\theta+\phi)$

and $D^2(\phi-\theta)=-\frac{2g}{a}(\phi-\theta)$

Taking 2θ + ϕ = X and ϕ – θ = Y, we get

$$D^2X=-\frac{g}{2a}X \text{ and } D^2Y=-\frac{2g}{a}Y;$$

which obviously represent two independent simple harmonic motions. Thus X and Y are principal co-ordinates, that is (2θ + ϕ) and (ϕ – θ) are principal co-ordinates.

Also, periods of small oscillations are given by

$$2\pi\sqrt{\left(\frac{2a}{g}\right)} \text{ and } 2\pi\sqrt{\left(\frac{a}{2g}\right)}.$$

Example 28:

A square ABCD formed by four rods each of length 2l and mass hinged at their ends, rests on a horizontal frictionless table. An impulsive of magnitude I is applied to the vertex A in the direction AD Find the equations of motion, and prove that the K.E. developed immediately after the application of the impulsive forces is

$$T = (I^2/2m)$$

Solution:

When the square is struck, its shape will in general be a rhombus. Suppose that at any time t, the angels made by sides AD (or BC) and BA (or CD) with the x-axis are θ_1 and θ_2 respectively, while the co-ordinates of the centre M are (x, y).

*Suppose that the force F_V acting on a system are such that $\lim_{\tau\to 0}\int_0^{\tau} F_V dt = I_v$

where t represents a time interval, Then we call F_v impulsive forces and I_v are called Impulse.

** If we call P_α to be the generalised impulse, (3) can be written as

Hence x, y, θ_1, θ_2, are the generalised co-ordinates.

From the adjoining diagram, we see that the position vectors of the centre E,F, G.H of the rods are given respectively by

$r_E = (x - l\cos\theta_1)i + (\dot{y} - l\sin\theta_1)j,$

$r_F = (x + l\cos\theta_2)i + (y - l\sin\theta_2)j,$

$r_G = (x + l\cos\theta_2)i + (y - l\sin\theta_1)j,$

and $r_H = (x - l\cos\theta_2)i + (y + l\sin\theta_1)j.$

$\because v_E = \dot{r}_E = (\dot{x} - l\sin\theta_1\dot{\theta}_1)i + (\dot{y} - l\cos\theta_1\dot{\theta}_1)j,$

$v_F = \dot{r}_F = (\dot{x} - l\sin\theta_2\dot{\theta}_2)i + (\dot{y} - l\cos\theta_2\dot{\theta}_2)j,$

$v_G = \dot{r}_G = (\dot{x} - l\sin\theta_1\dot{\theta}_1)i + (\dot{y} + l\cos\theta_1\dot{\theta}_1)j,$

$v_H = \dot{r}_H = (\dot{x} + l\sin\theta_2\dot{\theta}_2)i + (\dot{y} + l\cos\theta_2\dot{\theta}_2)j.$

Now, K.E. of the rod $AB = \frac{1}{2}m\dot{r}_E^2 + \frac{1}{3}ml^2\dot{\theta}_1^2 \frac{1}{2} = T_{AB}$ say

K.E. of the rod $CB = \frac{1}{2}m\dot{r}_F^2 + \frac{1}{3}ml^2\dot{\theta}_1^2 \frac{1}{2} = T_{BC}$ say

K.E. of the rod $CD = \frac{1}{2}m\dot{r}_G^2 + \frac{1}{3}ml^2\dot{\theta}_2^2 \frac{1}{2} = T_{CD}$ say

K.E. of the rod $DA = \frac{1}{2}m\dot{r}_H^2 + \frac{1}{3}ml^2\dot{\theta}_1^2 \frac{1}{2} = T_{DA}$ say

$\therefore$ K.E. of system $\left(\frac{\partial T}{\partial}\right)_2 = (4m\dot{x})_2 = 4m\dot{x}, \left(\frac{\partial T}{\partial \dot{y}}\right)_2 = (4m\dot{y})_2 = 4m,$

$\frac{1}{2}m(4\dot{x}^2 + 4\dot{y}^2 + 2l^2\dot{\theta}_1^2 + 2l^2\dot{\theta}_2^2) + \frac{ml^2}{3}(\dot{\theta}_1^2 + \dot{\theta}_2^2).$

$$= 2m(\dot{x}^2 + \dot{y}^2) + \frac{4}{3}ml^2(\dot{\theta}_1^2 + \dot{\theta}_2^2). \qquad ...(1)$$

Let us now assume that initially the rhombus is a square at rest with its sides parallel to the co-ordinate axes and its centre located at the origin.

Then, we get $x = 0 = y,\ \theta_1 = \frac{\pi}{2}, \theta_2 = 0, \dot{x} = 0,\ \dot{y} = 0, \dot{\theta}_1 = 0, \dot{\theta}_2 = 0.$

Now if we use the notation, $(\ )_1$ and $(\ )_2$ to denote quantities before and after the impulse is applied, we have

$$\left(\frac{\partial T}{\partial \dot{x}}\right)_1 = (4m\dot{x})_1 = 0 \left(\frac{\partial T}{\partial \dot{y}}\right)_1 = (4m\dot{y})_1 = 0$$

$$\left(\frac{\partial T}{\partial \dot{\theta}}\right)_1 = \left(\frac{8}{3} ml^2 \dot{\theta}_1\right) = 0, \left(\frac{\partial T}{\partial \dot{\theta}_2}\right) = \left(\frac{8}{3} ml^2 \dot{\theta}_2\right) = 0$$

and $\left(\frac{\partial T}{\partial \dot{x}}\right)_2 = (4m\dot{x})_2 = 4m\dot{x}, \left(\frac{\partial T}{\partial \dot{y}}\right)_2 = (4m\dot{y})_2 = 4m,$

$$\left(\frac{\partial T}{\partial \dot{\theta}_1}\right)_2 = \left(\frac{8}{3} ml^2 \dot{\theta}_1\right)_2 = \frac{8}{3} ml^2 \dot{\theta}_1 \left(\frac{\partial T}{\partial \dot{\theta}_2}\right)_2 = \frac{8}{3} ml^2 \dot{\theta}_2.$$

Hence Lagrange's equations are

$$\left(\frac{\partial T}{\partial \dot{x}}\right)_2 - \left(\frac{\partial T}{\partial \dot{x}}\right)_1 = p_x \Rightarrow 4m\dot{x} = P_x \qquad ...(2)$$

$$\left(\frac{\partial T}{\partial \dot{y}}\right)_2 - \left(\frac{\partial T}{\partial \dot{y}}\right)_1 = p_y \Rightarrow 4m\dot{y} = P_y \qquad ...(3)$$

$$\left(\frac{\partial T}{\partial \dot{\theta}_1}\right)_2 - \left(\frac{\partial T}{\partial \dot{\theta}_1}\right)_1 = p_{\theta_1} \Rightarrow \frac{8}{3} ml^2 \dot{\theta}_1 = P_{\theta_1}. \qquad ...(4)$$

$$\left(\frac{\partial T}{\partial \dot{\theta}_1}\right)_2 - \left(\frac{\partial T}{\partial \dot{\theta}_2}\right)_1 = p_{\theta_1} \Rightarrow \frac{8}{3} ml^2 \dot{\theta}_2 = P_{\theta 2}. \qquad ...(5)$$

Now we shall find the value of P_x, P_y, p_{θ_1} and p_{θ_2}

we have $P_\alpha = \sum I_v \bullet \frac{\partial r_v}{\partial q_\alpha}$ where I_v are the impulsive forces.

$$\therefore P_x = I_A \bullet \frac{\partial r_A}{\partial x} + I_B \bullet \frac{\partial r_B}{\partial x} + I_c \bullet \frac{\partial r_c}{\partial x} + I_D \bullet \frac{\partial r_D}{\partial x} \qquad ...(6)$$

$$P_y = I_A \bullet \frac{\partial r_A}{\partial y} + I_B \bullet \frac{\partial r_B}{\partial y} + I_c \bullet \frac{\partial r_c}{\partial y} + I_D \bullet \frac{\partial r_D}{\partial y} \qquad ...(7)$$

$$P_{\theta_1} = I_A \bullet \frac{\partial r_A}{\partial \theta_1} + I_B \bullet \frac{\partial r_B}{\partial \theta_1} + I_c \bullet \frac{\partial r_c}{\partial \theta_1} + I_D \bullet \frac{\partial r_D}{\partial \theta_1} \qquad ...(8)$$

$$P_{\theta 2} = I_A \bullet \frac{\partial r_A}{\partial \theta_2} + I_B \bullet \frac{\partial r_B}{\partial \theta_2} + I_c \bullet \frac{\partial r_c}{\partial \theta_2} + I_D \bullet \frac{\partial r_D}{\partial \theta_1} \qquad ...(9)$$

where

$$r_A = (x - l\cos\theta_1 - l\cos\theta_2)\,i + (y - l\sin\theta_1 + l\sin\theta_2)\,j \qquad ...(10)$$

$$r_B = (x - l\cos\theta_1 + l\cos\theta_2)\,i + (y - l\sin\theta_1 - l\sin\theta_2)\,j \qquad ...(11)$$

$$r_C = (x + l\cos\theta_1 + l\cos\theta_2)\,i + (y + l\sin\theta_1 - l\sin\theta_2)j \qquad ...(12)$$

$$r_D = (x + l\cos\theta_1 - l\cos\theta_2)\,i + (y + l\sin\theta_1 + l\sin\theta_2)j \qquad ...(13)$$

But initially the impulsive force at A is in the direction of the positive y-axis, we have

$$I_A = Ij. \qquad ...(14)$$

$\therefore$ Equation (6), (7), (8) and (9) $\Rightarrow$

$$P_x = 0, P_y = I, P_{\theta_1} = -Il\cos\theta_1 \text{and} P\theta_2 = Il\cos\theta_2 \qquad ...(15)$$

Thus equation (2) and (3),(4) and (5) give

$$4m\dot{x} = 0, 4m\dot{y} = I, \frac{8}{3}ml^2\dot{\theta}_1 = -Il\cos\theta_1, \frac{8}{3}ml^2\dot{\theta}_2 = Il\cos\theta_2 \qquad ...(16)$$

Second Part

We have $\dot{x} = 0, \dot{y} = \frac{1}{4m'}\dot{\theta}_1 = -\frac{3I}{8ml}\cos\theta_1$ and $\dot{\theta}_2 = \frac{3I}{8ml}\cos\dot{\theta}_2$

$$\therefore T = 2m(x^2 + y^2) + \frac{4}{3}ml^2(\dot{\theta}^2{}_1 + \dot{\theta}_2^2)$$

$$= 2m\left(0 + \frac{I^2}{16m^2}\right) + \frac{4}{3}ml^2\left[\frac{9I^2}{64m^2l^2}\cos^2\theta_1 + \frac{9I^2}{64m^2l^2}\cos^2\theta_2\right] \qquad ...(17)$$

But immediately after the application of the impulsive forces,

$\theta_1 = \frac{1}{2}\pi$ and $\theta_2 = 0$ approximately, so (17) gives $T = \frac{I^2}{2m}\cdot$

Example 29:

Three equal uniform rods, AB, BC, CD are freely joined at B and C and the ends A and B are fastened to smooth, fixed pivots whose distance apart is equal to the length of either rod. The frame being at rest in the form of the square, a blow J is given perpendicular to AB at its middle point and in the plane of square. Show that the energy set up is $\frac{3J^2}{40m}$ *where m is the mass of each rod.*

Find also the blows at the joints B and C.

Solution:

the blow J is given at G_1 the C.G. of AB, at right angles to AB

The rods AB and CD will turn through the same angle θ but the rod BC will remain parallel to AD.

Now let T be the K.E. of the system, then we have

$$T = 2.\frac{1}{2}m\left[\frac{a^2}{3}\dot{\theta}^2 + a^2\dot{\theta}^2\right] + \frac{1}{2}m(2a\dot{\theta})^2 = \frac{10ma^2}{3}\dot{\theta}^2.$$

∴ Lagrange's θ-equation gives

$$\left(\frac{\partial T}{\partial \dot{\theta}}\right) - \left(\frac{\partial T}{\partial \dot{\theta}}\right) = J \Rightarrow \frac{20ma^2}{3}\dot{\theta} = Ja$$

$$i.e.\ \dot{\theta} = \frac{3J}{20ma} \therefore T = \frac{10ma^2}{3}.\frac{9J^2}{400m^2a^2} = \frac{3J^2}{40m}.$$

SECOND PART

Let P and Q be the impulses at B and C respectively. With directions, as shown in the figure, then considering the motion of AB and taking moments about A, we get

Change in the angular momentum about the axis through A

= moments of the impulses about this axis

$$i.e.\ m\frac{4a^2}{3}\dot{\theta} = Ja - P2a \Rightarrow P = \frac{1}{2}J - \frac{2ma}{3}$$

$$\therefore P = \frac{1}{2}J - \frac{2ma}{3}\frac{3J}{20ma} = \frac{2J}{5}$$

Again considering the motion of CD and taking moments about D, we obtain

$$m = \frac{4a^2}{3}\dot{\theta} = Q2a \Rightarrow Q = \frac{2ma}{3}\dot{\theta} = \frac{3J}{20ma} = \frac{1}{10}J.$$

Example 30:

Three equal uniform rods AB, BC, CD, each of mass m and length 2a, are at rest in a straight line smoothly jointed at B and C. A blow I is given to the middle rod at a distance c from the centre O in a direction perpendicular to it, show that the initial velocity of O is $\frac{2I}{3T'}$ *and that the initial angular velocities of the rods are*

$$\frac{(5a+9c)I}{10ma^2}, \frac{6cI}{5ma^2} \text{ and } \frac{(5a-9c)I}{10ma^2}$$

Solution:

Just after the blow let $\dot{y}$ be the linear velocity of O, and let θ, φ, ψ be the inclination of BC, CD and AB to their initial positions. Now let T be K.E. of the system, then we have

$$T=\frac{1}{2}\left(\frac{a^2}{3}\dot{\theta}^2+\dot{y}^2\right)+\frac{1}{2}m\left\{\frac{a^2}{3}\dot{\psi}^2+(\dot{y}-a\dot{\theta}+a\dot{\psi})^2\right\}$$
$$+\frac{1}{2}m\left\{\frac{a^2}{3}\dot{\phi}^2+(\dot{y}-a\dot{\theta}+a\dot{\phi})^2\right\}$$
$$=\frac{1}{2}m\left[3\dot{y}^2+\frac{7a^2}{3}\dot{\theta}^2+\frac{4a^2}{3}\dot{\psi}^2+\frac{4}{3}a^2\dot{\phi}^2+2a\dot{y}\dot{\psi}+2a\dot{y}\dot{\phi}+2a\dot{\theta}\dot{\phi}-2a\dot{\theta}\dot{\psi}\right]$$

Also just before the action of the impulse, we have T = 0

∴ Lagrange's equation for blows gives

$$\left(\frac{\partial T}{\partial \dot{y}}\right)_2-\left(\frac{\partial T}{\partial \dot{y}}\right)_1=I\Rightarrow 3m\dot{y}+ma\dot{\phi}+ma\dot{\psi}=I$$

$$\Rightarrow 3\dot{y}+a\dot{\theta}+a\dot{\psi}=\frac{1}{m}.$$

Similarly Lagrange's θ, ϕ, ψ equations give

$$\frac{7}{3}a\dot{\theta}+a\dot{\phi}-a\dot{\psi}=\frac{Ic}{ma'}\dot{y}+\frac{4}{3}a\dot{\phi}+a\dot{\theta}=0,\dot{y}-a\dot{\theta}+\frac{4}{3}a\dot{\psi}=0$$

Last two equations $\Rightarrow 3\dot{y}+2a\dot{\phi}+2a\dot{\psi}=03\dot{y}=\frac{2I}{m'}$ or $\dot{y}=\frac{2I}{3m}$

Also $2a\dot{\theta}=\frac{4a}{3}(\dot{\psi}-\dot{\phi})$, i.e. $\frac{3}{2}a\dot{\theta}+a\dot{\phi}-\dot{\psi}=0$

$$\therefore \frac{5a}{6}\dot{\theta}=\frac{Ic}{ma}\text{ i.e. }\dot{\theta}=\frac{6Ic}{5ma^2}$$

Substituting now the values of $\dot{y}$ and $\dot{\theta}$ in the equation

$\dot{y}+\frac{4}{3}a\dot{\phi}+a\dot{\theta}=0$ we get

$$\frac{2I}{3m}+\frac{4}{3}a\dot{\phi}+\frac{6Ic}{5ma}=0 \text{ or } \dot{\phi}=\frac{(5a-9c}{10ma^2}.$$

Also substituting values of $\dot{y}$ and $\dot{\theta}$ in equation $\dot{y}-a\dot{\theta}+\frac{4}{3}a\dot{\psi}=0,$

we get $\frac{I}{3m}-\frac{6Ic}{5ma}+\frac{4}{3}a\dot{\psi}=0$ or $\dot{\psi}=\frac{(5a-9c)I}{10ma^2}.$

Example 31:

Six equal uniform rods form a regular hexagon, loosely jointed at the angular points, and rest on a smooth table; a blow is given perpendicular to one of them at its middle point, find the resulting motion and show that the opposite rod begins to move with one tenth of the velocity of the rod that is struck.

Solution:

The impulse is given at the middle point of AB in the direction as shown in the figure so the ensuing motion of BC, AF and CD, EF will be symmertrical.

Therefore they have after impulse same angular velocity say ω.

Also the rods AB and DE will begin to move at right angles to themselves. Now let $\dot{x}$ and $\dot{y}$ be their velocities just after the action of impulse.

$\therefore$ Velocity of C in the direction of $\dot{x}$

= velocity of B + velocity of C relative to B

$$= \dot{x} + 2a\omega \sin 30^\circ = \dot{x} + a\omega \qquad ...(1)$$

and velocity of C in the same direction (with respect to the rod CD)

= velocity of D + velocity of C relative to D,

$$= \dot{y} - 2a\omega \sin 30^\circ = \dot{y} - a\omega \qquad ...(2)$$

$$\therefore (1) \text{ and } (2) \Rightarrow \dot{x} + a\omega = \dot{y} - a\omega \Rightarrow a\omega = \frac{1}{2}(\dot{y} - \dot{x})$$

Now actual velocity of the C.G of the rod BC is $a\omega\cos 30^\circ$ parallel to AB and $\dot{x} + a\omega \sin 30^\circ$ at right angles to AB

$\therefore$ (velocity)^2 of $G_1 = (a\omega\cos 30^\circ)^2 + (\dot{x} + a\omega\sin 30^\circ)^2$

Thus K.E.of BC

$$= \frac{1}{2}m[\frac{1}{3}a^2\omega^2 + \{(a\omega\cos 30)^2 + (\dot{x} + a\omega\sin 30)^2\}]$$

$$= \frac{1}{2}m(\frac{4}{3}a^2\omega^2 + \dot{x}^2 + a\omega\dot{x}) = \frac{1}{12}m[5\dot{x}^2 + 2\dot{y}^2 - \dot{x}\dot{y}] = T_1 \text{say}$$

$$[\because a\omega = \frac{1}{2}(\dot{y} - \dot{x})]$$

Also Kinetic energy of CD

$$= \frac{1}{2}m[\frac{1}{3}a^2\omega^2 + \{(a\omega\cos 30)^2 + (\dot{y} - a\omega\sin 30)^2\}]$$

$$= \frac{1}{2}m(\frac{4}{3}a^2\omega^2 + \dot{y}^2 - a\omega\dot{y}] = \frac{1}{12}m[2\dot{x}^2 + 5\dot{y}^2 - \dot{x}\dot{y}] = T_2 \text{say}$$

$\therefore$ Total kinetic energy T is given by $T = 2T_1 + 2T_2 + T_3 + T_4$

where T_3 and T_4 are K.E.'s of AB and DE respectively.

$$\therefore T = \frac{1}{2}m\dot{x}^2 + \frac{1}{2}m\dot{y}^2 + 2.(m/12)[5\dot{x}^2 + 2\dot{y}^2 - \dot{x}\dot{y}]$$

$$+ 2.(m/12)[5\dot{y}^2 + 2\dot{x}^2 - \dot{x}\dot{y}]$$

or $T = \frac{1}{3}m(5\dot{x}^2 + 5\dot{y}^2 - \dot{x}\dot{y})$ $\quad\therefore$ Lagrange's x-equation is

$$\left(\frac{\partial T}{\partial \dot{x}}\right)_2 - \left(\frac{\partial T}{\partial \dot{x}}\right)_1 = I \Rightarrow \frac{1}{3}m(10\dot{x} - \dot{y}) = I$$

Lagrange's y-equation gives $\frac{1}{3}m(10\dot{y} - \dot{x}) = 0$

Solving (1) and (2), we get $\dot{x} = \frac{10}{33}\frac{1}{m}$ and $\dot{y} = \frac{1}{33m}$

Also $\dot{y} = (x/10)$ and $a\omega = \frac{1}{2}(\dot{y} - \dot{x}) \Rightarrow \omega = -\frac{31}{22m}a$

EXERCISES

1. A homogeneous rod OA, of mass m_1 and length 2a, is freely at O to a fixed point, as its other end is freely attached another homogeneous rod AB, of mass m_2 and length 2b; the system moves under gravity, find equations to determine motion.

2. A rhombus of smoothly jointed equal uniform rods of length 2a lies on a smooth horizontal table. It is struck by a blow J perpendicular to one of its sides at a point distant c from the middle point, show that the angular velocity of the side instantly becomes $\frac{3jc}{8ma^2}$, where m is the mass of each side.

3. A square ABCD formed of four freely jointed uniform rods, is rotating about its centre of gravity on smooth horizontal table with angular velocity ω when the joint A is suddenly fixed. Show that the initial angular velocities of the rods are equal and that 3/5 of the kinetic energy immediately lost.

4. A thin circular ring of a radius a and M lies on a smooth horizontal plane and two light elastic strings are attached to it at ends of a diameter. The other ends of the string being fastened fixed points in the diameter produced. Show that for small oscillations in the plane of the ring the periods are the values of $2\pi/p$ given by

$$\frac{Mlp^2}{2T} - = 0 \text{ or } \frac{b}{l-b} \text{ or } \frac{1}{a}$$

where b is the natural length l the equilibrium length and T the equilibrium tension of each string.

5. A heavy uniform rod of mass m and length 2a rotating in a vertical plane falls and strikes a smooth horizontal plane. If u and ω $\beta\varepsilon$ its linear and angular velocities and θ the inclination to the vertical just before impact, prove that the impulse J is given by

$$(1 + 3\sin^2\theta)j = m(u + a\omega\sin\theta).$$

6

Composite Materials

INTRODUCTION

Till now, we have been considering materials that are homogeneous and isotropic. These qualifications are with respect to their elastic properties. If the properties are the same at every point in the body, then it is said to be homogeneous. Isotropy implies that the properties are independent of directions. Materials like steel, copper, aluminium, etc. are both homogeneous and isotropic. However, wood is homogeneous but is not isotropic since its strength along the fibres is greater than its strength in a direction transverse to the fibres. Materials that are not isotropic are called anisotropic materials. At any point of such a body the elastic properties are different for different directions. Directions for which the elastic properties are the same are said to be elastically equivalent. Generally speaking, one would like to use materials that are suitable to specific applications. For example, a cable wire or rope that is used for hauling purposes needs to have the required tensile strength in the direction of the cable. A structure built up of bricks is good to carry compressive loads. A reinforced concrete beam with steel reinforcement at the bottom is good to carry bending loads which will induce compressive stresses in concrete and tensile stresses in the steel reinforcement. This is an example of a composite material that is designed for a specific operation. On the other hand, consider a bundle of glass fibres or carbon filaments, which is useless when used as an engineering structure. It has no shape and no defined hard surface for machining purposes. The bundle can resist tensile forces, but it is useless for compressive, bending and torsional forces. But, when the same bundle of filaments or fibres is dipped into a bath of resin, drained and allowed to harden, it behaves as a new material possessing properties that are comparable to those of steel or other metals, and can resist forces in tension, comparable to those of steel or other metals, and can resist forces in tension, compression and bending. It has a definite shape, a durable surface and it can be machined. Such a material is called a composite material.

Generally speaking, composites are produced when two or more materials are joined to give a combination of properties that cannot be attained in the original materials. Composites can be placed into three categories–particulate, fibre and laminar–based on the shapes of the constituent materials. Concrete, a mixture of cement and gravel, is a particulate composite; fiberglass, containing glass fibres embedded in a polymer, is a fibre–reinforced composite; and plywood, having alternating layers of wood veneer, is a laminar composite. In this chapter, we shall focus our attention mainly on fibre–reinforced composites and laminates.

STRESS–STRAIN RELATIONS

In Chapter 3, it was stated that for a linearly elastic body, the stresses are linearly related to the strains and are given by

$$\begin{aligned}
\sigma_x &= a_{11}\,\varepsilon_x + a_{12}\,\varepsilon_y + a_{13}\,\varepsilon_z + a_{14}\,\gamma_{xy} + a_{15}\,\gamma_{yz} + a_{16}\,\gamma_{zx}\\
\sigma_y &= a_{21}\,\varepsilon_x + a_{22}\,\varepsilon_y + a_{23}\,\varepsilon_z + a_{24}\,\gamma_{xy} + a_{25}\,\gamma_{yz} + a_{26}\,\gamma_{zx}\\
\sigma_z &= a_{31}\,\varepsilon_x + a_{32}\,\varepsilon_y + a_{33}\,\varepsilon_z + a_{34}\,\gamma_{xy} + a_{35}\,\gamma_{yz} + a_{36}\,\gamma_{zx}\\
\tau_{xy} &= a_{41}\,\varepsilon_x + a_{42}\,\varepsilon_y + a_{43}\,\varepsilon_z + a_{44}\,\gamma_{xy} + a_{45}\,\gamma_{yz} + a_{46}\,\gamma_{zx}\\
\tau_{yz} &= a_{51}\,\varepsilon_x + a_{52}\,\varepsilon_y + a_{53}\,\varepsilon_z + a_{54}\,\gamma_{xy} + a_{55}\,\gamma_{yz} + a_{56}\,\gamma_{zx}\\
\tau_{zx} &= a_{61}\,\varepsilon_x + a_{62}\,\varepsilon_y + a_{63}\,\varepsilon_z + a_{64}\,\gamma_{xy} + a_{65}\,\gamma_{yz} + a_{66}\,\gamma_{zx}
\end{aligned} \tag{1}$$

Assuming that the sixth–order determinant of the coefficients a_{ij}s in Eq. (1) is not zero, one can solve for $\varepsilon_x, \varepsilon_y, \ldots, \gamma_{zx}$ in terms of $\sigma_x, \sigma_y, \ldots, \tau_{zx}$. The expressions for the strain components will then be

$$\begin{aligned}
\varepsilon_x &= b_{11}\,\sigma_x + b_{12}\,\sigma_y + b_{13}\,\sigma_z + b_{14}\,\tau_{xy} + b_{15}\,\tau_{yz} + b_{16}\,\tau_{zx}\\
\varepsilon_y &= b_{21}\,\sigma_x + b_{22}\,\sigma_y + b_{23}\,\sigma_z + b_{24}\,\tau_{xy} + b_{25}\,\tau_{yz} + b_{26}\,\tau_{zx}\\
\varepsilon_z &= b_{31}\,\sigma_x + b_{32}\,\sigma_y + b_{33}\,\sigma_z + b_{34}\,\tau_{xy} + b_{35}\,\tau_{yz} + b_{36}\,\tau_{zx}\\
\gamma_{xy} &= b_{41}\,\sigma_x + b_{42}\,\sigma_y + b_{43}\,\sigma_z + b_{44}\,\tau_{xy} + b_{45}\,\tau_{yz} + b_{46}\,\tau_{zx}\\
\gamma_{yz} &= b_{51}\,\sigma_x + b_{52}\,\sigma_y + b_{53}\,\sigma_z + b_{54}\,\tau_{xy} + b_{55}\,\tau_{yz} + b_{56}\,\tau_{zx}\\
\gamma_{zx} &= b_{61}\,\sigma_x + b_{62}\,\sigma_y + b_{63}\,\sigma_z + b_{64}\,\tau_{xy} + b_{65}\,\tau_{yz} + b_{66}\,\tau_{zx}
\end{aligned} \tag{2}$$

where the coefficients b_{ij}s are related to a_{ij}s. The stress–strain relations given by Eq. (1) contain altogether 36 elastic constants. However, this number can be reduced based on the material properties. Let us assume that there exists an elastic potential V such that

$$\sigma_x = \frac{\partial V}{\partial \varepsilon_x}, \quad \sigma_y = \frac{\partial V}{\partial \varepsilon_y}, \quad \ldots, \quad \tau_{zx} = \frac{\partial V}{\partial \gamma_{zx}} \tag{3}$$

The physical meaning of the potential V will become clear soon. Assuming the existence of such a potential, from Eq. (3), one obtains

$$\frac{\partial \sigma_x}{\partial \varepsilon_y} = \frac{\partial^2 V}{\partial \varepsilon_x \partial \varepsilon_y} = \frac{\partial \sigma_y}{\partial \varepsilon_x}, \quad \frac{\partial \sigma_x}{\partial \gamma_{xy}} = \frac{\partial^2 V}{\partial \varepsilon_x \partial \gamma_{xy}} = \frac{\partial \tau_{xy}}{\partial \varepsilon_x}, \text{ etc.} \tag{4}$$

From Eqs. (4) and (1), one immediately gets

$$a_{12} = a_{21;}\, a_{31} = a_{13}, \ldots, a_{65} = a_{56}$$

And, in general, based on the existence of such a potential,

$$a_{ij} = a_{ji}\ (i, j = 1, 2, 3, \ldots, 6) \tag{5a}$$

Consequently, in Eq. (11.2)

$$b_{ij} = b_{ji}\ (i, j = 1, 2, 3, \ldots, 6) \tag{5b}$$

As a result of this, i.e. $a_{ij} = a_{ji}$ and $b_{ij} = b_{ji}$, the 36 elastic constants a_{ij}s in Eq. (1) or the b_{ij}s in Eq. (2) get reduced to 21. In other words, since $a_{12} = a_{21}$, $a_{13} = a_{31}, \ldots, a_{16} = a_{61}$ and similarly, $b_{12} = b_{21}$, $b_{13} = b_{31}, \ldots, b_{16} = b_{61}$, the number of independent elastic constants in Eqs (1) and (2) gets reduced by 15, resulting in 21 constants. So, for a general case of anisotropy, the number of indep ndent elastic constants is 21. However, because of the symmetry of the ma erial properties, it is claimed by material scientists that the number of independent elastic constants does not exceed 18.

Integrating Eq. (3), one gets

$$\begin{aligned} V = {} & \frac{1}{2} a_{11}\, \varepsilon_x^2 + a_{12}\, \varepsilon_x \varepsilon_y + a_{13}\, \varepsilon_x \varepsilon_z + a_{14}\, \varepsilon_x \gamma_{xy} + a_{15}\, \varepsilon_x \gamma_{yz} + a_{16}\, \varepsilon_x \gamma_{zx} \\ & + \frac{1}{2} a_{22}\, \varepsilon_y^2 + a_{23} \varepsilon_y \varepsilon_z + a_{24} \varepsilon_y \gamma_{xy} + a_{25} \varepsilon_y \gamma_{yz} + a_{26} \varepsilon_y \gamma_{zx} \\ & + \frac{1}{2} a_{33}\, \varepsilon_z^2 + a_{34} \varepsilon_z \gamma_{xy} + a_{35} \varepsilon_z \gamma_{yz} + a_{36} \varepsilon_z \gamma_{zx} \\ & + \frac{1}{2} a_{44}\, \gamma_{xy}^2 + a_{45} \gamma_{xy} \gamma_{yz} + a_{46} \gamma_{xy} \gamma_{zx} \\ & + \frac{1}{2} a_{55}\, \gamma_{yz}^2 + a_{56} \gamma_{yz} \gamma_{zx} \\ & + \frac{1}{2} a_{66}\, \gamma^2 zx \end{aligned} \tag{6}$$

By differentiating Eq. (6) with respect to ε_x, ε_y, ε_z, etc., one gets expressions for σ_x, σ_y, σ_z, etc., thus verifying Eq. (3). The terms of Eq. (6) can be grouped to give for V and equivalent expression as

$$V = \frac{1}{2}(a_{11}\varepsilon_x + a_{12}\varepsilon_y + a_{13}\varepsilon_z + a_{14}\gamma_{xy} + a_{15}\gamma_{yz} + a_{16}\gamma_{zx})\varepsilon_x$$

$$+\frac{1}{2}(a_{12}\varepsilon_x + a_{22}\varepsilon_y + a_{23}\varepsilon_z + \ldots + a_{16}\gamma_{zx})\varepsilon_y$$
$$+\frac{1}{2}(a_{13}\varepsilon_x + a_{23}\varepsilon_y + a_{33}\varepsilon_z + \ldots + a_{36}\gamma_{zx})\varepsilon_z \tag{7}$$
$$+\frac{1}{2}(a_{14}\varepsilon_x + a_{24}\varepsilon_y + a_{34}\varepsilon_z + a_{44}\gamma_{xy} + \ldots)\gamma_{xy} + \ldots$$

Using Eq. (1), the elastic potential assumes the form

$$V = \frac{1}{2}(\sigma_x\varepsilon_x + \sigma_y\varepsilon_y + \sigma_z\varepsilon_z + \tau_{xy}\gamma_{xy} + \tau_{yz}\gamma_{yz} + \tau_{xz}\gamma_{zx}) \tag{8}$$

This is nothing but the elastic energy per unit volume at any point of the body, when forces are applied uniformly, i.e. without dynamic effects.

Consider a body for which Eqs (1) and (2) are valid and let uniform stress $\sigma_x = P$ be applied along the x–axis. The remaining stress components are zero. Then, the strain components as given by Eq. (2) are constant for every point, i.e.

$$\varepsilon_x = b_{11}P \qquad \varepsilon_y = b_{21}P \qquad \varepsilon_z = b_{31}P \tag{9}$$
$$\gamma_{xy} = b_{41}P \qquad \gamma_{yz} = b_{51}P \qquad \gamma_{zx} = b_{61}P$$

It becomes obvious that small segments passing through different points and parallel to x–axis get extended by the same amount. In general, all segments parallel to a given direction n and drawn through different points, undergo equal elongations. A homogeneous anisotropic body exhibiting this property is said to be rectilinearly anisotropic. For such a body, all parallel directions are elastically equivalent. Equations (9) show that elements of the same size in the form of rectangular parallelepipeds with respective parallel faces deform identically, no matter where they are located. However, it should be noted that, in general, rectangular parallelepipeds deform into oblique parallelepipeds having no right angles between the faces. The constants a_{ij}s appearing in Eq. (1) are sometimes called material constants or components of modulus. The constants b_{ij}s appearing in Eq. (2) are called components of compliance. In other words, material constants or components of modulus are used to determine stresses from strains, and components of compliance are used to get strains from stresses.

BASIC CASES OF ELASTIC SYMMETRY

The number of elastic constants involved either in Eq. (1) or Eq. (2) is 36, and these get reduced to 21 independent elastic constants under the assumption that an elastic potential V (equivalent to strain energy density function) exists. It was also stated that even in the most general case, the number of independent elastic constants (according to material scientists) does not exceed 18. For an isotropic body, it has been shown earlier that the number of independent elastic constants is only 2, these being the Lame's constants λ and μ, Eq. (4); or the engineering constants, E the Young's modulus and μ the poisson's ratio. For an anisotropic body, the number of independent elastic constants get reduced depending on the type of symmetry that exists.

When a surface of revolution is rotated through any angle about the axis of revolution, the position of every point on the surface, but not on the axis, gets changed, but the position of the figure as a whole is not changed. In other words, the surface can be made to coincide with itself after an operation which changes the positions of some of its point. Any geometrical figure which can be brought into coincidence with itself after an operation which changes the position of any of its points is said to possess symmetry. A body which can be brought into coincidence with itself by a rotation about an axis, is said to possess an axis of symmetry. A body which after rotation can be brought into coincidence with itself by reflection in a plane, is said to possess a *plane of symmetry.*

(i) Transversely isotropic: Consider a fibre–reinforced body in which the filaments are fairly long and are all oriented in the same direction, Fig. 1. Let the z–axis be parallel to the fibre elements and let the x and y axes lie in a plane perpendicular to the element orientation. If the fibres are uniformly distributed in the matrix, then it is obvious that the elastic properties at any point in the $x = y$ plane, which is the plane of symmetry, are independent of the directions in that plane. A body of this nature is said to be transversely isotropic. The axis normal to the plane of elastic symmetry is some times called the principal direction. It is assumed that no debonding between the fibres and matrix occurs when the body is stressed.

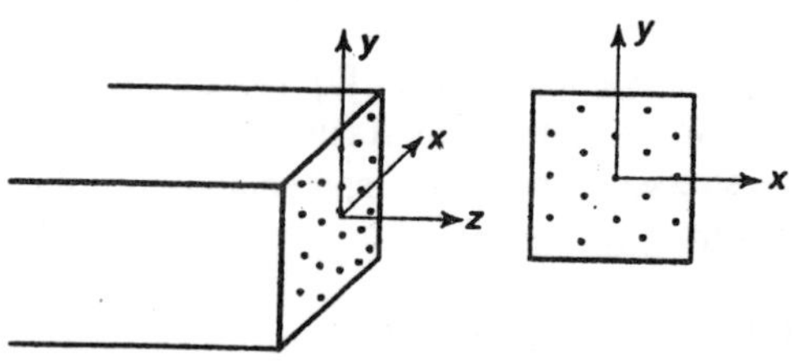

Fig. 1

If a uniform force P is applied in the z direction, such that $\sigma_z = P$ and all other stress components are zero, then any rectangular parallelepiped with faces parallel to x, y and z planes will deform into a rectangular parallelepiped with equal lateral contraction (or extension) in x and y directions. Equations (2) can therefore be written as (using double subscripts for ε_x, ε_y and ε_z),

$$\varepsilon_{zz} = b_{11}\sigma_z, \qquad \varepsilon_{xx} = b_{13}\sigma_z, \qquad \varepsilon_{yy} = b_{23}\sigma_z b_{13}\sigma_z, \qquad \gamma_{xy} = \gamma_{yz} = \gamma_{zx} = 0$$

Because of transverse isotropy, $b_{13} = b_{23}$ (i.e. transverse strains are equal). For a uniform stress σ_x, with the remaining stress components being zero, the strain components will be

$$\varepsilon_{xx} = b_{11}\sigma_x, \qquad \varepsilon_{yy} = b_{21}\sigma_x, \qquad \varepsilon_{zz} = b_{31}\sigma_x, \qquad \gamma_{xy} = \gamma_{yz} = \gamma_{zx} = 0$$

One should observe that the transverse strains ε_{yy} and ε_{zz} will not be equal.

A similar set of equations can be written with stress σ_y, and other stress components being zero. Now consider the shearing stresses. For a shearing stress τ_{xy} in the xy plane, the deformation of a rectangular parallelepiped will only be in the xy plane. Thus, for τ_{xy} alone (with other stress components being zero),

$$\varepsilon_{xx} = \varepsilon_{yy} = \varepsilon_{zz} = 0 \qquad \gamma_{xy} = b_{44}\tau_{xy}, \qquad \gamma_{yz} = \gamma_{zx} = 0$$

Now consider with τ_{yz} alone. The strain components will be

$$\varepsilon_{xx} = \varepsilon_{yy} = \varepsilon_{zz} = 0 \qquad \gamma_{yz} = b_{55}\tau_{yz}, \qquad \gamma_{zx} = \gamma_{xy} = 0$$

Similarly, with τ_{zx} alone, all strain components other than $\gamma_{zx} = b_{66}\,\tau_{zx}$ will be zero. In general, with all stress components σ_{ij} and τ_{ij} acting, the strain components ε_{ij} can be obtained by superposition of all the above expressions. Thus,

$$\begin{aligned}
\varepsilon_{xx} &= b_{11}\sigma_x + b_{12}\sigma_y + b_{13}\sigma_z \\
\varepsilon_{yy} &= b_{12}\sigma_x + b_{11}\sigma_y + b_{13}\sigma_z \\
\varepsilon_{zz} &= b_{31}\sigma_x + b_{31}\sigma_y + b_{33}\sigma_z \\
\gamma_{xy} &= b_{44}\tau_{xy} \\
\gamma_{yz} &= b_{55}\tau_{yz} \\
\gamma_{zx} &= b_{66}\tau_{zx}
\end{aligned} \qquad (10)$$

In the above set of equations, there are eight constants. However, because of the reciprocal relations, $b_{13} = b_{31}$, and as a result, the number gets reduced to seven elastic constants. Also, as a result of the plane of symmetry, $b_{55} = b_{66}$. Further, the elastic constants in the plane of isotropy, i.e. b_{11}, b_{12} and b_{44} are also related. One can see this if Eq. (10) are written using the familiar engineering constants E, v and G, as follows.

$$\varepsilon_{xx} = \frac{1}{E_{xx}}\sigma_x - \frac{V_{xy}}{E_{yy}}\sigma_y - \frac{V_{xz}}{E_{zz}}\sigma_{zz}$$

$$\varepsilon_{yy} = \frac{V_{yx}}{E_{xx}}\sigma_x - \frac{1}{E_{yy}}\sigma_y - \frac{V_{yz}}{E_{zz}}\sigma_{zz}$$

$$\varepsilon_{zz} = \frac{V_{zx}}{E_{xx}}\sigma_x - \frac{V_{zy}}{E_{yy}}\sigma_y - \frac{1}{E_{zz}}\sigma_{zz} \qquad (11)$$

$$\gamma_{xy} = \frac{1}{G_{xy}}\tau_{xy}$$

$$\gamma_{yz} = \frac{1}{G_{yz}}\tau_{yz}$$

$$\gamma_{zx} = \frac{1}{G_{yz}}\tau_{zx}$$

In these equations, v_{xy} is the Poisson's ratio in x direction due to a stress in y direction, i.e. the ratio of lateral contraction in x direction to axial extension in y direction. This contraction is indicated by a negative sign. Slight variations exist in the subscript representation of poisson's ratio from book to book. Also, the strain–stress equations are written in different ways. What is important is the reciprocal identity, i.e. $b_{ij} = b_{ji}$.

Since xy plane is a plane of isotropy, $E_{xx} = E_{yy}$; $V_{xy} = V_{zx} = V_{zy}$; $G_{yz} = G_{zx}$. Also, because of reciprocal identity,

$$\frac{V_{zy}}{E_{zz}} = \frac{V_{zy}}{E_{yy}} \quad \text{and} \quad \frac{V_{xz}}{E_{zz}} = \frac{V_{zx}}{E_{xx}} \qquad (12a)$$

Further, in the plane of isotropy, one has from Eq. (14)

$$G = \frac{E}{2(1+v)}$$

i.e.

$$G_{xy} = \frac{E_{xx}}{2(1+v_{xy})} = \frac{1}{\left(\dfrac{1}{E_{xx}} + \dfrac{V_{xy}}{E_{xx}}\right)} \qquad (12b)$$

In Eq. (11)

$$E_{xx} = \frac{1}{b_{11}}, \quad -\frac{V_{xy}}{E_{xx}} = b_{12}, \ G_{xy} = \frac{1}{b_{44}}$$

Substituting these in Eq. (12b), one gets

$$\frac{1}{b_{44}} = \frac{1}{2}.\frac{1}{(b_{11} - b_{12})}$$

i.e.

$$b_{44} = 2(b_{11} - b_{12}) \qquad (13)$$

As a result of this, the number of independent elastic constants in Eq. (11.11) are only five; these being E_{xx}, E_{zz}, V_{xy}, E_{xz}, G_{yz}.

(ii) ***Orthotropic body:*** Let the fibres in a composite be aligned along the x and y axes and let these be uniformly distributed. z–axis is taken normal to this plane, Fig. 2. The planes normal to x, y and z axes are planes of symmetry (by reflection) and the body is said to be orthogonally anisotropic or orthotropic. The axes x, y and z are the principal directions.

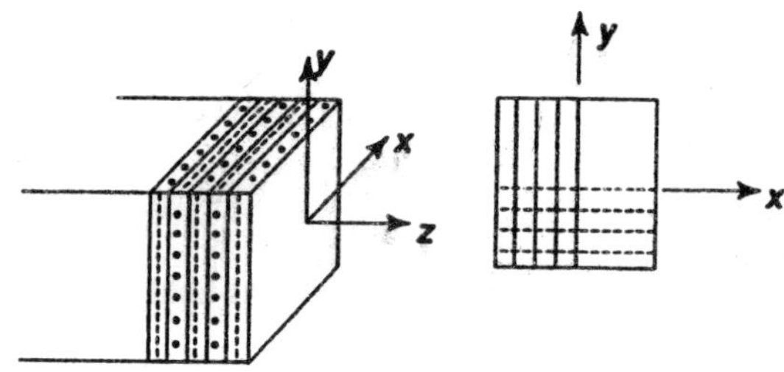

Fig. 2

To be quite general, the fibres parallel to x–axis may be different from the fibres parallel to y–axis.

Consequently, for an axial force in z direction, the lateral contractions (or extensions) in x and y directions will be different. Similarly, the elasticity moduli corresponding to these two directions will also be different. But, when the thickness in z direction is large, the properties in x and y directions tend to become equal. However, retaining the difference, the strain–stress relations will be (recalling that $b_{ij} = b_{ji}$):

$$\begin{aligned}
\varepsilon_{xx} &= b_{11}\sigma_x + b_{12}\sigma_y + b_{13}\sigma_z \\
\varepsilon_{yy} &= b_{12}\sigma_x + b_{22}\sigma_y + b_{23}\sigma_z \\
\varepsilon_{zz} &= b_{13}\sigma_x + b_{23}\sigma_y + b_{33}\sigma_z \\
\gamma_{xy} &= b_{44}\tau_{xy} \\
\gamma_{yz} &= b_{55}\tau_{yz} \\
\gamma_{zx} &= b_{66}\tau_{zx}
\end{aligned} \tag{14}$$

Observe that because of orthogonal symmetries, the shear stresses cause deformations only in their respective planes. There are thus altogether nine independent elastic constants. One can rewrite Eq. (14) using the familiar engineering constants (v_{ij} is the poisson's effect in i direction due to a force in j direction):

$$\varepsilon_{xx} = \frac{1}{E_{xx}}\sigma_x - \frac{V_{xy}}{E_{yy}}\sigma_y - \frac{V_{xz}}{E_{zz}}\sigma_z$$

$$\varepsilon_{yy} = -\frac{V_{yx}}{E_{xx}}\sigma_x + \frac{1}{E_{yy}}\sigma_y - \frac{V_{yz}}{E_{zz}}\sigma_z$$

$$\varepsilon_{zz} = -\frac{V_{zx}}{E_{xx}}\sigma_x - \frac{V_{zy}}{E_{yy}}\sigma_y + \frac{1}{E_{zz}}\sigma_z \tag{15}$$

$$\gamma_{xy} = \frac{1}{G_{xy}}\tau_{xy}$$

$$\gamma_{yz} = \frac{1}{G_{yz}}\tau_{yz}$$

$$\gamma_{zx} = \frac{1}{G_{zx}}\tau_{zx}$$

In the above equations, the following conditions hold good because of the reciprocal identity:

$$\frac{V_{xy}}{E_{yy}} = \frac{V_{yx}}{E_{xx}} \tag{16}$$

i.e. $E_{xx}v_{xy} = E_{yy}v_{yx}$

and, $E_{xx}v_{xz} = E_{zz}v_{zx}$, $E_{yy}v_{yz} = E_{zz}v_{zy}$

It should be observed that $v_{yx} \neq v_{xy}$, $v_{zx} \neq v_{zx}$ and $v_{zy} \neq v_{yz}$, unlink in the case of an isotropic body. If the fibres in the x and y directions are identical in their elastic properties and assuming that they are uniformly interwoven, the following additional relations hold good among the elastic constants.

$$b_{11} = b_{22}, \quad b_{13} = b_{23}, \quad b_{55} = b_{66} \tag{17}$$

Consequently, the number of elastic constants gets reduced to six.

LAMINATES

So far, our discussion has been quite general, in the sense that the composite element that was being considered was an element in a three–dimensional or a bulk material having special properties. However, composites are manufactured keeping in view certain specific applications. For example, a plywood with veneers oriented in different directions is essentially a laminate designed to meet specific requirements.

Laminates, which are essentially thick sheets, are produced not only to be used as such or in a moulded form as corrugated sheets but also to produce bulk materials (by cementing one sheet on top of another, or wrapping one sheet after another about a mandrel). So, an analysis of composite laminates becomes important. Let xy plane represent the midplane of a composite laminate, with z–axis normal to the plane.

(i) Unidirectional Laminates: Let the composite consist of fibres all aligned parallel to x–axis. Such a composite will obviously be stronger in the x direction than in the y direction. We assume that the laminate will be subjected to a plane state of stress. At any point, the rectangular stress components will be σ_x, σ_y and $\tau_{xy.}$ The strain components will be ε_x, ε_y, ε_z, and γ_{xy}, their values being (remembering that $b_{12} = b_{21}$) :

$$\begin{aligned} \varepsilon_{xx} &= b_{11}\sigma_{xx} + b_{12}\sigma_{yy} \\ \varepsilon_{yy} &= b_{12}\sigma_{xx} + b_{22}\sigma_{yy} \\ \varepsilon_{zz} &= b_{31}\sigma_{xx} + b_{32}\sigma_{yy} \\ \gamma_{xy} &= b_{44}\tau_{xy} \end{aligned} \tag{18}$$

We shall be using double subscripts for stresses and strains in order to be consistent with a_{ij}s, b_{ij}s and τ_{ij}s.

There are six independent elastic constants called compliance coefficients. Using engineering constants E and V,

$$\begin{aligned} \varepsilon_{xx} &= \frac{1}{E_{xx}}\sigma_{xx} - \frac{V_{xy}}{E_{yy}}\sigma_{yy} \\ \varepsilon_{yy} &= -\frac{V_{yx}}{E_{xx}}\sigma_{xx} + \frac{1}{E_{yy}}\sigma_{yy} \\ \varepsilon_{zz} &= -\frac{V_{zx}}{E_{xx}}\sigma_{xx} - \frac{V_{zy}}{E_{yy}}\sigma_{yy} \\ \gamma_{xy} &= \frac{1}{G_{xy}}\tau_{xy} \end{aligned} \tag{19}$$

v_{xy} is the Poisson's effect (ratio) in x direction due to a stress in y direction, v_{zx} is the Poisson's ratio in z direction due to a stress in x direction, etc. A negative sign is used to indicate lateral contraction for an extensional stress at right angles. As in Eq. (16), because of reciprocal identity,

$$\frac{V_{xy}}{E_{yy}} = \frac{V_{yx}}{E_{xx}} \tag{20}$$

Table 1 gives typical values of Young's moduli in x and y directions, shear modulus, volume fraction of fibre V_f, Poisson's ratio, and specific gravity for selected unidirectional (along x–axis) composites.

Table 1

Material	E_{xx} *(GPa)*	E_{yy} *(GPa)*	G_{xy} *(GPa)*	V_{yx}	V_f	*Sp. gravity*
Graphite + Epoxy	181	10.3	7.17	0.28	0.70	1.6
Boron + Epoxy	204	18.5	5.59	0.23	0.5	2.0
Graphite + Epoxy	138	8.96	7.1	0.3	0.6	1.6
Glass + Epoxy	38.6	8.27	4.14	0.26	0.45	1.8
Kevlar + Epoxy	76	5.5	2.3	0.34	0.60	1.46

Similar to Eq. (18), one can express the stress components in terms of strain components. Thus, for a plane state of stress with $\sigma_{zz} = 0$,

$$\begin{aligned}\sigma_{xx} &= b_{11}\varepsilon_{xx} + b_{12}\varepsilon_{yy}\\ \sigma_{yy} &= b_{12}\varepsilon_{xx} + b_{22}\varepsilon_{yy}\\ \tau_{xy} &= b_{44}\gamma_{xy}\end{aligned} \tag{21}$$

Or, one can solve Eq. (19) to obtain expressions for σ_{xx}, σ_{yy} and τ_{xy} in terms of the engineering constants and the strain components. These come out as

$$\begin{aligned}\sigma_{xx} &= \frac{E_{xx}}{1-V_{xy}V_{yx}}(\varepsilon_{xx} + V_{xy}\varepsilon_{yy})\\ \sigma_{yy} &= \frac{E_{yy}}{1-V_{xy}V_{yx}}(V_{yx}\varepsilon_{xx} + \varepsilon_{yy})\\ \tau_{xy} &= G_{xy}\gamma_{xy}\end{aligned} \tag{22}$$

The reciprocal identity given by Eq. (20) holds good.

(a) Off–axis Loading: In writing the strain–stress equations, the axes x and y were chosen along the principal directions, i.e. along and perpendicular to the fibre directions. If the laminate is stressed such that the rectangular stress components for these axes can easily be determined, then one can directly use Eq. (19), and in practice, through experiments or otherwise, the elastic constants (like E_{xx}, E_{yy} V_{yx}, etc.) along the principal directions can be determined. However, if the loading is in an arbitrary direction, say x' and y' directions that are oriented at an angle θ to x and y axes, then, it is desirable to get $\varepsilon_{x'x'}$, $\varepsilon_{y'y'}$, etc. using the elastic constants transformed to these new axes. Consider Fig. 3, where the principal directions are x and y and the arbitrary loading directions are x' and y'. The axes x' and y' are rotated through an angle θ counter–clockwise.

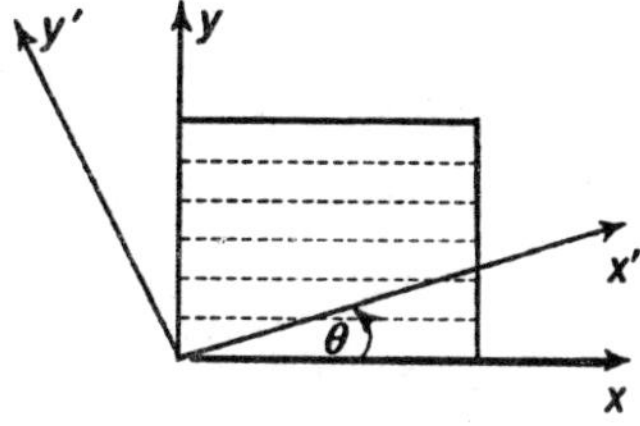

Fig. 3

The positive x'–axis makes an angle θ with fibre direction (i.e. x–axis) and the angle is positive when it is measured counter–clockwise.

Let the stresses applied be $\sigma_{x'x'}$, $\sigma_{y'y'}$ and $\tau_{x'y'}$ and let the corresponding strain components be, $\varepsilon_{x'x'}$, $\varepsilon_{y'y'}$ and $\gamma_{x'y'}$. The procedure to get $\varepsilon_{i'j'}$ in terms of $\sigma_{i'j'}$ is as follows:

$$\sigma_{ij} = f_1(\sigma_{i'j'}, \theta) \quad \text{according to Eqs (59) and (60)}$$
$$\varepsilon_{ij} = f_2(\sigma_{ij}, b_{ij}) \quad \text{according to Eq. (18)}$$
$$\varepsilon_{i'j'} = f_3(\varepsilon_{ij}, \theta) \quad \text{according to Eqs (20) and (36a)}$$
$$\therefore \quad \varepsilon_{i'j'} = f_3[f_2(\sigma_{ij}, b_{ij}), \theta]$$
$$= f_3\{f_2[f_1(\sigma_{i'j'}, \theta), b_{ij}, \theta]\}$$

The above transformations are illustrated in Fig. 4.

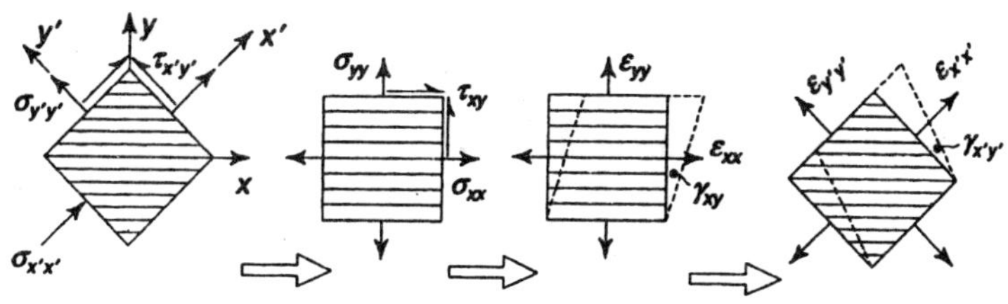

Fig. 4

To transform $\sigma_{i'j'}$ into σ_{ij}, we make use of Eqs (59) and (60). σ_{xx} is the normal stress on the x plane, and the normal to this plane which is the x–axis, makes angles $-\theta$ and $(90 - \theta)$ with x' and y' axes. Hence, $n_x' = \cos\theta$ and $n_y' = -\sin\theta$. Similarly, for the y–axis, $n_x' = \cos(90 - \theta) = \sin\theta$ and $n_y' = \cos\theta$.

$$\sigma_{xx} = \sigma_{x'x'}\cos^2\theta + \sigma_{y'y'}\sin^2\theta - 2\tau_{x'y'}\sin\theta\cos\theta$$
$$\sigma_{yy} = \sigma_{x'x'}\sin^2\theta + \sigma_{y'y'}\cos^2\theta + 2\tau_{x'y'}\sin\theta\cos\theta \quad (11.23)$$

$$\tau_{xy} = +\frac{1}{2}(\sigma_{x'x'} - \sigma_{y'y'}) \sin 2\theta + \tau_{x'y'} \cos 2\theta$$

From Eq. (18)

$$\begin{aligned}
\varepsilon_{xx} &= b_{11} (\sigma_{x'x'} \cos^2\theta + \sigma_{y'y'} \sin^2\theta - \tau_{x'y'} \sin 2\theta) \\
&\quad + b_{12} (\sigma_{x'x'} \sin^2\theta + \sigma_{y'y'} \cos^2\theta + \tau_{x'y'} \sin 2\theta) \\
\varepsilon_{yy} &= b_{12} (\sigma_{x'x'} \cos^2\theta + \sigma_{y'y'} \sin^2\theta - \tau_{x'y'} \sin 2\theta) \\
&\quad + b_{22} (\sigma_{x'x'} \sin^2\theta + \sigma_{y'y'} \cos^2\theta + \tau_{x'y'} \sin 2\theta) \qquad (24) \\
\varepsilon_{zz} &= b_{31} (\sigma_{x'x'} \cos^2\theta + \sigma_{y'y'} \sin^2\theta - \tau_{x'y'} \sin 2\theta) \\
&\quad + b_{32} (\sigma_{x'x'} \sin^2\theta + \sigma_{y'y'} \cos^2\theta + \tau_{x'y'} \sin 2\theta) \\
\gamma_{xy} &= \frac{1}{2} b_{44}[\sigma_{x'x'} - \sigma_{y'y'}) \sin 2\theta + \tau_{x'y'} \cos 2\theta]
\end{aligned}$$

To obtain $\varepsilon_{i'j'}$ in terms of ε_{ij} we make use of Eqs (20) and (36). In using Eq. (36a) for the shear strain, one ignores in the denominator, quantities of higher order compared to unity. For x'–axis, $n_x = \cos\theta$, $n_y = \sin\theta$, and for y'–axis, $n_x = \cos(90 + \theta) = -\sin\theta$, $n_y = \cos\theta$.

$$\begin{aligned}
\varepsilon_{x'x'} &= \varepsilon_{xx} \cos^2\theta + \varepsilon_{yy} \sin^2\theta + \gamma_{xy} \sin\theta \cos\theta \\
\varepsilon_{y'y'} &= \varepsilon_{xx} \sin^2\theta + \varepsilon_{yy} \cos^2\theta + \gamma_{xy} \sin\theta \cos\theta \qquad (25) \\
\gamma_{x'y'} &= -2\varepsilon_{xx} \cos\theta \sin\theta + 2\varepsilon_{yy} \sin\theta + \gamma_{xy} (\cos^2\theta - \sin^2\theta)
\end{aligned}$$

Substituting for ε_{xx}, ε_{yy} and γ_{xy} from Eq. (24),

$$\begin{aligned}
\varepsilon_{x'x'} &= \cos^2\theta\, [b_{11}(\sigma_{x'x'} \cos^2\theta + \sigma_{y'y'} \sin^2\theta - \tau_{x'y'} \sin 2\theta) \\
&\quad + b_{12} (\sigma_{x'x'} \sin^2\theta + \sigma_{y'y'} \cos^2\theta + \tau_{x'y'} \sin 2\theta)] \\
&\quad + \sin^2\theta\, [b_{12} (\sigma_{x'x'} \cos^2\theta + \sigma_{y'y'} \sin^2\theta - \tau_{x'y'} \sin 2\theta) \\
&\quad + b_{22} (\sigma_{x'x'} \sin^2\theta + \sigma_{y'y'} \cos^2\theta + \tau_{x'y'} \sin 2\theta)] \\
&\quad + 1/4 \sin 2\theta\, [b_{44} (\sigma_{x'x'} - \sigma_{y'y'}) \sin 2\theta + 2b_{11}\, \tau_{x'y'} \cos 2\theta]
\end{aligned}$$

Grouping the coefficients of $\sigma_{x'x'}$, $\sigma_{y'y'}$ and $\tau_{x'y'}$, we get

$$\begin{aligned}
\varepsilon_{x'x'} &= \sigma_{x'x'} \left(b_{11} \cos^4\theta + 2b_{12} \cos^2\theta \sin^2\theta + b_{22} \sin^4\theta + \frac{1}{4} b_{44} \sin^2 2\theta \right) \\
&\quad + \sigma_{y'y'} \Big(b_{11} \cos^2\theta \sin^2\theta + b_{12} \cos^4\theta + b_{12} \sin^4\theta \\
&\quad + b_{22} \cos^2\theta \sin^2\theta - \frac{1}{4} b_{44} \sin^2 2\theta \Big) \\
&\quad + \tau_{x'y'} \Big(-b_{11} \cos^2\theta \sin 2\theta + b_{12} \cos^2\theta \sin 2\theta - b_{12} \sin^2\theta \sin 2\theta \\
&\quad + b_{22} \sin 2\theta \sin^2\theta + \frac{1}{4} b_{44} \sin 2\theta \cos 2\theta \Big)
\end{aligned}$$

Using the notation $n = \cos\theta$ and $m = \sin\theta$, one can rewrite the expression for $\varepsilon_{x'x'}$ as

$$\varepsilon_{x'x'} = \left(b_{11}n^4 + 2b_{12}n^2m^2 + b_{22}m^4 + b_{44}n^2m^2 2\theta\right)\sigma_{x'x'}$$
$$+\left(b_{11}n^2m^2 + b_{12}n^4 + b_{12}m^4 + b_{22}n^2m^2 - b_{44}n^2m^2\right)\sigma_{y'y'}$$
$$+\left[-2b_{11}n^3m + 2b_{12}n^3m - 2b_{12}nm^3 + 2b_{22}nm^3 + b_{44}nm(n^2 - m^2)\right]\tau_{x'y'}$$

Substituting for b_{11}, b_{22} and b_{44} from Eqs (18) and (19),

$$\varepsilon_{x'x'} = \left[\frac{1}{E_{xx}}n^4 + \frac{1}{E_{yy}}m^4 - \left(\frac{2V_{xy}}{E_{yy}} - \frac{1}{G_{xy}}\right)n^2m^2\right]\sigma_{x'x'}$$
$$+\left[\left(\frac{1}{E_{xx}} + \frac{1}{E_{yy}} - \frac{1}{G_{xy}}\right)n^2m^2 - \frac{V_{xy}}{E_{yy}}(n^4 + m^4)\right]\sigma_{y'y'}$$
$$+\left[-2\left(\frac{1}{E_{xx}}n^2 - \frac{1}{E_{yy}}m^2\right)nm - (n^2 - m^2)nm\left(\frac{2V_{xy}}{E_{yy}} - \frac{1}{G_{xy}}\right)\right]\tau_{x'y'}$$

Observing that

$$n^4 + m^4 = n^2 + m^4 + 2n^2m^2 - 2n^2m^2$$
$$= (n^2 + m^2)^2 - 2n^2m^2 = 1 - 2n^2m^2,$$

the expression for $\varepsilon_{x'x'}$ can be simplified as

$$\varepsilon_{x'x'} = \left[\frac{n^4}{E_{xx}} + \frac{m^4}{E_{yy}} + \left(\frac{1}{G_{xy}} - \frac{2V_{xy}}{E_{yy}}\right)n^2m^2\right]\sigma_{x'x'}$$
$$+\left[\left(\frac{1}{E_{xx}} + \frac{1}{E_{yy}} - \frac{1}{G_{xy}} + \frac{2V_{xy}}{E_{yy}}\right)n^2m^2 - \frac{V_{xy}}{E_{yy}}\right]\sigma_{y'y'} \tag{26}$$
$$+\left[2\left(\frac{m^2}{E_{yy}} - \frac{n^2}{E_{xx}}\right) + \left(\frac{1}{G_{xy}} - \frac{2V_{xy}}{E_{yy}}\right)(n^2 - m^2)\right]nm\tau_{x'y'}$$

Similarly, expressions for $\varepsilon_{y'y'}$ and $\varepsilon_{x'x''}$ can be obtained.

For an anisotropic body, it is preferable to use compliance coefficients while expressing strains in terms of stresses. Thus, one can write

$$\varepsilon_{x'x'} = b'_{11}\sigma_{x'x'} + b'_{12}\sigma_{y'y'} + b'_{14}\tau_{x'y'}$$
$$\varepsilon_{y'y'} = b'_{12}\sigma_{x'x'} + b'_{22}\sigma_{y'y'} + b'_{24}\tau_{x'y'}$$
$$\varepsilon_{z'z'} = b'_{31}\sigma_{x'x'} + b'_{32}\sigma_{y'y'} + b'_{34}\tau_{x'y'} \tag{27}$$
$$\gamma_{x'y'} = b'_{14}\sigma_{x'x'} + b'_{24}\sigma_{y'y'} + b'_{44}\tau_{x'y'}$$

From Eq. (26),

$$b'_{11} = \frac{n^4}{E_{xx}} + \frac{m^4}{E_{yy}} + \left(\frac{1}{G_{xy}} - \frac{2V_{xy}}{E_{yy}}\right)n^2m^2$$

$$b'_{12} = \left(\frac{1}{E_{xx}} + \frac{1}{E_{yy}} - \frac{1}{G_{xy}} + \frac{2V_{xy}}{E_{yy}}\right)n^2m^2 - \frac{V_{xy}}{E_{yy}} \tag{28}$$

$$b'_{14} = \left[2\left(\frac{m^2}{E_{yy}} - \frac{n^2}{E_{xx}}\right) + \left(\frac{1}{G_{xy}} - \frac{2V_{xy}}{E_{yy}}\right)(n^2 - m^2)\right]nm$$

Proceeding on the lines for getting $\varepsilon_{x'x'}$ one can get expressions for $\varepsilon_{y'y'}$ and $\gamma_{x'y'}$. The compliance coefficients for these will come out as

$$b'_{22} = \frac{m^4}{E_{xx}} + \frac{n^4}{E_{yy}} + \left(\frac{1}{G_{xy}} - \frac{2V_{xy}}{E_{yy}}\right)n^2m^2$$

$$b'_{24} = \left[2\left(\frac{n^4}{E_{yy}} - \frac{m^2}{E_{xx}}\right) - \left(\frac{1}{G_{xy}} - \frac{2V_{xy}}{E_{yy}}\right)(n^2 - m^2)\right]nm \tag{29}$$

$$b'_{44} = 4\left(\frac{1}{E_{xx}} + \frac{1}{E_{yy}} + \frac{2V_{xy}}{E_{yy}} - \frac{1}{G_{xy}}\right)n^2m^2 + \frac{1}{G_{xy}}$$

Since changes in thickness of the laminate are not of much concern, the compliance coefficients for $\varepsilon_{z'z'}$, have not been written. Hence, for a unidirectionally reinforced laminate, if the material constants (i.e., *Es* and *vs*) along the principal directions are known, one can get from Eqs (28) and (29), the compliance coefficients for any arbitrarily oriented axes x' and y'. One should remember that positive x'–axis makes an angle $+\theta$ counterclockwise with the fibre axis, i.e. $+x$ axis. Equations (28) and (59) can be written in a compact from using the notations of Eqs (18) and (19). Thus,

$$\begin{aligned}
b'_{11} &= b_{11}n^4 + b_{22}m^4 + 2b_{12}n^2m^2 + b_{44}n^2m^2 \\
b'_{12} &= b_{11}n^2m^2 + b_{22}n^2m^2 + b_{12}(n^4 + m^4) - b_{44}n^2m^2 \\
b'_{14} &= -2b_{11}n^3m + 2b_{22}nm^3 + 2b_{12}(n^2 - m^2)nm + b_{44}(n^2 - m^2)nm \\
b'_{22} &= b_{11}m^4 + b_{22}n^4 + 2b_{12}n^2m^2 + b_{44}n^2m^2 \\
b'_{24} &= -2b_{11}nm^3 + 2b_{22}n^3m - 2b_{12}(n^2 - m^2)nm - b_{44}(n^2 - m^2)nm \\
b'_{44} &= 4b_{11}n^2m^2 + 4b_{22}n^2m^2 - 8b_{12}n^2m^2 + b_{44}(n^2 - m^2)^2
\end{aligned} \tag{30a}$$

where

$$b_{11} = \frac{1}{E_{xx}},\ b_{22} = \frac{1}{E_{yy}},\ b_{12} = -\frac{V_{xy}}{E_{yy}} = -\frac{V_{yx}}{E_{xx}},\ b_{44} = \frac{1}{G_{xy}} \tag{30b}$$

Equations (30a) can be further modified for ease of applications when we take up for analysis multidirectional composites. Towards this, consider the following trigonometric identities:

$$n^4 = \frac{1}{8}(3+4\cos 2\theta+\cos 4\theta),\quad n^3m = \frac{1}{8}(2\sin 2\theta+\sin 4\theta) \qquad (31a)$$

$$n^2m^2 = \frac{1}{8}(1-\cos 4\theta),\quad nm^3 = \frac{1}{8}(2\sin 2\theta-\sin 4\theta) \qquad (31b)$$

$$m^4 = \frac{1}{8}(3-4\cos 2\theta+\cos 4\theta) \qquad (31c)$$

Substituting these for b'_{11} in Eq. (30a)

$$b'_{11} = \frac{1}{8}(3+4\cos 2\theta+\cos 4\theta)b_{11} + \frac{1}{8}(3-4\cos 2\theta+\cos 4\theta)b_{22}$$

$$+\frac{1}{8}(1-\cos 4\theta)(2b_{12}+b_{44})$$

$$+\frac{1}{8}(1-\cos 4\theta)(2b_{12}+b_{44})+\frac{1}{2}(b_{11}-b_{22})\cos 2\theta$$

$$=\frac{1}{8}(b_{11}+b_{22}-2b_{12}-b_{44})\cos 4\theta$$

$$= p_{11} + p_2 \cos 2\theta + p_3 \cos 4\theta \qquad (32)$$

Similarly, substituting for their components in Eq. (30a),

$$b'_{12} = \frac{1}{8}(b_{11}+b_{22}+6b_{12}-b_{44}) - \frac{1}{8}(b_{11}+b_{22}-2b_{12}-b_{44})\cos 4\theta$$

$$= p_4 - p_3 \cos 4\theta$$

$$b'_{14} = -p_2 \sin 2\theta - 2p_3 \sin 4\theta$$

$$b'_{22} = p_1 - p_2 \cos 2\theta + p_3 \cos 4\theta \qquad (32b)$$

$$b'_{24} = -p_2 \sin 2\theta + 2p_3 \sin 4\theta$$

$$b'_{44} = \frac{1}{2}(b_{11}+b_{22}-2b_{12}+b_{44}) - \frac{1}{2}(b_{11}+b_{22}-2b_{12}-b_{44})\cos 4\theta$$

$$= p_5 - 4p_3 \cos 4\theta$$

In the above expressions for b'_{ij}

$$p_1 = \frac{1}{8}\left(3b_{11}+3b_{22}+2b_{12}+b_{44}\right)$$

$$p_2 = \frac{1}{2}(b_{11}-b_{22})$$

$$p_3 = \frac{1}{8}(b_{11}+b_{22}-2b_{12}-b_{44}) \qquad (33)$$

$$p_4 = \frac{1}{8}(b_{11} + b_{22} + 6b_{12} - b_{44})$$

$$p_5 = \frac{1}{2}(b_{11} + b_{22} - 2b_{12} - b_{44})$$

These equations are useful in two ways. Firstly, the quantities p_is are material properties of the composite.

Once these quantities are determined, they can be used for any off–axis loading direction. Secondly, as mentioned before, when we take up multidirection composites, these equations become useful.

Example 1:

Consider a graphite–epoxy laminate whose elastic constants along and perpendicular to the fibres are as follows.

E_{xx} = 181 GPa, E_{yy} = 10.3 GPa,

G_{xy} = 7.17 GPa, V_{yx} = 0.28,

V_{xy} = 0.01594

Obtain the compliance coefficients appropriate to $x'y'$ axes which are at

(a) + 30° (counter–clockwise) to xy axes and (b) +90° to xy axes

(a) θ = 30°, n = cos 30° = 0.866, m = sin 30° = 0.5

From Eqs. (28) and (29),

b'_{11} = [3.107 + 6.068 + (139.5 – 3.094) × 0.1875] × 10^{-12}

= (34.75) × 10^{-3} (GPa)$^{-1}$

b'_{12} = [(5.525 + 97.09 – 139.47 + 3.094) × 0.1875 – 1.547] × 10^{-12}

= – (7.88 × 10^{-3}) (GPa)$^{-1}$

b'_{14} = [2(24.3 – 4.143) + (139.47 – 3.094) (0.75 – 0.25)] × 0.433 × 10^{-12}

= (40.3 + 68.188) × 0.433 × 10^{-12} = 46.98 × 10^{-3} (GPa)$^{-1}$

b'_{22} = [0.345 + 54.61 + (139.47 – 3.094) × 0.1875] × 10^{-12}

= 80.53 × 10^{-3} (GPa)$^{-1}$

b'_{24} = [2 (72.81 – 1.381) – (139.47 – 3.094) (0.75 – 0.25)] × 0.433 × 10^{-12}

= (142.86 – 68.188) × 0.433 × 10^{-12} = 32.33 × 10^{-3} GPa)$^{-1}$

b'_{44} = [4(5.525 – 97.09) + 3.094 – 139.47)×0.1875+139.47] × 10^{-12}

= 114.15 × 10^{-3} (GPa)$^{-1}$

(b) $\theta = 90°$, $n = 0$, $m = 1$,

From Eqs (27) and (28),

$$b'_{11} = \frac{1}{E_{yy}} = 97.09 \times 10^{-3}\ (\text{GPa})^{-1}$$

$$b'_{12} = -\frac{V_{xy}}{E_{xx}} = -1.22 \times 10^{-3}\ (\text{GPa})^{-1}$$

$$b'_{14} = 0$$

$$b'_{22} = \frac{1}{E_{xx}} = 1.55 \times 10^{-3}\ (\text{GPa})^{-1}$$

$$b'_{24} = 0$$

$$b'_{44} = \frac{1}{G_{xy}} = 139.5 \times 10^{-3}\ (\text{GPa})^{-1}$$

It should be observed that $x'y'$ frame is obtained through rotation of the xy frame by 90° counter–clockwise. Consequently, the values of the elastic constants get switched since the x'–axis will be along the y–axis, and the y'–axis will be along the x–axis (but in the opposite direction). Thus, $E_{x'x'} = E_{y'y'} = E_{y'y'} = E_{xx}$ and $G_{x'y'} = G_{xy}$ as the results show.

Example 2:

At a point in a laminate the following stress state exists:

$\sigma_{x'x'} = 100$ MPa, $\sigma_{y'y'} = 30$ MPa,

$\tau_{x'y'} = 30$ MPa

The laminate is unidirectionally reinforced and the fibre orientation is 30° to x'–axis, as shown in Fig. 5. The elastic constants along the principal directions of the laminate are

$E_{xx} = 100$ GPa, $E_{yy} = 10$ GPa,

$G_{xy} = 5$ GPa, $V_{yx} = 0.25$

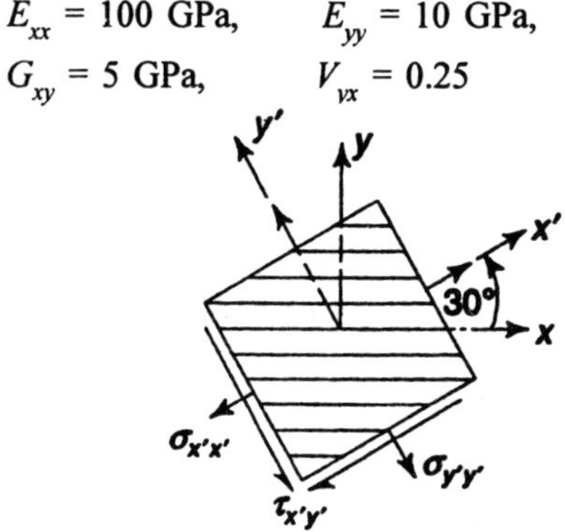

Fig. 5

Determine the principal stresses, principal strains and their orientations in the plane of the laminate.

From Eq. (61), the principal stresses are

$$\sigma_{1,2} = \frac{1}{2}(\sigma_{x'x'} + \sigma_{y'y'}) \pm \sqrt{\left[\left(\frac{\sigma_{x'x'} - \sigma_{y'y'}}{2}\right)^2 + \tau_{x'y'}^2\right]}$$

$$= \frac{1}{2}(130) \pm \sqrt{\left[(35)^2 + 30^2\right]} = 111 \text{ or } 19$$

$\therefore$ σ_1 = 111 MPa and σ_2 = 19 MPa (check: $\sigma_{x'x'} + \sigma_{y'y'} = \sigma_1 + \sigma_2$)

From Eq. (62),

$$\tan 2\phi = \frac{2\tau_{x'y'}}{\sigma_{x'x'} - \sigma_{y'y'}}$$

$$= \frac{60}{70} = 0.8571$$

$\therefore$ $\phi_1' = 40.6°$ and $\phi_2' = 130.6°$

From Strength of Materials, the algebraically maximum principal stress, which in our present case is σ_1 = 111 MPa, lies within the principal 45° angle. Thus, σ_1 = 111 MPa makes an angle of 40.6° with x'–axis, and σ_2 = 19 MPa makes an angle of 130.6° with x'–axis (counter clockwise).

To determine the principal strains, the required rectangular components can be obtained either with respect to $x'y'$ or with respect to xy axes. To determine the components with respect to $x'y'$ axes, we need the corresponding compliance coefficients. To obtain the strain components with respect to xy axes, we need to transform the given stress components to these axes, and then use Eq. (19). Let us transform the given stress components to xy axes. From Eq. (23),

$$\sigma_{xx} = (100\times 0.75) + (30 \times 0.25) - (2 \times 30 \times 0.433) = 56.52 \text{ MPa}$$

$$\sigma_{yy} = (100.0 \times 0.25) + (30 \times 0.75) + (2 \times 30 \times 0.433) = 73.48 \text{ MPa}$$

(check $\sigma_{xx} + \sigma_{yy} = \sigma_{x'x'} + \sigma_{y'y'}$)

$$\tau_{xy} = \frac{1}{2}(70)\times 0.866 + (30\times 0.5) = 45.31 \text{ MPa}$$

From Eq (09) and using the reciprocal identity, Eq. (20),

$$\varepsilon_{xx} = \left(\frac{56.52}{100} - \frac{0.25}{100}\times 73.48\right)\times 10^{-3} = 0.3815\times 10^{-3}$$

$$\varepsilon_{yy} = \left(-\frac{0.25}{100}\times 56.52+\frac{73.48}{8}\right)\times 10^{-3} = 7.207\times 10^{-3}$$

$$\gamma_{xy} = \frac{45.31}{5}\times 10^{-3} = 9.062\times 10^{-3}$$

The principal strains corresponding to these are, from Eq. (50),

$$\varepsilon_{1,2} = \frac{\varepsilon_{xx}+\varepsilon_{yy}}{2} \pm \sqrt{\left[\left(\frac{\varepsilon_{xx}-\varepsilon_{yy}}{2}\right)^2+\left(\frac{\gamma_{xy}}{2}\right)^2\right]}$$

$$= \frac{1}{2}(0.3815+7.207)\times 10^{-3} \pm \sqrt{\left[(3.413)^2+(4.531)^2\right]}\times 10^{-3}$$

$$= (3.7942 \times 10^{-3}) \pm (5.6726)$$

$$= 9.4668 \times 10^{-3} \quad \text{or} \quad -1.8784 \times 10^{-3}$$

(check: $\varepsilon_{xx} + \varepsilon_{yy} = \varepsilon_1 + \varepsilon_2$)

From Eq. (51), the directions of these principal strains are

$$\tan 2\phi^* = \frac{\gamma_{xy}}{\varepsilon_{xx}-\varepsilon_{yy}} = \frac{9.062}{0.3815-7.207} = -1.3277$$

$\therefore \qquad \phi_1 = -26.5^\circ \qquad \text{and} \qquad \phi_2 = +63.5^\circ$

These angles are with respect to the x–axis. By subtracting 30°, we get the orientations of the principal strain axes with respect to x'–axis. Thus,

$$\phi_1 = -26.5^\circ \qquad \text{and} \qquad \phi_2 = 33.5^\circ.$$

Unlike an isotropic case, in general, the principal stress axes do not coincide with the principal strain axes in an anisotropic body. In this example, the principal stress axes are at 40.6° and 130.6°, while the principal strain axes are at 33.5° and 123.5° (i.e. – 56.5°) to x'–axis.

Example 3

In Example 2, the directions of the principal strains were obtained by transforming the applied stresses to the principal direction axes. Show that the same results can be obtained by using the loading or the stress axes as reference and obtaining the corresponding compliance coefficients.

$\theta = 30^\circ, \qquad n \cos 30^\circ = 0.866, \qquad m = \sin 30^\circ = 0.5$

From Eqs. (28) and (29)

$b'_{11} = [5.6243 + 6.25 + (200 - 5) \times 0.1875] \times 10^{-12} = 48.4368 \times 10^{-12}$

$b'_{12} = [(10 + 100 - 200 + 5) \times 0.1875 - 2.5] \times 10^{-12} = -18.475\times 10^{-12}$

$b'_{14} = [2(25 - 7.4996) + (200 - 5) \times 0.5]\times 0.433\times 10^{-12} = 57.3728\times 10^{-12}$

$$b'_{22} = [0.625 + 56.24 + (200 - 5) \times 0.1875] \times 10^{-12} = 93.4275 \times 10^{-12}$$

$$b'_{24} = [2(74.996 - 2.5) - (200 - 5) \times 0.5] \times 0.433 \times 10^{-12}$$

$$= 20.564 \times 10^{-12}$$

$$b'_{44} = [4(10 + 100 + 5 - 200) \times 0.1875 + 200] \times 10^{-12}$$

$$= 136.25 \times 10^{-12}$$

From Eq. (25)

$$\varepsilon_{x'x'} = [(48.4368 \times 100) - (18.4375 \times 30) + (57.3728 \times 30)] \times 10^{-6}$$

$$= 0.006012$$

$$\varepsilon_{y'y'} = [-(18.4375 \times 100) + (93.4275 \times 30) + (20.564 \times 30)] \times 10^{-6}$$

$$= 0.001576$$

(check : $\varepsilon_{x'x'} + \varepsilon_{y'y'} + \varepsilon_{xx} + \varepsilon_{yy}$)

$$\gamma_{x'y'} = [(57.3728 \times 100) + (20.564 \times 30) + (136.25 \times 30)] \times 10^{-6}$$

$$= 0.010434$$

The principal strains are

$$\varepsilon_{1,2} = \frac{\varepsilon_{x'x'} + \varepsilon_{y'y'}}{2} \pm \sqrt{\left[\left(\frac{\varepsilon_{x'x'} - \varepsilon_{y'y'}}{2}\right)^2 + \left(\frac{\gamma_{x'y'}}{2}\right)^2\right]}$$

$$= \frac{1}{2}(6.012 + 1.576) \times 10^{-3} \pm \sqrt{\left[\left(\frac{6.012 - 1.576}{2}\right)^2 + (5.217)^2\right]} \times 10^{-3}$$

$$= (3.794 \pm 5.669) \text{ ' } 10^{-3} = 0.00946; -0.00187$$

(b) Off–axis Components of Modulus: In the previous discussions we obtained the off–axis components of compliances b'_{ij}s from Eqs (28) and (29). The motivation for considering this first is that in practice, composites are used to comply with situations where stresses or loads are prescribed which usually are not along principal directions. To estimate the deformations using Eq. (27), one needs compliance coefficients from known elastic constants along principal directions. On the other hand, when we need to analyse multidirectional fibre composites we need to know the stress values for given off–axis strain values. To get this, one can follow a similar procedure as was adopted earlier, i.e. obtain $\sigma_{i'j'}$ in terms of $\varepsilon_{i'j'}$s.

$\varepsilon_{i'j'} = f_1\ (\varepsilon_{i'j'}, \theta)$ from Eqs (20) and (36a)

$\sigma_{ij} = f_2\ (\varepsilon_{ij}, a_{ij})$ from Eq. (21)

$\sigma_{i'j'} = f_3\ (\sigma_{ij}, \theta)$ from Eqs (56) and (60)

$= f_3\ [f_2\ (\varepsilon_{ij}, a_{ij}), \theta]$

$$= f_3 \{f_2 [f_1 (\varepsilon_{i'j'}, \theta), a_{ij}], \theta\}$$

Alternatively, one can solve for $\sigma_{x'x'}$, $\sigma_{y'y'}$ and $\tau_{x'y'}$, from Eq. (27). For this, we need expressions for $\varepsilon_{x'x'}$, $\varepsilon_{y'y'}$ and $\gamma_{x'y'}$.

$$\begin{aligned}\varepsilon_{x'y'} &= b'_{11}\sigma_{x'x'} + b'_{12}\sigma_{y'y'} + b'_{14}\tau_{x'y'} \\ \varepsilon_{y'y'} &= b'_{12}\sigma_{x'x'} + b'_{22}\sigma_{y'y'} + b'_{24}\tau_{x'y'} \\ \gamma_{x'y'} &= b'_{41}\sigma_{x'x'} + b'_{42}\sigma_{y'y'} + b'_{44}\tau_{x'y'}\end{aligned} \qquad \text{(a)}$$

Inn Eq. (a), $b'_{41} = b'_{14}$ and $b'_{42} = b'_{24}$. The determinant of the coefficients in Eq. (a) is

$$\begin{aligned}\Delta &= b'_{11}(b'_{22}b'_{44} - b'^{2}_{24}) - b'_{12}(b'_{12}b'_{44} - b'_{24}b'_{14}) + b'_{14}(b'_{12}b'_{24} - b'_{22}b'_{14}) \\ &= b'_{11}b'_{22}b'_{44} + 2b'_{12}b'_{24}b'_{14} - b'_{22}b'^{2}_{14} - b'_{11}b'^{2}_{24} - b'_{44}b'^{2}_{12}\end{aligned}$$

Hence, the solutions for $\sigma_{x'x'}$, $\sigma_{y'y'}$ and $\tau_{x'y'}$, from Eq. (a), are

$$\begin{aligned}\sigma_{x'x'} &= \frac{1}{\Delta}[(b'_{22}\, b'_{44} - b'^{2}_{24})\varepsilon_{x'x'} - (b'_{12}\, b'_{44} - b'_{14}b'_{24})\varepsilon_{y'y'} \\ &\quad + (b'_{12}b'_{24} - b'_{14}b'_{22})\gamma_{x'y'}] \\ \sigma_{y'y'} &= \frac{1}{\Delta}[(b'_{12}b'_{44} - b'_{24}b'_{14})\varepsilon_{x'x'} - (b'_{11}b'_{44} - b'^{2}_{14})\varepsilon_{y'y'} \\ &\quad + (b'_{11}b'_{24} - b'_{14}b'_{12})\gamma_{x'y'}] \\ \tau_{x'y'} &= \frac{1}{\Delta}[(b'_{12}b'_{42} - b'_{22}b'_{14})\varepsilon_{x'x'} - (b'_{11}b'_{24} - b'_{12}b'_{14})\varepsilon_{y'y'} \\ &\quad + (b'_{11}b'_{22} - b'^{2}_{12})\gamma_{x'y'}]\end{aligned} \qquad (34)$$

From the general stress–strain equations, for a laminate under plane state of stress with an off–axis coordinate system, one has, similar to Eq. (27),

$$\begin{aligned}\sigma_{x'x'} &= a'_{11}\ \varepsilon_{x'x'} + a'_{12}\ \varepsilon'_{y'y'} + a'_{14}\ \gamma_{x'y'} \\ \sigma_{y'y'} &= a'_{12}\ \varepsilon_{x'x'} + a'_{22}\ \varepsilon'_{y'y'} + a'_{24}\ \gamma_{x'y'} \\ \tau_{x'y'} &= a'_{14}\ \varepsilon_{x'x'} + a'_{24}\ \varepsilon'_{y'y'} + a'_{44}\ \gamma_{x'y'}\end{aligned} \qquad (35)$$

Comparing the coefficients $\varepsilon_{i'j'}$ in Eqs (34) and (35), one gets

$$\begin{aligned}a'_{11} &= \frac{1}{\Delta}(b'_{22}b'_{44} - b'^{2}_{24}) \\ a'_{12} &= -\frac{1}{\Delta}(b'_{12}b'_{44} - b'_{14}b'_{24}) \\ a'_{14} &= \frac{1}{\Delta}(b'_{12}b'_{24} - b'_{14}b'_{22}) \\ a'_{22} &= -\frac{1}{\Delta}(b'_{11}b'_{44} - b'^{2}_{14})\end{aligned} \qquad (36)$$

$$a'_{24} = \frac{1}{\Delta}(b'_{11}b'_{24} - b'_{14}b'_{12})$$

$$a'_{44} = \frac{1}{\Delta}(b'_{11}b'_{22} - b'^{2}_{12})$$

Application of Eq. (36) to get a'_{ij}s involves calculations of b'_{ij}s. Instead, one can follow the procedure adopted earlier. This is shown schematically in Fig. 6.

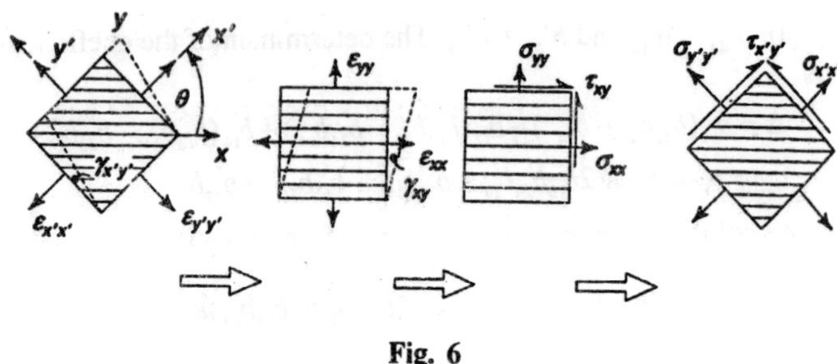

Fig. 6

The results are the following:

$$a'_{11} = a_{11}n^4 + a_{22}m^4 + 2a_{12}n^2m^2 + 4a_{44}n^2m^2$$

$$a'_{12} = a_{11}n^2m^2 + a_{22}n^2m^2 + a_{12}(n^4 + m^4) - 4a_{44}\, n^2m^2$$

$$a'_{14} = -a_{11}n^3m + a_{22}nm^3 + a_{12}(n^2 - m^2)nm + 2a_{44}\,(n^2 - m^2)nm \quad (37a)$$

$$a'_{24} = -a_{11}nm^3 + a_{22}n^3m - a_{12}(n^2 + m^2)nm - 2a_{44}\,(n^2 - m^2)nm$$

$$a'_{22} = a_{11}m^4 + a_{22}n^4 + 2a_{12}n^2m^2 + 4a_{44}n^2m^2$$

$$a'_{44} = a_{11}n^2m^2 + a_{22}n^2m^2 - 2a_{12}n^2m^2 + a_{44}\,(n^2 - m^2)^2$$

where, from Eqs (21) and (22),

$$a_{11} = \frac{E_{xx}}{1 - V_{yx}V_{xy}}, \quad a_{22} = \frac{E_{yy}}{1 - V_{yx}V_{xy}}, \quad a_{12} = +\frac{V_{yx}\,E_{xx}}{1 - V_{yx}V_{xy}}, \quad a_{44} = G_{xy}$$

Equations (37a) can be recast as was done in the case of compliance coefficients, i.e. Eqs (32) and (33). Using the trigonometric identities given by Eq. (31),

$$a'_{11} = \frac{1}{8}(3 + 4\cos 2\theta + \cos 4\theta)a_{11} + \frac{1}{8}(3 - 4\cos 2\theta + \cos 4\theta)a_{22}$$

$$+\frac{1}{8}(1 - \cos 4\theta)\,(2a_{12} + 4a_{44})$$

$$= \frac{1}{8}(3a_{11} + 3a_{22} + 2a_{12} + 4a_{44}) + \frac{1}{2}(a_{11} - a_{22})\cos 2\theta$$

$$+\frac{1}{8}(a_{11}+a_{22}-2a_{12}-4a_{44})\cos 4\theta$$

$$= Q_1 + Q_2 \cos 2\theta + Q_3 \cos 4\theta \tag{38a}$$

Similarly, for other components we get,

$$a'_{22} = Q_1 - Q_2 \cos 2\theta + Q_3 \cos 4\theta$$

$$a'_{12} = Q_4 - Q_3 \cos 4\theta$$

$$a'_{14} = -\frac{1}{2}Q_2 \sin 2\theta - Q_3 \sin 4\theta \tag{38b}$$

$$a'_{24} = -\frac{1}{2}Q_2 \sin 2\theta + Q_3 \sin 4\theta$$

$$a'_{44} = Q_5 - Q_3 \cos 4\theta$$

where,

$$Q_1 = \frac{1}{8}(3a_{11}+3a_{22}+2a_{12}+4a_{44})$$

$$Q_2 = \frac{1}{8}(a_{11}-a_{22})$$

$$Q_3 = \frac{1}{8}(a_{11}+a_{22}-2a_{12}-4a_{44}) \tag{39}$$

$$Q_4 = \frac{1}{8}(a_{11}+a_{22}+6a_{12}-4a_{44})$$

$$Q_5 = \frac{1}{8}(a_{11}+a_{22}-2a_{12}+4a_{44})$$

The Q_is involve only material properties and once they are determined, the compliance coefficients for any off–axis direction can be determined using Eq. (38b).

Example 4:

A unidirectionally reinforced composite of 'Toray' filament and 'Nameo' resin has the following moduli and poisson's ratio.

E_{xx} = 181 GPa, E_{yy} = 10.3 GPa, V_{xy} = 0.0159, G_{xy} = 7.17 GPa, $(1 - V_{xy}V_{yx})^{-1}$ 1.0045

Estimate the components of moduli for an off–axis orientation of

(a) $\theta = +30°$, $n = \cos\theta = 0.866$ and $m = \sin\theta = 0.5$.

From Eq. (37b),

$$a_{11} = \frac{E_{xx}}{1-V_{yx}V_{xy}} = 181\times 1.0045 = 181.8 \text{ GPa}$$

$$a_{22} = \frac{E_{yy}}{1 - V_{yx}V_{xy}} = 10.3 \times 1.0045 = 10.34 \text{ GPa}$$

$$a_{12} = +\frac{V_{yx}E_{xx}}{1 - V_{yx}V_{xy}} = 0.0159 \times 181 \times 1.0045 = 2.891 \text{ GPa}$$

$$a_{44} = G_{xy} = 7.17 \text{ GPa}$$

Further, $n = 0.866$, $m = 0.5$, $nm = 0.433$

$n^2 = 0.750$, $n^4 = 0.562$, $m^2 = 0.25$, $m^2 = 0.0625$

$n^3m = 0.3248$, $nm^3 = 0.1083$, $n^2m^2 = 0.1875$

Substituting in Eq. (37a),

$$a'_{11} = (181.8 \times 0.562) + (10.34 \times 0.625) + (2 \times 2.891 \times 0.1875)$$
$$+ (4 \times 7.17 \times 0.1875) = 109.2 \text{ GPa}$$

$$a'_{12} = (181.8 \times 0.1875) + (10.34 \times 0.1875) + (2.891)(0.6245)$$
$$- (4 \times 7.17 \times 0.1875) = 32.45 \text{ GPa}$$

$$a'_{14} = -(181.8 \times 0.3248) + (10.34 \times 0.1083) + (2.891 \times 0.5 \times 0.433)$$
$$+ (2 \times 7.17 \times 0.5 \times 0.433) = -54.19 \text{ GPa}$$

$$a'_{22} = 23.64 \text{ GPa}, \quad a'_{24} = -20.05 \text{ GPa}, \quad a'_{44} = 36.78 \text{ GPa};$$

Similarly, for (b) with $\theta = +45°$

$a'_{11} = 56.6$ GPa, $a'_{12} = 42.32$ GPa, $a'_{14} = -42.87$ GPa, $a'_{22} = 4659$ GPa;

$a'_{24} = -42.87$ GPa; $a'_{44} = 46.59$ GPa

(i) Multidirectional Laminates: Multidirectional laminates can be formed by cementing plies with different fibre orientations. The effective in–plane modulus of laminate plies is found to be simply the arithmetic mean of the moduli of the constituent plies. Laminates with midplane symmetry will behave like homogeneous anisotropic plates. A multidirectional composite laminate is defined by a code which describes the stacking sequence of the ply groups. For example, the code

$$[0_2/90_2/45/-45_3]_s \tag{40}$$

means the following:

The thickness of the laminate is h. Starting from the bottom of the laminate, at $z = -\frac{h}{2}$, the first ply group has tow plies of 0° orientation, followed by the next group with two 90° plies, followed by one 45° ply, and finally the last group with three – 45° plies. For a symmetric laminate, the ascending order from the bottom face is identical to the descending order from the top face, i.e $z = +\frac{h}{2}$. The subscript *S* denotes that the laminate is symmetric with respect to the midplane, i.e. $z = 0$. The upper half of the laminate is the same as the lower half except that the stacking sequence is reversed to maintain midplane symmetry.

A subscript *T* is used to describe the total laminate without resorting to describe symmetry or otherwise. For example, the laminate described by the code given in Eq. (40) can be written with the following code also:

$$[0_2/90_2/45/-45_3/45_3/45/90_2/0_2]_T \tag{41a}$$

or

$$[0_2/90_2/45/-45_6/45/90_2/0_2]_T \tag{41b}$$

where the middle six ply groups with the same orientations have been grouped together. Figure 7 shows the laminate schematically.

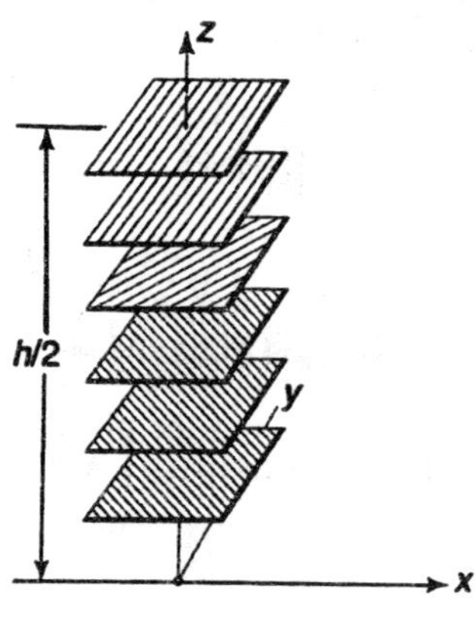

Fig. 7

Inplane Stress–Strain Relations: In deriving the stress–strain relations for a multidirectional laminate, the following assumptions are made:

(i) The laminate is symmetric, i.e.

$$\theta(z) = \theta(-z) \tag{42a}$$

and

$$a_{ij}(z) = a_{ij}(-z) \tag{42b}$$

Hence, both the ply orientation and the ply material modulus are symmetric with respect to the midplane of the laminate.

(ii) The strain is uniformly the same across the thickness of the laminate, i.e.

$$\begin{aligned} \varepsilon_{xx}(z) &= \varepsilon_{xx}^{*} \\ \varepsilon_{yy}(z) &= \varepsilon_{yy}^{*} \\ \gamma_{xy}(z) &= \gamma_{xy}^{*} \end{aligned} \tag{43}$$

The above assumption is fairly reasonable when the total laminate thickness is small and bonding between plies is good. x and y axes are arbitrary axes with reference to which the strains are prescribed. These axes may not in general coincide with any fibre axes.

Because of different orientations of the plies, the components of moduli for any given direction are not the same for each ply. Hence, for a given uniform strain, the stresses vary from ply to ply, and it is useful to discuss in terms of average stresses across the thickness of the laminate. Thus,

$$\bar{\sigma}_{xx} = \frac{1}{h}\int_{-h/2}^{h/2} \sigma_{xx}\, dz$$

$$\bar{\sigma}_{yy} = \frac{1}{h}\int_{-h/2}^{h/2} \sigma_{xx}\, dz \tag{44}$$

$$\bar{\tau}_{xy} = \frac{1}{h}\int_{-h/2}^{h/2} \tau_{xy}\, dz$$

Now, from Eq. (35) for any ply, remembering that x and y are arbitrary axes,

$$\sigma_{xx} = a_{11}\varepsilon_{xx} + a_{12}\varepsilon_{yy} + a_{14}\gamma_{xy}$$

$$\sigma_{yy} = a_{12}\varepsilon_{xx} + a_{22}\varepsilon_{yy} + a_{24}\gamma_{xy}$$

$$\tau_{xy} = a_{14}\varepsilon_{xx} + a_{24}\varepsilon_{yy} + a_{44}\gamma_{xy}$$

Since the strains are the same in the plies,

$$\bar{\sigma}_{xx} = \frac{1}{h}\int_{-h/2}^{h/2}\left[a_{11}\varepsilon^*_{xx} + a_{12}\varepsilon^*_{yy} + a_{14}\gamma^*_{xy}\right] dz$$

$$= \frac{1}{h}\left[\varepsilon^*_{xx}\int a_{11}\, dz + \varepsilon^*_{yy}\int a_{12}\, dz + \gamma^*_{xy}\int a_{14}\, dz\right]$$

$$= \frac{1}{h}\left[A_{11}\varepsilon^*_{xx} + A_{12}\varepsilon^*_{yy} + A_{14}\gamma^*_{xy}\right] \tag{45a}$$

where,

$$A_{11} = \int_{-h/2}^{h/2} a_{11}\, dz; \quad A_{12} = \int_{-h/2}^{h/2} a_{12}\, dz; \quad A_{14} = \int_{-h/2}^{h/2} a_{14}\, dz \tag{45b}$$

Similarly,

$$\bar{\sigma}_{yy} = \frac{1}{h}\left[A_{11}\varepsilon^*_{xx} + A_{22}\varepsilon^*_{yy} + A_{24}\gamma^*_{xy}\right] \tag{45c}$$

$$\bar{\tau}_{xy} = \frac{1}{h}\left[A_{14}\varepsilon^*_{xx} + A_{24}\varepsilon^*_{yy} + A_{44}\gamma^*_{xy}\right] \tag{45d}$$

where,

$$A_{22} = \int_{-h/2}^{h/2} a_{22}\, dz; \quad A_{24} = \int_{-h/2}^{h/2} a_{24}\, dz; \quad A_{44} = \int_{-h/2}^{h/2} a_{44}\, dz \qquad (45e)$$

$\bar{\sigma}_{xx}, \bar{\sigma}_{yy}$ and $\bar{\tau}_{xy}$, are the average stresses across the thickness of the laminate, i.e. stresses per unit thickness. Hence, for a laminate of thickness h, the stress resultants are

$$N_{xx} = h\bar{\sigma}_{xx}, \qquad N_{yy} = h\bar{\sigma}_{yy}, \qquad N_{yy} = h\bar{\tau}_{xy}$$

Substituting for $\bar{\sigma}_{xx}, \bar{\sigma}_{yy}$ and $\bar{\tau}_{xy}$

$$\begin{aligned} N_{xx} &= A_{11}\varepsilon^{*}_{xx} + A_{12}\varepsilon^{*}_{yy} + A_{14}\gamma^{*}_{xy} \\ N_{yy} &= A_{12}\varepsilon^{*}_{xx} + A_{22}\varepsilon^{*}_{yy} + A_{24}\gamma^{*}_{xy} \\ N_{xy} &= A_{14}\varepsilon^{*}_{xx} + A_{24}\varepsilon^{*}_{yy} + A_{44}\gamma^{*}_{xy} \end{aligned} \qquad (46)$$

The quantities A_{11}, A_{22}, etc. are the integrated values (across the thickness) of the off–axis components of moduli of the laminate.

Evaluation of Inplane Moduli: The resultant values of the moduli components are obtained by integrating the moduli component values across the thickness. In practice, when different plies of finite thicknesses are bonded to get the laminate, the values of a_{ij}s change in discrete steps from ply to ply and they are not continuous functions of z as indicated by Eqs (45b and e). However, continuing the integration sign as used earlier,

$$A_{11} = \int_{/h/2}^{h/2} a_{11}\, dz$$

$$= \int (Q_1 + Q_2 \cos 2\theta + Q_3 \cos 4\theta)\, dz$$

$$= Q_1 \int dz + Q_2 \int \cos 2\theta\, dz + Q_3 \int \cos 4\theta\, dz$$

The Qs for a laminate are constant because of our assumption that the laminate consists of plies of the same kind having identical material constants, but in the bonding process, the plies are put with their fibre axes oriented differently, i.e. θ changes from ply to ply, but the Qs are the same for each ply. Thus,

$$A_{11} = Q_1 h + Q_2 V_1 + Q_3 V_2 \qquad (47a)$$

where

$$V_1 = \int_{-h/2}^{h/2} \cos 2\theta\, dz, \quad \text{and} \quad V_2 = \int_{-h/2}^{h/2} \cos 4\theta\, dz \qquad (47b)$$

Similarly, from Eqs (45 b and e),

$$A_{22} = Q_1 h - Q_2 V_1 + Q_3 V_2$$

$$A_{12} = Q_4 h - Q_3 V_2$$
$$A_{14} = -\frac{1}{2} Q_2 V_3 - Q_3 V_4 \qquad (11.47c)$$
$$A_{24} = -\frac{1}{2} Q_2 V_3 - Q_3 V_4$$
$$A_{44} = Q_5 h - Q_3 V_2$$

where $V_3 = \int_{-h/2}^{h/2} \sin 2\theta \, dz,$ and $V_4 = \int_{-h/2}^{h/2} \sin 4\theta \, dz$ (47d)

It was assumed that the laminate consists of even number of plies and are symmetrically positioned, i.e. the positioning sequence from the bottom of the laminate, i.e. from $z = -\frac{h}{2}$ to $z = 0$ is the same as from $z = +\frac{h}{2}$ to $z = 0$, the midplane. Hence the limits for the integrals can be changed to $z = 0$ and $z = +\frac{h}{2}$, and the quantities multiplied by 2. Also, it was explicitly stated that the orientations θ change in finite steps, from ply to ply. So, the integration sign can be changed to summation sign, i.e.

$$V_1 = 2\sum_{i=1}^{n} \cos 2\theta_i h_i \qquad (48)$$

where θ_i is the fibre orientation of the ith ply whose thickness is h_i, and the total number of plies in the laminate is $2n$, so that the number of plies from $z = 0$ to $z = \frac{h}{2}$ is n. The summation in Eq. (48) is over all the plies from $z = 0$ to $z = \frac{h}{2}$, i.e. n. Let the laminate be composed of $2k_1$ number of plies with fibre orientation θ_1, $2k_2$ number of plies with fibre orientation θ_2, and $2k_i$ number with θ_i orientation. Then,

$$V_1 = 2k_1 h_1 \cos 2\theta_1 + 2k_2 h_2 \cos 2\theta_2 + \ldots + 2k_i h_i \cos 2\theta_i + \ldots \qquad (49)$$

Also, $2k_1 + 2k_2 + \ldots 2k_i + \ldots = 2n$

and $2k_1 h_1 + 2k_2 h_2 + \ldots + 2k_i h_i + \ldots = h$

or $2k_1 \frac{h_1}{h} + 2k_2 \frac{h_2}{h} + \ldots + 2k_i \frac{h_i}{h} + \ldots = 1$

It is easily seen that $(2k_i h_i/h)$ is the volume fraction of plies with fibre orientation i in the laminate. If the volume fractions are indicated by v_is, then Eq. (49) can be written as

$$V_1^* = \frac{V_1}{h} = v_1 \cos 2\theta_1 + v_2 \cos 2\theta_2 + \ldots + v_i \cos 2\theta_i + \ldots \qquad (50)$$

where, $v_1 + v_2 + \ldots + v_i + \ldots = 1$.

Equation (50) is simply the rule mixtures equation which will be discussed later. Thus, Eqs (47b and d) can be rewritten as

$$
\begin{aligned}
V_1^* &= \frac{V_1}{h} = v_1 \cos 2\theta_1 + v_2 \cos 2\theta_2 + \ldots \\
V_2^* &= \frac{V_2}{h} = v_1 \cos 4\theta_1 + v_2 \cos 4\theta_2 + \ldots \\
V_3^* &= \frac{V_3}{h} = v_1 \sin 2\theta_1 + v_2 \sin 2\theta_2 + \ldots \\
V_4^* &= \frac{V_4}{h} = v_1 \sin 4\theta_1 + v_2 \sin 4\theta_2 + \ldots
\end{aligned}
\tag{51}
$$

If the thickness of each ply is the same, say t, then on the basis of Eq. (49), one can write

$$
\begin{aligned}
V_1 &= t(2k_1 \cos 2\theta_1 + 2k_2 \cos 2\theta_2 + \ldots) \\
V_2 &= t(2k_1 \cos 4\theta_1 + 2k_2 \cos 4\theta_2 + \ldots) \\
V_3 &= t(2k_1 \sin 2\theta_1 + 2k_2 \sin 2\theta_2 + \ldots) \\
V_4 &= t(2k_1 \sin 4\theta_1 + 2k_2 \sin 4\theta_2 + \ldots)
\end{aligned}
\tag{52}
$$

where, as mentioned earlier, $2k_1$ is the number of plies in the laminate with θ_1 orientation of fibres, $2k_2$ is the number of plies in the laminate with θ_2 orientations, etc.

Using either Eq. (51) or Eq. (52), one can easily compute the inplane moduli of multidirectional laminates with any ply orientation. The information needed is orientation and volume fraction (or the number of plies) of each ply group. Using Eqs (51) or (52), the values of V_i^*s (or V_is) can be determined. Since the plies are identical, the values of Q_is are the same for each ply and these can be evaluated from Eq. (39). The values of a_{ij}s needed in Eq. (39) are obtained from Eq. (37b). Finally, Eq. (47) give the values of A_{ij}s. The units of A_{ij}s are Pa m or Nm^{-1}.

As an illustration of the steps involved consider the following case:

Cross–ply composites are commonly used in practice when uniform strength is required in both x and y directions. The laminate consists of plies with fibres oriented at $\theta = 0°$ and $\theta = 90°$ The laminate is symmetric. Let v_0 be the volume fraction of plies with $\theta = 0°$ orientation, and v_{90} be the volume fraction of plies with fibres at $\theta = 90°$ orientation. In general, v_0 and v_{90} are not equal.

$\theta_1 = 0°$, $\cos 2\theta = \cos 4\theta = 1$, $\sin 2\theta = \sin 4\theta = 0$

$\theta_2 = 90°$, $\cos 2\theta = -1$, $\cos 4\theta = +1$, $\sin 2\theta = \sin 4\theta = 0$

From Eq. (51),

$$V_1 = v_0 h - v_{90}\, h = (v_0 - v_{90})h$$
$$V_2 = (v_0 + v_{90})h = h$$
$$V_3 = V_4 = 0$$

From Eqs (47a and c),

$$\begin{aligned} A_{11} &= [Q_1 + Q_2(v_0 - v_{90}) + Q_3]h \\ A_{22} &= [Q_1 - Q_2(v_0 - v_{90}) + Q_3]h \\ A_{44} &= [Q_5 - Q_3]h \\ A_{12} &= [Q_4 - Q_3]h \\ A_{14} &= A_{24} \equiv 0 \end{aligned} \tag{53}$$

A laminate of this type is called an orthotopic laminate, Fig. 8. A three-dimensional body formed by cross–ply laminates was said to be orthogonally anisotropic or orthotropic.

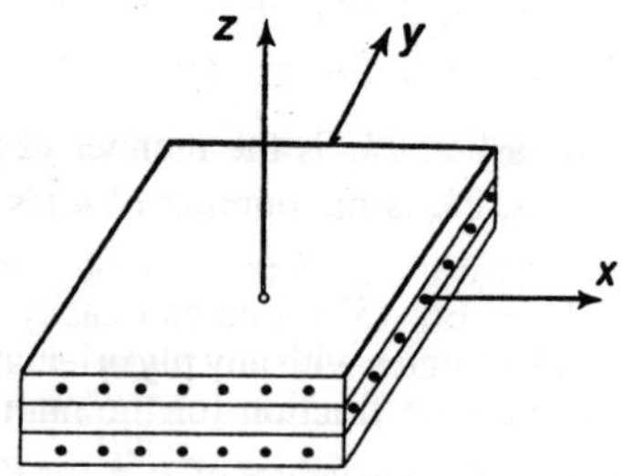

Fig. 8

It can easily be seen from the figure that because of the difference in fibre densities in x and y directions, E_{xx} and E_{yy} can be different. Further, a shear stress τ_{xy} produces only shear strain in the xy plane, and does not cause any linear strain either in x direction or in y direction. These are being reflected in A_{14} and A_{24}, both being zero. Further, the moduli components A_{11} and A_{22} vary linearly with $(v_0 - v_{90})$. When v_0 or v_{90}, is zero, we get unidirectional composite laminates. When $v_0 = v_{90}$, the volume fractions are equal and $A_{11} = A_{22}$.

Example 5

Estimate the in–plane moduli and compliances for a cross–ply laminate formed by using unidirectional composites with 'Toray' filament and 'Namco' resin. The modulus data for this ply was given in Example 4 and is repeated here:

$E_{xx} = 181$ GPa, $E_{yy} = 10.3$ GPa, $V_{xy} = 0.0159$, $G_{xy} = 7.17$ GPa,

$(1 - V_{yx}V_{xy})^{-1} = 1.0045$

From Example 4, the values of a_{ij}s for any ply are

$$a_{11} = 181.8 \text{ GPa}, \quad a_{22} = 10.34 \text{ GPa}, \quad a_{12} = 2.891 \text{ GPa},$$

$$a_{44} = 7.17 \text{ GPa}$$

From Eq. (39),

$$Q_1 = \frac{1}{8}(3a_{11} + 3a_{22} + 2a_{12} + 4a_{44})$$

$$= \frac{1}{8}[(3 \times 181.8) + (3 \times 10.34) + (2 \times 2.891) + (4 \times 7.17)]$$

$$= 76.36 \text{ GPa}$$

$$Q_2 = \frac{1}{2}(a_{11} - a_{22}) = \frac{1}{2}(181.8 - 10.34) = 85.73 \text{ GPa}$$

$$Q_3 = \frac{1}{8}(a_{11} + a_{22} - 2a_{12} - 4a_{44}) \qquad \text{(b)}$$

$$= \frac{1}{8}[181.8 + 10.34 - (2 \times 2.891) - (4 \times 7.17)] = 19.71 \text{ GPa}$$

$$Q_4 = \frac{1}{8}(a_{11} + a_{22} + 6a_{12} - 4a_{44})$$

$$= \frac{1}{8}[181.8 + 10.34 + (6 \times 2.891) - (4 \times 7.17)] = 22.6 \text{ GPa}$$

$$Q_5 = \frac{1}{8}(a_{11} + a_{22} - 2a_{12} + 4a_{44}) = 26.88 \text{ GPa}$$

Substituting these in the expressions for A_{ij}s from Eq. (53),

$$\frac{1}{h}A_{11} = 76.36 + (v_0 - v_{90})85.73 + 19.71$$

$$\frac{1}{h}A_{22} = 76.36 - (v_0 - v_{90})85.73 + 19.71 \qquad \text{(c)}$$

$$\frac{1}{h}A_{44} = 26.88 - 19.71 = 7.17$$

$$\frac{1}{h}A_{12} = 22.60 - 19.71 = 2.89$$

$$A_{14} = A_{24} = 0$$

The average values of the compliance coefficients are obtained by the inversion of Eq. (46). If Δ is the determinant of the A_{ij}s in Eq. (46), then

$$\Delta = A_{11}(A_{22}A_{44} - A^2_{24}) - A_{12}(A_{12}A_{44} - A_{24}A_{14}) + A_{14}(A_{12}A_{24} - A_{22}A_{14})$$

Corresponding to Eq. (46), one can write for ε^*_{ij}

$$\begin{aligned}
\varepsilon^*_{xx} &= \bar{b}_{11}N_{xx} + \bar{b}_{12}N_{yy} + \bar{b}_{14}N_{xy} \\
\varepsilon^*_{yy} &= \bar{b}_{12}N_{xx} + \bar{b}_{22}N_{yy} + \bar{b}_{24}N_{xy} \\
\gamma^*_{xy} &= \bar{b}_{41}N_{xx} + \bar{b}_{42}N_{yy} + \bar{b}_{44}N_{xy}
\end{aligned} \tag{54}$$

Solving Eq. (46) for ε^*_{ij} s and comparing with the coefficients in Eqs (54), one gets

$$\begin{aligned}
\bar{b}_{11} &= \frac{1}{\Delta}(A_{22}A_{44} - A_{24}^2) \\
\bar{b}_{12} &= -\frac{1}{\Delta}(A_{12}A_{44} - A_{14}A_{24}) \\
\bar{b}_{14} &= \frac{1}{\Delta}(A_{12}A_{24} - A_{14}A_{22}) \\
\bar{b}_{22} &= \frac{1}{\Delta}(A_{11}A_{44} - A_{14}^2) \\
\bar{b}_{24} &= \frac{1}{\Delta}(A_{11}A_{24} - A_{14}A_{12}) \\
\bar{b}_{44} &= \frac{1}{\Delta}(A_{11}A_{22} - A_{12}^2)
\end{aligned} \tag{55}$$

To find the values of A_{ij}s we need the values of v_0 and v_{90}. Assuming $v_0 = v_{90} = 0.5$,

$A_{11} = 96.07h,\quad A_{22} = 96.07h,\quad A_{44} = 7.17h,\quad A_{12} = 2.89h,$

$A_{14} = A_{24} = 0$

Substituting these,

$\Delta = 96.07\ (96.07 \times 7.17) - 2.89\ (2.89 \times 7.17) = 66.12 \times 10^{30}h^3\ (Pa)^3$

$\frac{1}{\Delta} = 0.0151 \times 10^{-30}h^{-3}$

$$\therefore \quad \bar{b}_{11}h = 0.0151\times10^{-30}(96.07\times7.17)\times10^{18} = 10.40\times10^{-12}(Pa)^{-1}$$

$$\begin{aligned}
\bar{b}_{12}h &= -0.0151\times10^{-30}(2.89\times7.17)\times10^{18} \\
&= -\ 0.313 \times 10^{-12}\ (Pa)^{-1}
\end{aligned} \tag{d}$$

$$\bar{b}_{22}h = 0.0151\times10^{-30}(96.07\times7.17)\times10^{18} = 10.40\times10^{-12}(Pa)^{-1}$$

$$\begin{aligned}
\bar{b}_{44}h &= 0.0151\times10^{-30}(96.07\times96.07 - 2.89^2)\times10^{18} \\
&= 138.3 \times 10^{-12}\ (Pa)^{-1}
\end{aligned}$$

Angle–ply Laminates: Another class of laminates that are commonly used in practice are the angle–ply laminates. In these laminates, there are only two ply orientations which have the same magnitudes but are of opposite sings. The laminate is said to be balanced when there are equal number of plies with positive and negative orientations. Hence, for a balanced angleply laminate assuming complete symmetry, one has, from Eqs (42a and b), the following:

$$\theta_1 = +\beta, \qquad \theta_2 = -+\beta, \qquad v_1 = v_2 = \frac{1}{2} \tag{56}$$

Substituting these in Eq. (51),

$$\begin{aligned} V_1^* &= \frac{1}{2}(\cos 2\beta + \cos 2\beta) = \cos 2\beta \\ V_2^* &= \cos 4\beta \\ V_3^* &= V_4^* = 0 \end{aligned} \tag{57}$$

From Eqs (47a and c),

$$\begin{aligned} A_{11} &= [Q_1 + Q_2 \cos 2\beta + Q_3 \cos 4\beta]h \\ A_{22} &= [Q_1 - Q_2 \cos 2\beta + Q_3 \cos 4\beta]h \\ A_{12} &= [Q_4 - Q_3 \cos 4\beta]h \\ A_{44} &= [Q_5 - Q_3 \cos 4\beta]h \\ A_{14} &= A_{24} = 0 \end{aligned} \tag{58}$$

As an example, consider a balanced symmetric angleply laminate with $\beta = 45°$. For such a laminate,

$$A_{11} = A_{22} = (Q_1 - Q_3)\,h, \quad A_{12} = (Q_4 + Q_3)\,h, \quad A_{44} = (Q_5 + Q_3)\,h$$

$$A_{14} = A_{24} = 0$$

For the composite considered in Example 5, the values of Q_is are,

$$Q_1 = 76.36 \text{ GPa}, \quad Q_2 = 85.73 \text{ GPa}, \quad Q_3 = 19.71 \text{ GPa},$$

$$Q_4 = 22.6 \text{ GPa}, \quad Q_5 = 26.28 \text{ GPa}$$

For a laminate formed from these composites,

$$\begin{aligned} A_{11} &= A_{22} = (76.36 - 19.71)h = 56.65\,h \text{ GPa} \\ A_{12} &= (22.6 + 19.71)h = 42.31\,h \text{ GPa} \\ A_{44} &= (26.88 + 19.71)h = 46.59\,h \text{ GPa} \\ A_{14} &= A_{24} = 0 \end{aligned}$$

The components of compliance are obtained from Eq. (55)

$$\bar{b}_{11}h = \bar{b}_{22}h = 39.91 \times 10^{-12}(Pa)^{-1}$$

$$\bar{b}_{12}h = -29.81\times10^{-12}(Pa)^{-1}$$

$$\bar{b}_{44}h = 21.46\times10^{-12}(Pa)^{-1}$$

$$\bar{b}_{44}h = \bar{b}_{24} = 0$$

The corresponding Engineering constants are

$$\bar{E}_{xx} = \frac{1}{\bar{b}_{11}h} = 25.05\text{ GPa}$$

$$\bar{E}_{yy} = \frac{1}{\bar{b}_{22}h} = 25.05\text{ GPa}$$

$$\bar{G}_{xy} = \frac{1}{\bar{b}_{44}h} = 46.59\text{ GPa}$$

$$\bar{v}_{xy} = -\bar{b}_{12}\bar{E}_{xx}\ h = 0.746$$

The reason for working out the values of the Engineering constants is to show that in the case of a composite, the value of Poisson's ratio can be greater than 0.5. In this particular case, the Poisson's ratio in the x direction is 0.746.

PLY STRESS AND PLY STRAIN

The stress analysis of symmetrical laminates discussed in the previous section was based on the fundamental assumption that all the plies in the laminate experienced uniform strains. As the fibre orientations in the plies are different, for a given strain, the stresses induced in individual plies will be different. Also, under the same assumption of uniform strain, a given load or stress gets distributed according to the stiffness of each ply. As an example, consider a laminate having the code $[0_4/90_4]$s. This is a symmetric cross–ply laminate having a total of 16 plies. Let the plies be of the same composite material that we have been discussing so far, i.e. the values of $h\bar{b}_{ij}$s are as given in Eq. (d). Let the thickness of each ply be 130 × 10^{-6} m, and let the laminate be subjected to a uniaxial stress resultant N_{xx} = 1 MN/m.

Thickness of laminate = h = 16 × 130 × 10^{-6} = 2.08 × 10^{-3}m

The compliance coefficients for the laminate from Eq. (d) are

$$\bar{b}_{11} = \frac{1}{h}\times10.40\times10^{-12} = \frac{1}{2.08}\times10.40\times10^{-9} = 5\times10^{-9}\ (\text{N/m})^{-1}$$

$$1\,\bar{b}_{12} = -\frac{1}{h}\times0.313\times10^{-12} = -\frac{1}{2.08}\times0.313\times10^{-9} = -0.15\times10^{-9}\ (\text{N/m})^{-1}$$

$$\bar{b}_{14} = 0$$

From Eq. (54), the strains are

$$\varepsilon_{xx}^{*} = \bar{b}_{11} N_{xx} = 5\times10^{-9}\times10^{6} = 5\times10^{-3}$$

$$\varepsilon_{yy}^{*} = \bar{b}_{12} N_{xx} = -0.15\times10^{-9}\times10^{6} = -0.15\times10^{-3}$$

These are the strains experienced by each ply. For the ply, the stress–strain equations are

$$\sigma_{xx} = a_{11}\varepsilon_{xx} + a_{12}\varepsilon_{yy}$$

$$\sigma_{yy} = a_{21}\varepsilon_{xx} + a_{22}\varepsilon_{yy}$$

From Example 4, for 0° fibre orientation,

$$a_{11} = 181.8 \text{ GPa}, \quad a_{22} = 10.34 \text{ GPa}, \quad a_{21} = 2.891 \text{ GPa}$$

$$\therefore \quad \sigma_{xx} = (181.8 \times 5 \times 10^{6}) - (2.891 \times 0.15 \times 10^{6}) = 908.6 \text{ MPa.}$$

$$\sigma_{yy} = (2.891 \times 5 \times 10^{6}) - (10.34 \times 0.15 \times 10^{6}) = 10.9 \text{ MPa}$$

$$\tau_{xy} = 0$$

For the 90° fibre orientation,

$$a_{11} = 10.34 \text{ GPa}, \quad a_{22} = 181.8 \text{ GPa}, \quad a_{21} = 2.891 \text{ GPa}$$

$$\therefore \quad \sigma_{xx} = (10.34 \times 5 \times 10^{6}) - (2.891 \times 0.15 \times 10^{6}) = 51.3 \text{ MPa}$$

$$\sigma_{yy} = (2.891 \times 5 \times 10^{6}) - (181.8 \times 0.15 \times 10^{6}) = -12.9 \text{ MPa}$$

$$\tau_{xy} = 0$$

It should be observed that for 0° fibre orientation ply group, σ_{xx} and σ_{yy} are the stresses along and perpendicular to the fibres, whereas for 90° ply group, σ_{xx} and σ_{yy} are stresses perpendicular and parallel to the fibres.

From these results, the average stresses in x and y directions are

$$\bar{\sigma}_{xx} = \frac{1}{2}(908.6 + 51.3) = 479.95 \text{ MPa}$$

$$\bar{\sigma}_{y} = 0, \qquad \bar{\tau}_{xy} = 0$$

The resultant stress in x direction is

$\bar{\sigma}_{xx} h =$ (908.6 + 10^{6} Pa) × (2.08 + 10^{-3} m) = 998 × 10^{3} N/m = 1 MN/m

This checks with the resultant applied stress N_{xx}.

One of the reasons in estimating the stresses and strains in individual plies is to check whether they meet the failure criteria. Failure criteria will be discussed in the next section. For example, in the present case if the maximum strain criterion is applied with the limit that

$$\varepsilon_{\max} \text{ alone fibre} \le 10 \times 10^{-3}$$

ε_{max} perpendicular to fibre $\leq 4.5 \times 10^{-3}$

then, for the 90° ply group, $\varepsilon^{*}_{xx} = 5\times10^{-3}$ is the strain perpendicular to the fibres and this is greater than 4.5×10^{-3}, which is the limit. Hence, based on the maximum strain criterion, failure would have occurred in the 90° ply group, when the resultant applied stress reached a value

$$N_{xx(max)} = \frac{4.5}{5} 0.9 \text{ MN/m}$$

FAILURE CRITERIA OF COMPOSITE MATERIALS

It is obvious from the discussions so far that the failure theories for composite materials would be quite different and more complex compared to theories of failure for an isotropic solid. In this section, we shall briefly consider some of the failure criteria found suitable for composites. We shall restrict our discussion to orthotropic materials in general and to laminates in particular. As in the case of isotropic materials, the maximum stress theory and the maximum strain theory are the basic theories that are considered first.

(a) Maximum Stress Theory: The maximum stress theory assumes that failure occurs when any of the stresses in the principal material axes reach a critical value. There are three possible modes of failure, and the conditions for these are

$$\begin{aligned} \sigma_{11} &= \sigma^{*}_{11} \\ \sigma_{22} &= \sigma^{*}_{22} \\ \tau_{12} &= \tau^{*}_{12} \end{aligned} \qquad (59)$$

The stresses σ_{ij} are referred to the principal directions 1 and 2. σ^{*}_{11} is the ultimate tensile or compressive stress in direction 1, σ^{*}_{22} is the ultimate tensile or compressive stress in direction 2 and τ^{*}_{12} is the ultimate shear stress acting on plane 1 (i.e. plane with normal in direction 1) in direction 2. The values of $\sigma^{*}_{11}, \sigma^{*}_{22}$ and τ^{*}_{12} are obtained experimentally for a given composite.

If the load were to be applied at an angle θ to the fibre axis direction 1, Fig. 5, and if x', y' are the corresponding frame of reference for the applied stresses, then, from the transformation equations

$$\begin{aligned} \sigma_{11} &= \sigma_{x'x'} \cos^2 \theta \\ \sigma_{22} &= \sigma_{x'x'} \sin^2 \theta \\ \tau_{12} &= + \sigma_{x'x'} \sin \theta \cos \theta \end{aligned} \qquad (60)$$

Combining Eqs (59) and (60), according to the maximum stress theory, failure occurs when σ_{x} assumes the smallest of the following three values

$$\sigma_{x'x'} = \frac{\sigma_{11}^*}{\cos^2\theta}$$
$$\sigma_{x'x'} = \frac{\sigma_{22}^*}{\sin^2\theta} \qquad (61)$$
$$\sigma_{x'x'} = \frac{\tau_{12}^*}{\sin\theta\cos\theta}$$

Instead of $\sigma_{x'x'}$ alone, if the stresses acting are $\sigma_{x'x'}$, $\sigma_{y'y'}$ and $\tau_{x'y'}$ then from Eq. (23), the stresses in the principal directions 1 and 2 are

$$\sigma_{11} = \sigma_{x'x'}\cos^2\theta + \sigma_{y'y'}\sin^2\theta - \tau_{x'y'}\sin 2\theta$$
$$\sigma_{22} = \sigma_{x'x'}\sin^2\theta + \sigma_{y'y'}\cos^2\theta + \tau_{x'y'}\sin 2\theta \qquad (62)$$
$$\tau_{12} = \frac{1}{2}(\sigma_{x'x'} - \sigma_{y'y'})\sin 2\theta + \tau_{x'y'}\cos 2\theta$$

If the applied stresses $\sigma_{x'x'}$, $\sigma_{y'y'}$ and $\tau_{x'y'}$ either individually or in combination cause σ_{11} or σ_{22} or σ_{12} to exceed their maximum allowable values, failure occurs.

(b) Maximum Strain Theory: According to this theory, failure occurs when the strain along any principal direction assumes a critical value, i.e. when

$$\varepsilon_{11} = \varepsilon_{11}^*$$
$$\varepsilon_{22} = \varepsilon_{22}^* \qquad (63)$$
$$\gamma_{12} = \gamma_{12}^*$$

where ε_{11}^* is the maximum tensile or compressive strain in direction 1, ε_{22}^* is the maximum tensile or compressive strain in direction 2, and γ_{12}^* is the maximum shear strain in plane 1–2. If E_{11}, E_{22} and G_{12} are the material constants, then, according to the maximum strain theory, failure occurs when any of the following conditions hold:

$$\varepsilon_{11} = \frac{\sigma_{11}}{E_{11}} - v_{12}\frac{\sigma_{22}}{E_{22}} \geq \varepsilon_{11}^*$$
$$\varepsilon_{22} = -v_{12}\frac{\sigma_{11}}{E_{11}} + \frac{\sigma_{22}}{E_{22}} \geq \varepsilon_{22}^* \qquad (64)$$
$$\gamma_{12} = \frac{\tau_{11}}{G_{12}} \geq \gamma_{12}^*$$

If $\sigma_{x'x'}$, $\sigma_{y'y'}$ and $\tau_{x'y'}$ are the stresses applied, then the values of σ_{11}, σ_{22} and, τ_{12} are obtained form Eq. (23), which can then be substituted into Eq. (62).

(c) Distortion Energy Theory: While the maximum stress and maximum strain theories are easy to apply, they have limitations since experiments do not completely support them. Another theory which is commonly used in design processes is the energy of distortion theory, which sometimes is called the Tsai–Hill theory. This theory is similar to the distortion energy or the deviatoric stress theory applied to isotropic solids. For an isotropic solid, Eq. (12) gives the distortion energy as

$$U^* = \frac{1}{12G}\left[(\sigma_1 - \sigma_2)^2 + (\sigma_2 - \sigma_3)^2 + (\sigma_3 - \sigma_1)^2\right]$$

where σ_1, σ_2 and σ_3 are the principal stresses and G is the shear modulus. For an orthotropic solid, this expression is generalised and written as

$$2f(\sigma_{ij}) = F(\sigma_{11} - \sigma_{22})^2 + G(\sigma_{22} - \sigma_{33})^2 + H(\sigma_{33} - \sigma_{11})^2 + 2L\,\tau_{12}^2 + 2M\,\tau_{23}^2 + 2N\tau_{31}^2 = 1 \tag{65}$$

where 1, 2 and 3 are the principal directions of symmetry and *F, G, H, L, M* and *N* are parameters characterising the anisotropy of the material. In the stress–space, Eq. (65) represents a six–dimensional surface. The critical values of σ_{ij}s and τ_{ij}s will give the limits to this yield surface. If the applied stresses lie within the surface, then no failure occurs. The values of the parameters are obtained from tests conducted on a sample of the composite. Let $\sigma_{11}^*, \sigma_{22}^*, \sigma_{23}^*$ be the normal or yield strengths in the directions of anisotropic symmetry. Then, with σ_{11}^* alone, Eq. (65) gives

$$F\sigma_{11}^{*2} + H\sigma_{11}^{*2} = 1$$

$$F + H = \frac{1}{\sigma_{11}^{*2}} \tag{66a}$$

Similarly, with $\sigma_2^*$and σ_3^* individually applied, one gets

$$F + G = \frac{1}{\sigma_{22}^{*2}} \tag{66b}$$

$$G + H = \frac{1}{\sigma_{33}^{*2}} \tag{66c}$$

On the same lines, for τ_{12}, τ_{23} and τ_{31}, one gets

$$2L = \frac{1}{\tau_{12}^{*2}}, \quad 2M = \frac{1}{\tau_{23}^{*2}}, \quad 2N = \frac{1}{\tau_{31}^{*2}} \tag{66d}$$

where τ_{ij}^*s are the yield strengths in shear.

From Eqs (66a–c), one can solve for *F, G* and *H*. The solutions are

$$F = \frac{1}{2}\left(\frac{1}{\sigma_{11}^{*2}} + \frac{1}{\sigma_{22}^{*2}} - \frac{1}{\sigma_{33}^{*2}}\right)$$

$$G = \frac{1}{2}\left(-\frac{1}{\sigma_{11}^{*2}} + \frac{1}{\sigma_{22}^{*2}} + \frac{1}{\sigma_{33}^{*2}} \right) \tag{67}$$

$$H = \frac{1}{2}\left(\frac{1}{\sigma_{11}^{*2}} - \frac{1}{\sigma_{22}^{*2}} + \frac{1}{\sigma_{33}^{*2}} \right)$$

Equation (65) with values for *F, G, H, L, M* and *N* substituted from Eqs (66d) and (67) describes a failure surface in a six–dimensional space. So long as the point (σ_{11}, σ_{22}, σ_{23}, τ_{12}, τ_{23}, τ_{31}) lies within this surface, no failure occurs. If the point happens to fall either on the surface or outside the surface, failure occurs.

Consider now a unidirectionally reinforced composite as shown in Fig. 1. Let *x*–axis be along the fibre direction instead of *z* as shown in that figure. Then, the plane *yz* will be a transverse plane of isotropy, and for this plane, the transverse yield strengths σ_{yy}^{*} and σ_{zz}^{*} will be equal to each other. In our notation, this means that σ_{22}^{*} and σ_{33}^{*} are equal. Also, for this body, τ_{12}^{*} and τ_{13}^{*} are equal, i.e. $L = N$. Hence, for an orthotropic body, substituting the present values for *F, G, H*, etc., Eq. (65) becomes

$$\frac{1}{2\sigma_{11}^{*2}}\left[(\sigma_{11}-\sigma_{22})^2-(\sigma_{22}-\sigma_{33})^2+(\sigma_{33}-\sigma_{11})^2\right]+\frac{1}{\sigma_{22}^{*2}}(\sigma_{22}-\sigma_{33})^2$$

$$2L+\frac{1}{\tau_{12}^{*2}}\left(\tau_{12}^2+\tau_{31}^2\right)+\frac{1}{\tau_{23}^{*2}}(\tau_{23})^2=1 \tag{68}$$

In the case of a laminate with unidirectional reinforcements, if the state of stress is a plane state, then one has $\sigma_{33} = \tau_{31} = \tau_{23} = 0$. Equation (68) then

$$\left(\frac{\sigma_1}{\sigma_{11}^{*}}\right)^2-\left(\frac{\sigma_{11}}{\sigma_{11}^{*}}\right)\left(\frac{\sigma_{22}}{\sigma_{11}^{*}}\right)+\left(\frac{\sigma_{22}}{\sigma_{22}^{*}}\right)^2+\left(\frac{\tau_{12}}{\tau_{12}^{*}}\right)^2=1 \tag{69}$$

Equation (69) describes a failure envelope and so long as the point (σ_{11}, σ_{22}, τ_{12}) lies within the surface no failure occurs. If the unidirectional laminate is subjected to a stress $\sigma_{x'x'}$ at an angle θ to *x*–axis, then from Eqs (60) and (69) failure occurs when

$$\sigma_{x'x'}=\left[\frac{\cos^4\theta}{\sigma_{11}^{*2}}+\left(\frac{1}{\tau_{12}^{*2}}-\frac{1}{\sigma_{11}^{*2}}\right)\sin^2\theta\cos^2\theta+\frac{\sin^4\theta}{\sigma_{22}^{*2}}\right]^{-\frac{1}{2}} \tag{70}$$

Example 6:

For a class of E–glass–epoxy composite with unidirectional reinforcement, the following data apply:

$E_{11} = 53.8$ GPa, $E_{22} = 17.9$ GPa

$\nu_{12} = 0.25$ $\quad G_{12} = 8.6$ GPa

σ_{11}^{*} (tens) = 1304 MPa $\quad \sigma_{11}^{*}$ (comp) = 1034 MPa

σ_{22}^{*} (tens) = 27.64 MPa $\quad \sigma_{22}^{*}$ (comp) = 138 MPa

$\tau_{12}^{*} = 55.2$ MPa

Determine the minimum value of $\sigma_{x'x'}$ applied at an angle of 30° to the fibre axis to cause failure according to (a) maximum stress theory (tension and compression), (b) maximum strain theory (tension) and (c) distortion energy theory (tension).

(a) Maximum Stress Theory

(i) Tension: From Eq. (61),

$$\sigma_{x'x'} = \frac{\sigma_{11}^{*}}{\cos^2\theta} = \frac{1304}{0.74} = 1378.7 \text{ MPa}$$

$$\sigma_{x'x'} = \frac{\sigma_{22}^{*}}{\sin^2\theta} = \frac{27.6}{0.25} = 110.4 \text{ MPa}$$

$$\sigma_{x'x'} = \frac{\tau_{12}^{*}}{\sin\theta\cos\theta} = \frac{55.2}{0.433} = 127.5 \text{ MPa}$$

Failure occurs when $\sigma_{x'x'} \geq 110.4$ MPa (tension).

(ii) Compression: From Eq. (61)

$$\sigma_{x'x'} = \frac{\sigma_{11}^{*}}{\cos^2\theta} = \frac{1034}{0.75} = 1378.7 \text{ MPa}$$

$$\sigma_{x'x'} = \frac{\sigma_{22}^{*}}{\sin^2\theta} = \frac{138}{0.25} = 552 \text{ MPa}$$

$$\sigma_{x'x'} = \frac{\tau_{12}^{*}}{\sin\theta\cos\theta} = \frac{55.2}{0.433} = 127.5 \text{ MPa}$$

Failure occurs when $\sigma_{x'x'} \geq 127.5$ MPa (Compression)

(b) Maximum Strain Theory: From Eq. (60)

$$\sigma_{11} = \sigma_{x'x'}\cos^2\theta, \qquad \sigma_{22} = \sigma_{x'x'}\sin^2\theta, \tag{e}$$

$$\tau_{12} = \frac{1}{2}\sigma_{x'x'}\sin 2\theta$$

Further, from Eq. (64), the maximum tensile strain in direction 1 when yield (or failure) stress σ_{11}^{*} is applied is

$$\varepsilon_{11}^{*} = \frac{\sigma_{11}^{*}}{E_{11}} = \frac{1.034}{23.8} = 0.01922$$

Similarly, the maximum tensile strain in direction 2 when yielding (or failure) occurs is

$$\varepsilon_{22}^{*} = \frac{\sigma_{22}^{*}}{E_{22}} = \frac{0.0276}{17.6} = 0.001542$$

Further, the shear strain at the time of yielding is

$$\gamma_{12}^{*} = \frac{\tau_{12}^{*}}{G_{12}} = \frac{0.0552}{8.6} = 0.00642$$

From Eqs (64) and (e)

$$\varepsilon_{11} = \frac{\sigma_{x'x'}}{E_{11}}\cos^2\theta - v_{12} = \frac{\sigma_{x'x'}}{E_{22}}\sin^2\theta$$

$$= \sigma_{x'x'}\left(\frac{\cos^2\theta}{E_{11}} - v_{12}\frac{\sin^2\theta}{E_{22}}\right)$$

$$\varepsilon_{22} = \sigma_{x'x'}\left(-v_{12}\frac{\cos^2\theta}{E_{11}} + \frac{\sin^2\theta}{E_{22}}\right)$$

$$\gamma_{12} = \sigma_{x'x'}\left(\frac{1}{2}\frac{\sin 2\theta}{G_{12}}\right)$$

Substituting the critical values for the strains and solving for

$$\sigma_{x'x'}\left(\text{with the reciprocal identity } \frac{v_{12}}{E_{22}} = \frac{v_{21}}{E_{11}}\right)$$

$$\sigma_{x'x'} = \varepsilon_{11}^{*}\left[\frac{\cos^2\theta}{E_{11}} - v_{12}\frac{\sin^2\theta}{E_{22}}\right]$$

$= 0.01922\ (0.01394 - 0.003492)^{-1}$

$= 1.8395$ GPa $= 1839$ MPa

$$\sigma_{x'x'} = \varepsilon_{22}^{*}\left[-v_{12}\frac{\cos^2\theta}{E_{11}} + \frac{\sin^2\theta}{E_{22}}\right]^{-1}$$

$= 0.001542\ (-0.01047 + 0.01397)^{-1}$

$= 0.44.6$ GPa $= 441$ MPa

$$\sigma_{x'x'} = \gamma_{12}^{*}\left(\frac{1}{2}\frac{\sin 2\theta}{G_{12}}\right)^{-1}$$

$= 0.00642\ (0.04511)^{-1}$

$= 0.1275$ GPa $= 142$ MPa

Based on the minimum of the three values, the critical stress value is 142 MPa.

(c) Distortion Energy Theory: From Eq. (70)

$$\sigma_{x'x'} = 10^8 \times [0.0052612 + (3.2819 - 0.00935) \times 0.1875 + 0.82047]^{-1/2}$$

$$= 83 \text{ MPa}$$

MICROMECHANICS OF COMPOSITES

In this section, the micromechanical aspects of fibrous composites based on the rule of mixtures, where the sharing of the loads by the matrix and the fibre is dependent on the volume–weighted averages of the component properties, will be examined.

Consider a composite of mass m_c and volume v_c. The total mass of the composites is the sum of the matrix mass m_m and the reinforcing fibre mass m_f, i.e.

$$m_c = m_m + m_f \tag{71}$$

The subscripts *c, m* and *f* refer to composite, matrix and fibre, respectively. The volume v_c of the composite is given by

$$v_c = v_m + v_f + v_v \tag{72}$$

where v_v is the volume of voids that the composite element may contain. Dividing Eq. (71) by m_c and Eq. (72) by v_c, one gets

$$M_m + M_f = 1 \tag{73}$$

and $$V_m + V_f + V_v = 1 \tag{74}$$

where the *M*s and *V*s stand for mass and volume fractions.

Consider a rectangular, unidirectional composite rod, Fig. 9, subjected to a force p_c in the direction of the fibres. Assume that the rod extends uniformly with no delaminations between the matrix and the fibres.

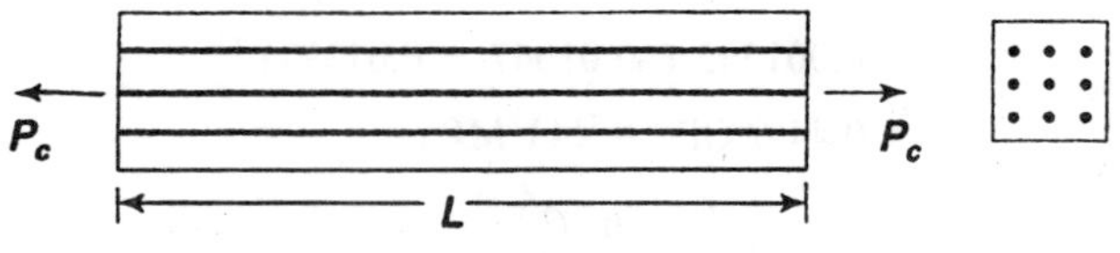

Fig. 9

Assume that transverse sections that were plane before loading remain plane after loading. This means that the strain in the matrix and the strains in the fibres are the same. Thus,

$$\varepsilon_{cl} = \varepsilon_m = \varepsilon_f = \frac{\Delta L}{L} \tag{75}$$

where ε_{cl} indicates the strain in the composite in the longitudinal direction.

The situation depicted by Eq. (75) is known as the isostrain situation. It is further assumed that the Poisson's ratios of the matrix and the fibres are equal. If E_m and E_f are the Young's moduli for the matrix and the fibre, then the stresses are

$$\sigma_m = E_m\, \varepsilon_{cl}, \qquad \sigma_f = E_f\, \varepsilon_{cl}$$

If A_c is the total cross–sectional area of the composite, then

$$P_c = P_m + P_f$$

i.e.

$$\sigma_c A_c = \sigma_m A_m + \sigma_f A_f \tag{76a}$$

$$= (A_m E_m + A_f E_f)\, \varepsilon_{cl} \tag{76b}$$

or

$$\frac{\sigma_c}{\varepsilon_{cl}} = E_{cl} = E_m \frac{A_m}{A_c} + E_f \frac{A_f}{A_c} \tag{77}$$

Since the lengths of the composite, the matrix and the fibres are all equal,

$$v_m = A_m L, \quad v_f = A_f L, \quad v_c = A_c L$$

and

$$\frac{A_m}{A_c} = \frac{v_m}{v_c} = V_m, \qquad \frac{A_f}{A_c} = \frac{v_f}{v_c} = V_f \tag{78}$$

Eq. (77) becomes

$$E_{cl} = E_m V_m + E_f V_f = E_{11} \tag{79}$$

E_{11} is the Young's modulus for the composite in the fibre direction. This is called the rule of mixtures for the Young's modulus in the fibre direction. From Eqs (76a) and (78), one can obtain an expression for the composite strength in the fibre direction as

$$\sigma_{cl} = \sigma_m V_m + \sigma_f A_f \tag{80}$$

If the composite is loaded in the transverse direction, and if it is assumed once again that there is no separation between the fibres and the matrix, then one can group the fibres together as one phase material that is continuous, and the matrix as one group, Fig. (10).

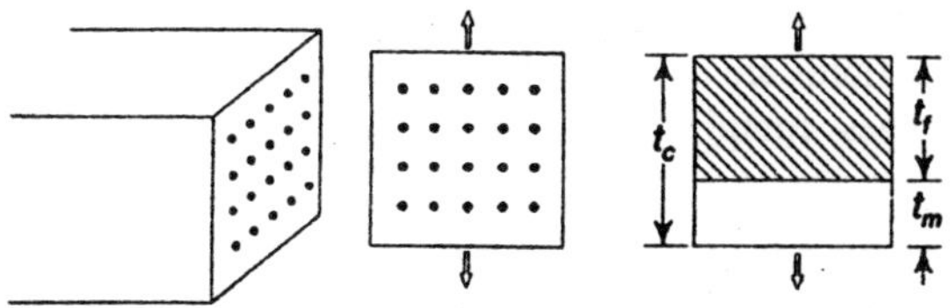

Fig. 10

If the applied load is uniformly distributed across the transverse faces, then the transverse stresses in the two phases are equal, i.e.

$$\sigma_{ct} = \sigma_m = \sigma_f \tag{81}$$

The total transverse displacement is

$$\Delta t_c = \Delta t_m + \Delta t_f \tag{82}$$

t_m and t_f are the equivalent gauge lengths of the matrix and the fibre respectively, when each is considered as one phase material. If L is the length of the member as shown in Fig. 9, then

$$Lt_m = v_m, \quad Lt_f = v_f, \quad Lt_c = v_c \tag{83}$$

Dividing Eq. (82) by t_c

$$\varepsilon_{ct} = \frac{\Delta t_c}{t_c} = \frac{\Delta t_m}{t_c} + \frac{\Delta t_f}{t_c}$$

$$= \frac{\Delta t_m}{t_m}.\frac{t_m}{t_c} + \frac{\Delta t_f}{t_f}.\frac{t_f}{t_c}$$

$$= \varepsilon_m \frac{v_m}{v_c} + \varepsilon_f \frac{v_f}{v_c}$$

i.e

$$\varepsilon_{ct} = \varepsilon_m V_m + \varepsilon_f V_f \tag{84}$$

Now, $\varepsilon_{ct} = \dfrac{\sigma_{ct}}{E_{ct}}, \quad \varepsilon_m = \dfrac{\sigma_m}{E_m}, \quad \varepsilon_f = \dfrac{\sigma_f}{E_f}$

Using Eq. (81), Eq. (84) becomes

$$\frac{\sigma_{ct}}{E_{ct}} = \frac{\sigma_{ct}}{E_m} V_m + \frac{\sigma_{ct}}{E_f} V_f$$

or,

$$\frac{1}{E_{ct}} = \frac{V_m}{E_m} + \frac{V_f}{E_f} \tag{85}$$

Equation (85) gives the Young's modulus for the composite in a direction transverse to the fibre direction according to the rules of mixtures. It should be observed that Eqs (79) and (85) for the values of the Young's moduli in the axial direction (i.e. in the direction of the fibres) and the transverse direction are obtained under the assumption that the Poisson's ratios for the matrix and the fibres are equal. If the Poisson's ratios are different, then the analysis becomes complicated. Some aspects of this will be discussed subsequently.

When a composite cylindrical rod of uniform cross–section is subjected to a force P_c, assuming that cross–sections remain plane, the stresses in the fibre and matrix, and the linear strain in the rod are given by

$$\varepsilon_{cl} = \frac{\sigma_f}{E_f} = \frac{\sigma_m}{E_m}$$

$$= \frac{p_c}{A_c \, E_{cl}}$$

$$= \frac{p_c}{A_c}\left[\frac{1}{E_m V_m + E_f V_f}\right], \text{ from Eq. (79)} \tag{86}$$

$$\sigma_f = E_f \, \varepsilon_{cl}, \quad \sigma_m = E_m \, \varepsilon_{cl} \tag{87}$$

The determination of the shear modulus G_c for the composite in terms of shear moduli for the fibre and matrix is not simple. However, under some simple assumptions, an expression can be obtained as indicated next.

Assume, as shown in Fig. 11, that the composite can be considered to be a combination of two continuous phase materials, one that of fibre and the other that of matrix.

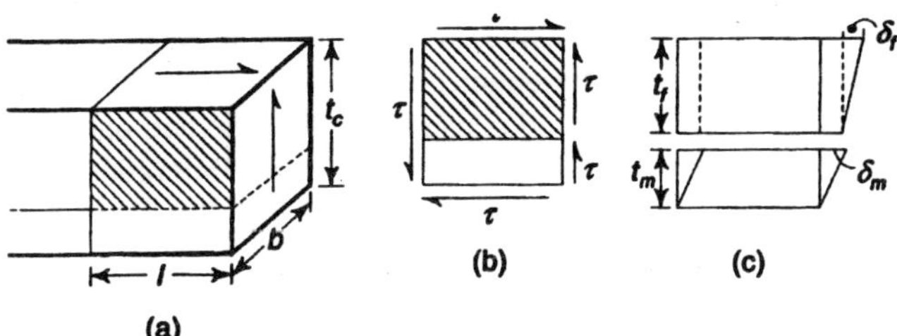

Fig 11

As shown in Fig. 11(b), the shear stresses on the complementary faces are equal. Consequently, if the shear moduli for the matrix and the composite are not equal, there will be some discontinuity in the shear strains as shown in Fig. 11(c). Ignoring this discontinuity,

the shear strain in the fibre is $\left(\dfrac{\delta_f}{t_f}\right)$

the shear strain in the matrix is $\left(\dfrac{\delta_m}{t_m}\right)$

and the total shear strain in the composite is $\left(\dfrac{\delta_f + \delta_m}{t_c}\right)$

The shear modulus for the composite is

$$G_c = \frac{\tau}{(\delta_f + \delta_m)/t_c} = \frac{\tau t_c}{\delta_f + \delta_m} \tag{88}$$

If G_f and G_m are the shear moduli for the fibre and matrix, then

$$G_f = \frac{\tau}{(\delta_f / t_f)} \quad \text{and} \quad G_m = \frac{\tau}{(\delta_m / t_m)}$$

i.e. $$d_f = \frac{\tau t_f}{G_f} \quad \text{and} \quad \delta_m = \frac{\tau t_m}{G_m}$$

Substituting these in Eq. (88),

$$G_c = \frac{\tau t_c}{(\tau t_f / G_f) + (\tau t_m / G_m)}$$

$$= \frac{t_c}{(t_f / G_f) + (t_m / G_m)}$$

or $$G_c = \frac{G_f\, G_m}{V_f G_m + V_m G_f} \tag{89}$$

Equation (89) gives the composite shear modulus in terms of the constituent shear moduli. In obtaining an expression for the composite elastic modulus E_{11} in the fibre direction, it was assumed that the Poisson's ratios for the fibre and matrix were equal. If the ratios happen to be different, one can get the composite Poisson's ratio in terms of the matrix and fibre ratios under some simple assumption. For this, consider Fig. 9 and Eq. (79). The longitudinal strain for the composite is

$$\varepsilon_{cl} = \frac{\sigma_{cl}}{E_{cl}} = \frac{\sigma_{cl}}{E_m V_m + E_f V_f}$$

If v_f and v_m are the Poisson's ratios for the constituents, then the change in the transverse dimension t_c is

$$\delta_{tc} = t_m v_m \varepsilon_{cl} + t_f v_f \varepsilon_{cl}$$

This is under the assumption that there are no transverse stresses when the bar is subjected to uniaxial tension. The transverse strain is therefore

$$\varepsilon_{ct} = \frac{\delta t_c}{t_c} = \frac{\varepsilon_{cl}}{t_c}(t_m V_m + t_f V_f)$$

$$= \varepsilon_{cl}\,(V_m v_m + V_f v_f) \tag{90a}$$

using Eq. (83). Hence, the Poisson's ratio for the composite is

$$v_c = \frac{\varepsilon_{ct}}{\varepsilon_{cl}} V_m v_m + V_f v_f \tag{90b}$$

It should be observed that the transverse strain ε_{ct} as given by Eq. (90a) is negative when ε_{cl}, the longitudinal strain, is positive.

Among the several important properties of composites, the specific strength and specific modulus are the special characteristics. These are defined as follows:

$$\text{Specific strength } = \frac{\sigma}{\rho} \tag{91}$$

$$\text{Specific modulus } = \frac{E}{\rho}$$

where σ is the yield or tensile strength, ρ is the density and E is the modulus of elasticity Properties of some typical fibres are given in Table 2. The highest specific modulus is usually found in materials having a low atomic number and covalent bonding, such as carbon and boron. One should be careful about the units involved in Eq. (91). In the metric system the yield strength σ will be expressed in kgf/cm^2 and the density in kgf/cm^3. Thus, the specific strength will be expressed in

$$\frac{\text{kgf}}{\text{cm}^2}.\frac{\text{cm}^3}{\text{kg}}: \quad \frac{\text{kg/cm}}{\text{s}^2\text{cm}^2}.\frac{\text{cm}^3}{\text{kg}}: \quad \frac{\text{cm}^2}{\text{s}^2}$$

In SI units also, the specific strength or the specific modulus will be expressed in m^2/s^2.

Table 2

Material	*Density(Mg/m³)*	*Tensile strength (MPa)*	*Elasticity modulus (GPa)*
Polymers			
Kevlar	1.44	4480	124
Polyethylene	1.14	3300	172
Metals			
Be	1.83	1275	303
Boron	2.36	3450	379
Glass			
E–glass	2.55	3450	72.4
S–glass	2.50	4480	86.9
R–glass	2.76	4137	85
Carbon			
High strength	1.75	5650	276
High modulus	1.90	1860	531

Example 7

One of the important light weight composites used for high temperature applications is borasic–reinforced aluminium containing 40% be volume of fibres. Estimate the density, modulus of elasticity and tensile strength parallel to the fibre axis. Also estimate the modulus of elasticity perpendicular to the fibres. The following data is given:

Table 3

Material	*Density(Mg/m³)*	*E (GPa)*	*Tensile strength (MPa)*
Fibres	2.36×10^3	380	2760
Aluminium	2.70×10^3	70	35

A cubic meter of composite consists of 0.4 m³ of fibres and 0.6 m³ of aluminium. Hence, from Eq. (71), the density ρ_c (kg/m³) of the composite is

$$\rho_c = 0.6(2.70 \times 10^3) + 0.4(2.36 \times 10^3)$$
$$= 2.56 \times 10^3 \text{ kg/m}^3$$

From Eq. (79),

$$E_{cl} = 70(0.6) + 380(0.4) = 194 \text{ GPa}$$

From Eq. (80),

$$\sigma_{cl} = 35(0.6) + 2760(0.4) = 1125 \text{ MPa}$$

In a direction perpendicular to the fibre axis, from Eq. (85),

$$\frac{1}{E_{ct}} = \frac{06}{70} + \frac{0.4}{380} = 9.624 \times 10^{-3}$$

$$\therefore \quad E_{ct} = 103.9 \text{ GPa}$$

Example 8:

A glass fibre reinforced nylon composite contains *E*–glass fibres 30% by volume. Calculate the percentage of load carried by the fibres when the composite is loaded. The moduli of elasticity of the constituents are *E* (glass) = 72 GPa, *E* (nylon) = 2.8 GPa.

Assuming isostrain condition,

$$\varepsilon_{cl} = \varepsilon_m = \varepsilon_f$$

But, $$\varepsilon_m = \frac{\sigma_m}{E_m} \quad \text{and} \quad \varepsilon_f = \frac{\sigma_f}{E_f}$$

$$\therefore \qquad \frac{\sigma_m}{E_m} = \frac{\sigma_f}{E_f}$$

i.e. $$\frac{\sigma_f}{\sigma_m} = \frac{E_f}{E_m} = \frac{72}{2.8} = 25.71$$

The load carried by the composite is

$$F_c = F_m + F_f$$

Hence, the fraction of the load carried by the fibre is

$$\frac{F_f}{F_m + F_f} = \frac{\sigma_f A_f}{\sigma_m A_m + \sigma_f A_f}$$

$$= \frac{\sigma_f v_f}{\sigma_m v_m + \sigma_f v_f}, \text{ using Eq. (78)}$$

$$= \frac{\sigma_f (0.3)}{\sigma_m (0.7) + \sigma_f (0.3)}$$

$$= \frac{(0.3)}{0.7(\sigma_m / \sigma_f) + 0.3}$$

$$= \frac{(0.3)}{0.7(1/25.71) + 0.3} = 0.92$$

Hence, the fibres carry 92% of the applied load.

Example 9:

An important part of a structure which currently is being made of an aluminium alloy having a modulus of elasticity of 60 GPa is to be replaced by a composite material consisting of E–glass fibres in nylon matrix. It is desired that while weight reduction is important, the specific modulus of the composite should not be lower than that of the current material. The direction of loading in the composite will be in the fibre direction. The density of aluminium alloy used is 2.8×10^6 kgf/m^3.

The specific modulus of the aluminium alloy is

$$\frac{60 \text{ GPa}}{2.8 \times 10^3 \text{ kg m}^{-3}} = \frac{60 \times 10^9 \text{ Nm}^3}{2.8 \times 10^3 \text{ kg m}^2}$$

From Table 2, the density of E–glass is 2.55×10^3 kg/m^3, and its modulus is 72 GPa. For nylon, the corresponding values are 1.14×10^3 kg/m^3 and 2.8 GPa. If we use 60% by volume of gauss fibres in the composite, then the density and modulus of the composite will be

$\rho_c = (0.6) \times 2.55 \times 10^3 + (0.4) \times 1.14 \times 10^3 = 1.986 \times 10^3$ kg/m^3

$E_c = (0.6) \times 72 + (0.4) \times 2.8$

$= 44.32$ GPa

$\therefore$ specific modulus of the composite is

$$\frac{44.32\times10^9 \text{ Nm}^{-2}}{1.986\times10^3 \text{ kg m}^{-3}} = 22.9\times10^6 \text{ Nkg}^{-1}\text{m} = 22.9\times10^6 \text{ m}^2 \text{ sec}^{-2}$$

While the specific modulus is marginally increased by 10%, the density is reduced by 29% of the original values.

Example 10

A microlaminate is produced using five sheets of 0.4 mm thick aluminium and four sheets of 0.2 mm epoxy which is reinforced with unidirectionally oriented Kevlar fibres. The volume fraction of Kevlar fibres in these intermediate epoxy sheet is 55%. Calculate the modulus of elasticity of the microlaminate parallel and perpendicular to the fibre alignment.

In each epoxy sheet of 0.2 mm thickness, the fibre content is 55%. Thus, in a 1 mm × 1 mm sheet size, the fibre content is

$(0.2 \times 0.55) = 0.11$ mm^3

and that of pure epoxy content is

$(0.2 \times 0.45) = 0.09$ mm^3

Since there are four such fibre reinforced epoxy sheets, the total fibre content is 0.44 mm^3, and that of pure epoxy is 0.36 mm^3.

A microlaminate of size 1 mm × 1 mm has a total volume equal to

$V_c = (5 \times 0.4) + (4 \times 0.2) = 2.88$ mm^3

Out of this, the aluminium content is 2 mm^3, the pure epoxy content is 0.36 mm^3, and that of fibres is 0.44 mm^3. Hence, the modulus along the fibre according to the rule of mixtures is

$$E_{cl} = \frac{1}{2.8}[(2\times70)+(0.36\times3)+(0.44\times124)]$$

$$= 69.87 \text{ GPa}$$

To evaluate the modulus perpendicular to the fibre orientation, we have to do it in two steps. The aluminium sheets being isotropic, its modulus will be direction independent. However, for the reinforced epoxy, we have to use Eq. (85). For each of the reinforced epoxy sheets, if E'_{cl} is the modulus in a direction transverse to fibre orientation, then

$$\frac{1}{E'_{cl}} = \frac{V_m}{E_m} + \frac{V_f}{E_f}$$

$$= \frac{1}{2.8}\left(\frac{0.09}{3} + \frac{0.11}{124}\right) = 0.1544\ (\text{GPa})^{-1}$$

$$\therefore \qquad E_{ct} = 6.477 \text{ GPa}$$

Now, the laminate consists of aluminium (volume content = 2 mm^3, E_{al} = 70 GPa), and fibre reinforced epoxy (volume content = 0.8 mm^3, E'_{ct} = 6.477 GPa). Hence, for the laminate, the modulus will be

$$\frac{1}{E_{ct}} = \frac{1}{2.8}\left[\frac{2}{70} + \frac{0.8}{6.477}\right] = 0.0543$$

$$\therefore \qquad E_{ct} = 18.4 \text{ GPa}$$

In getting the above answer, we have used Eq. (85) assuming isostress conditions, and this gives a low modulus value. However, for the reinforced epoxy, a modulus in the transverse direction has already been determined as E'_{ct}. So, if the bonding is good between the aluminium sheets and the reinforced epoxy sheets, one can use the isostrain condition and obtain a modulus value as

$$E_{ct} = \frac{1}{2.8}[(2 \times 70) + (0.8 \times 6.477)]$$

$$= 51.8 \text{ GPa}$$

The actual value will however be in between these two values.

Example 11

It is desired to design a tensile member made of a unidirectional composite material. The structure is to carry a load of 2.2 kN and is to be 3 m long having a circular cross–section. The matrix is to be epoxy with a yield strength of 80 MPa. The yield strength of the composite should not exceed the yield strength of the epoxy. This is to make sure that if the fibres break, the epoxy will be able to carry the load without any catastrophic failure. Assume a modulus of 3.5 GPa for the epoxy. It is also required that the composite member should not stretch more than 2.5 mm.

If the member is made entirely of epoxy without any fibres, then

$$\varepsilon_{\max} = \frac{25 \text{ mm}}{3000 \text{ mm}} = 0.88 \times 10^{-3}$$

$$\sigma_{\max} = \varepsilon_{\max} \times E$$

$$= 0.83 \times 10^{-3} \times 3.5 \times 10^{9}$$

$= 2.92 \times 10^6 \text{ Nm}^{-2}$

$$\text{Area of section} = \frac{2.2 \times 10^3}{2.92 \times 10^6 \text{ Nm}^{-2}}$$

$= 0.753 \times 10^{-3} \text{ m}^2$

$\therefore$ Diameter of the member = d = 31 mm

Assuming a specific weight of 1.25×10^3 kgf m–^{3}, the weight of the tensile member will be

$$W(\text{epoxy}) = (1.25 \times 10^3)(0.753 \times 10^{-3})(3) = 2.83 \text{ kgf}$$

For the composite, the maximum strain permitted is still 0.83×10^{-3}. The maximum yield strength for the composite is 80 MPa. Hence, the minimum modulus for the composite will be

$$E_c = (\text{minm}) = \frac{\sigma}{\varepsilon} = \frac{80 \times 10^6 \text{ Nm}^{-2}}{0.83 \times 10^{-3}} = 96.4 \times 10^9 \text{ Nm}^{-2}$$

From Table 2, the moduli of glass fibres are less than the minimum required. So one has to look for a fibre having a higher modulus. High modulus carbon having a modulus of 531 GPa, and a density of 1.90 Mgm^{-3} meets our requirement. If V_f is the volume fraction of the carbon fibre in the composite, the modulus of the composite will be

$$E_c = V_f(531) + (1 - V_f) \times 3.5 = 96.4$$

This should be equal to or greater than 96.4. Thus,

$$V_f(531) + (1 - V_f) \times 3.5 = 96.4$$

or $$V_f = 0.176$$

The volume fiction of the carbon is 0.176 and that of epoxy is 0.824. A composite of this nature will have a modulus not less than 96.4 GPa.

If the structure is made of such a composite, and if the fibres break when a load of 2.2 kN is applied, then the epoxy alone should be able to carry the load. If A_c is the total area of section, then 0.824 A_c is the area of epoxy and the stress on this should not exceed 80 MPa. Thus,

$$0.824\, A_c \times 80 \times 10^6 \text{ Nm}^{-2} = 2.2 \times 10^3 \text{ N}$$

or $$A_c = \frac{2.2 \times 10^3 \text{ N}}{0.824 \times 80 \times 10^6 \text{ Nm}^{-2}}$$

$= 0.0333 \times 10^{-3} \text{ m}^2 = 33.4 \text{ mm}^2$

The diameter of the composite is 6.5 mm.

$$\text{Volume} = 33.4 \times 10^{-6} \text{ m}^2 \times 3 \text{ m} = 10 \times 10^{-5} \text{ m}^3$$

Weight = [(1.9 × 0.176) + (1.25 × 0.824)] × 10 × 10^{-5} = 0.137 kgf

Therefore, the carbon fibre reinforced structure is less then one–quarter the diameter of pure epoxy structure, and one–twentieth the weight of pure epoxy.

PRESSURE VESSELS

Let the thickness of the vessel be small compared to the radius of the vessel, so that the curvature effects on the fibres can be neglected. The problem concerned is with the orientation of the fibres for optimum strength. In the netting theory, it is assumed that only the fibres take the load and that too in the directions of the fibres only. The strength of the fibre in its transverse direction is taken as zero. The contribution of the matrix to the strength is ignored.

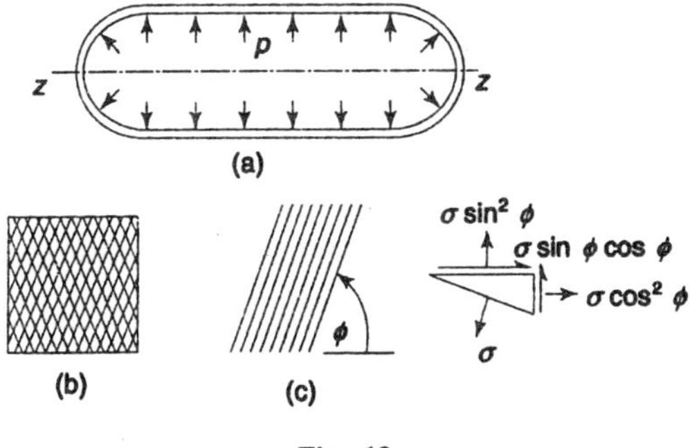

Fig. 12

Consider a cylindrical pressure vessel with closed ends, as shown in Fig, 12(a), subjected to an internal pressure p. The longitudinal and hoop stresses are

$$\sigma_z = \frac{pa}{2h}, \quad \sigma_\theta = \frac{pa}{h} \tag{92}$$

where a is the radius of the vessel and h is the thickness of the vessel. Assume a helical winding as shown in Fig. 12, and let us consider the stresses along the fibre orientation. If σ is the stress along the fibre orientation, then the stress in the z direction is $\sigma \cos^2 \phi$ and that in the hoop direction is $\sigma \sin^2 \phi$. For equilibrium,

$$\sigma \cos^2 \phi = \frac{pa}{2h} \quad and \quad \sigma \sin^2 \phi = \frac{pa}{h} \qquad (93)]$$

From these two,

$$\tan^2 \phi = 2 \quad \text{or} \quad \phi \cong 55° \tag{94}$$

Hence, the optimum orientation of the fibre does not coincide with the principal stress direction. The shear stress shown in Fig. 12(c) is balanced by the shear stress caused by the fibre in the $-\phi$ direction. In practice, the fibres are not made to run in the optimum directions as given by Eq. (94), because such a pattern cannot be used to form the end domes. Generally, a small winding angle ϕ is used to form both the cylindrical portion and the end domes, and then an overlay of fibres in the circumferential direction is put to resist the hoop stresses. Thus, in practice, the fibres run approximately in the principal stress directions.

TRANSVERSE STRESSES

In the previous sections, it was assumed that the Poisson's ratios of the matrix material and of the fibres were equal. When the ratios are different, one can expect forces between the surfaces of contact because of different contractile tendencies. To see this, consider a cylindrical composite member having a single fibre as a core, Fig. 13. Let a be the radius of the fibre

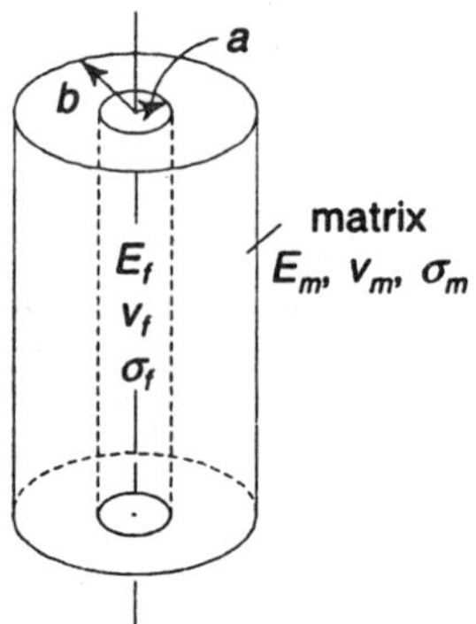

Fig. 13

and b the outer radius of the matrix. Let the composite cylinder be subjected to a uniaxial load in the z direction.

Let ε_r, ε_φ ε_z be the strains and σ_r, σ_φ σ_z the stress components in the polar coordinate system. Then, the general Hooke's law with Young's modulus in the longitudinal direction as E and Poisson's ratio as v, is

$$\begin{vmatrix} \varepsilon_r & 0 & 0 \\ 0 & \varepsilon_\theta & 0 \\ 0 & 0 & \varepsilon_z \end{vmatrix} = \frac{1-v}{E} \begin{vmatrix} \sigma_r & 0 & 0 \\ 0 & \sigma_\theta & 0 \\ 0 & 0 & \sigma_z \end{vmatrix} - \frac{v}{E}(\sigma_r + \sigma_\theta + \sigma_z) \begin{vmatrix} 1 & 0 & 0 \\ 0 & 1 & 0 \\ 0 & 0 & 1 \end{vmatrix} \tag{95}$$

The equation of equilibrium is

$$\frac{d\sigma_r}{dr}+\frac{\sigma_r-\sigma_\theta}{r}=0$$

This is the only equilibrium equation for the case under consideration. Let the strain in the z direction be constant. The strain–displacement equations are

$$\varepsilon_r=\frac{du_r}{dr},\qquad \varepsilon_\theta=\frac{u_r}{r},\qquad \varepsilon_z=\text{constant} \tag{96}$$

where u_r is the radial displacement. Equation (95) can be solved for σ_r and σ_θ in terms of ε_r, ε_θ, ε_z. The results are

$$\sigma_r = K[(1-v)\,\varepsilon_r + v(\varepsilon_\theta+\varepsilon_z)]$$

$$\sigma_\theta = K[(1-v)\,\varepsilon_\theta + v(\varepsilon_r+\varepsilon_z)] \tag{97a}$$

where $$K=\frac{E}{(1+v)(1-2v)} \tag{97b}$$

Substituting for ε_r and ε_θ from Eq. (96), one gets

$$\sigma_r=K\left[(1-v)\frac{du_r}{dr}+v\frac{u_r}{r}+v\varepsilon_z\right] \tag{98}$$

$$\sigma_\theta=k\left[v\frac{du_r}{dr}+(1-v)\frac{u_r}{r}+v\varepsilon_z\right]$$

Substituting into the equilibrium equation, the result appears as

$$\frac{d^2u_r}{dr^2}+\frac{1}{r}\frac{du_r}{dr}-\frac{u_r}{r^2}=0 \tag{99}$$

The solution of the above differential equation is

$$u_r=C_1\,r+\frac{C_2}{r} \tag{100}$$

where C_1 and C_2 are constants to be determined from the boundary conditions. Equation (100) is valid for both matrix and fibre. Representing the fibre equation by subscript f, and the matrix equation by m, Eq. (100) becomes

$$u_{rf}=C_{1f}\,r+\frac{C_{2f}}{r} \tag{101}$$

$$u_{rm}=C_{1m}\,r+\frac{C_{2m}}{r} \tag{102}$$

The boundary conditions to determine the constants are

(i) At the free surface $r=b$, $\sigma_{rm}=0$

(ii) At the interface $r = a$, because of continuity, $u_{rf} = u_{rm}$ and $\sigma_{rf} = \sigma_{rm}$

(iii) At $r = 0$, $u_{rf} = 0$ This gives $C_{2f} = 0$

Applying the above boundary conditions, the following equations are obtained:

(i) $$(1-v_m)\left(C_{1m}-\frac{C_{2m}}{b^2}\right)+v_m\left(C_{1m}-\frac{C_{2m}}{b^2}\right)+v_m\varepsilon_z=0$$

$$C_{1m}+\frac{1+2v_m}{b^2}C_{2m}=-v_m\varepsilon_z \qquad \text{(f)}$$

(ii) $$(1-v_m)\left(C_{1m}-\frac{C_{2m}}{a^2}\right)+v_m\left(C_{1m}-\frac{C_{2m}}{a^2}\right)+v_m\varepsilon_z$$

$$= (1 - v_f)\, C_{1f} + v_f\, C_{1f} + v_f\, \varepsilon_z$$

or, $$C_{1m}+\frac{1+2v_m}{a^2}C_{2m}+v_m\,\varepsilon_z=C_{1f}+v_f\,\varepsilon_z$$

i.e. $$C_{1m}+\frac{1+2v_m}{a^2}C_{2m}-C_{1f}=(v_f-v_m)\varepsilon_z \qquad \text{(g)}$$

Also, $$C_{1m}a+\frac{C_{2m}}{a^2}-C_{if}\;a=0$$

(h)

Equations (f)–(h) can be solved for the constants. The stress ($-P$), at the interface is obtained as

$$p=\frac{2\varepsilon_z(v_f-v_f)V_m}{(V_f/2K_m)+(V_m/2K_f)+(1/G_m)}$$

where K is given by Eq. (97b).

The stress components in the fibre are

$$\sigma_{rf} = \sigma_{\theta f} = -p$$

$$\sigma_{zf} = E_f\varepsilon_z - 2v_f p$$

The stress components in the matrix are

$$\sigma_{rm}=p\left(\frac{a^2}{b^2-a^2}\right)\left(1-\frac{b^2}{r^2}\right)$$

$$\sigma_{\theta m}=p\left(\frac{a^2}{b^2-a^2}\right)\left(1+\frac{b^2}{r^2}\right) \qquad (103)$$

$$\sigma_{zm}=E_m\varepsilon_z+2v_m\;p\left(\frac{a^2}{b^2-a^2}\right)$$

EXERCISES

1. A particular laminate has the following elastic constants along the principal axes *x–y*:

   $E_{xx} = 200$ GPa, $E_{yy} = 20$ GPa, $G_{xy} = 10$ GPa, $V_{yx} = 0.25$

   At a point in the laminate, the following state of stress exists:

   $\sigma_{x'x'} = 200$ MPa, $\sigma_{y'y'} = 20$ MPa, $\tau_{x'y'} = 20$ MPa

   The *x′*–axis makes an angle of 30° with the fibre axis, counterclockwise. Calculate the principal stresses, the principal strains and their orientations.

   $$\left[\begin{array}{l} Ans.\ \sigma_{1.2} = 202.2 \text{ MPa} \\ \phi' = 6.25^\circ \text{ and } 96.25^\circ \\ \varepsilon_{1.2} = 7.207\times10^{-3}; 2.255\times10^{-3} \\ \phi^* = -33.6^\circ \text{ and } 56.4^\circ \end{array}\right]$$

2. For a graphite–epoxy laminate having uniaxial reinforcements (parallel to *x*–axis), the following elastic constants apply:

   $E_{xx} = 181$ GPa, $E_{yy} = 10.3$ GPa, $G_{xy} = 7.17$ GPa, $V_{yx} = 0.28$

   Obtain the off–axis compliance coefficients when the axes are rotated by (a) +45° and (b) +60°. Express the results in results in $(10^{-12}\ \text{Pa})^{-1}$ units.

   $$\left[\begin{array}{l} Ans.\ (a)\ b'_{11} = 59.75; b'_{22} = 59.75; b'_{12} = -9.99 \\ b'_{44} = 105.7; b'_{14} = -45.78; b'_{24} = -45.78 \\ (b)\ b'_{11} = 80.53; b'_{22} = 34.75; b'_{12} = -7.88 \\ b'_{44} = 114.1; b'_{14} = -32.34; b'_{24} = -46.96 \end{array}\right]$$

3. Estimate the components of moduli and compliances for a cross–ply laminate formed from composites consisting of Toray filament and Namco resin. The modulus data are

   $E_{xx} = 181$ GPa, $E_{yy} = 10.3$ GPa, $V_{yx} = 0.159$ $G_{xy} = 7.17$GPa.

   $(1 - v_{yx}\, v_{xy})^{-1} = 1.0045$

   The laminate code is (a) $[0_2/90]$s (b) $[0_4/90]$. Assume that the composites are of uniform thicknesses.

$$\begin{bmatrix} \textit{Ans.}\ (a) A_{11} = 124.65h;\ A_{22} = 67.49h; \\ A_{44} = 7.17h;\ A_{12} = 2.89h; \\ b_{11}h = 8.03;\ b_{22}h = 14.82; \\ b_{44}h = 139.47;\ b_{12}h = -0.344; \\ (b) A_{11} = 147.51h;\ A_{22} = 44.63h; \\ A_{44} = 7.17h;\ A_{12} = 2.89h \\ b_{11}h = 6.78;\ b_{22}h = 22.43; \\ b_{44}h = 139.47;\ b_{12}h = -0.440; \end{bmatrix}$$

4. A laminate is formed from angleply composite plies having elastic constants given in Example 5. Estimate the components of moduli and compliances for the laminate described by the following codes: (a) $\phi = \pm 30°$ and (b) $\phi = \pm 60°$.

$$\begin{bmatrix} \textit{Ans.}\ (a) A_{11} = 109.3h;\ A_{22} = 23.6h; \\ A_{12} = 32.46\ h \\ A_{44} = 36.73h;\ A_{14} = A_{24} = 0 \\ h\bar{b}_{11} = 15.42;\ h\bar{b}_{22} = 71.36; \\ h\bar{b}_{12} = -21.8 \\ h\bar{b}_{44} = 27.22;\ h\bar{b}_{14} = \bar{b}_{24} = 0 \end{bmatrix}$$

$$\begin{bmatrix} \textit{Ans.}\ (b) A_{11} = 23.6h;\ A_{22} = 109.3h; \\ A_{12} = 32.46\ h \\ A_{44} = 36.73h;\ A_{14} = A_{24} = 0 \\ h\bar{b}_{11} = 71.36;\ h\bar{b}_{22} = 15.42; \\ h\bar{b}_{12} = -21.18;\ h\bar{b}_{44} = 27.22; \\ \bar{b}_{14} = \bar{b}_{24} = 0 \end{bmatrix}$$

5. For the laminates of Problem 4, estimate the average values of the engineering constants (*Es* & *vs*) corresponding to *x* and *y* axes.

$$\begin{bmatrix} \textit{Ans.}\ (a)\ \bar{E}_{xx} = 64.9\ \text{GPa}; \bar{E}_{yy} = 14\ \text{GPa}; \\ \bar{G}_{xy} = 36.7\ \text{GPa}; \bar{V}_{xy} = 1.37; \\ (b)\ \bar{E}_{xx} = 14\ \text{GPa}; \bar{E}_{yy} = 64.9\ \text{GPa}; \\ \bar{G}_{xy} = 36.7\ \text{GPa}; \bar{V}_{xy} = 0.297 \end{bmatrix}$$

6. For the laminate described in Example 6, determine the minimum failure stresses $\sigma_{x'x'}$ applied at $\theta = 45°$ and $\theta = 60°$ to the fibre axis according to

   (a) maximum stress theory in tension and compression;

   (b) maximum strain theory in tension only;

   (c) distortion energy theory in tension and compression.

   Use the data given in Example 6.

$$\left[\begin{array}{rl} \textit{Ans.}\text{ (a) Tension:} & \theta = 45°, \sigma_{x'x'} = 55.2 \text{ MPa} \\ & \theta = 60°, \sigma_{x'x'} = 36.8 \text{ MPa} \\ \text{Compression:} & \theta = 45°, \sigma_{x'x'} = 110.4 \text{ MPa} \\ & \theta = 60°, \sigma_{x'x'} = 127.5 \text{ MPa} \\ \text{(b) Tension:} & \theta = 45°, \sigma_{x'x'} = 73.6 \text{ MPa} \\ & \theta = 60°, \sigma_{x'x'} = 40 \text{ MPa} \\ \text{(c) Tension:} & \theta = 45°, \sigma_{x'x'} = 49.4 \text{ MPa} \\ & \theta = 60°, \sigma_{x'x'} = 35.3 \text{ MPa} \\ \text{Compression:} & \theta = 45°, \sigma_{x'x'} = 102 \text{ MPa} \\ & \theta = 60°, \sigma_{x'x'} = 105 \text{ MPa} \end{array}\right]$$

7. A cemented carbide cutting tool used for machining contains 75% by weight tungsten carbide (WC), 15% by weight titanium carbide (TiC), 5% by weight TaC, and 5% by weight cobalt (Co). Estimate the density of the composite, given the following densities for the constituents:

$$\rho_{wc} = 15.77 \text{ Mgm}^{-3}, \qquad \rho_{Tic} = 4.94 \text{ Mgm}^{-3},$$

$$\rho_{Tac} = 14.5 \text{ Mgm}^{-3}, \qquad \rho_{co} = 8.90 \text{ Mgm}^{-3},$$

[*Ans.* 11.50 Mgm^{-3}]

8. An electrical contact material is produced by infiltrating copper into a porous tungsten–carbide (WC) compact. The density of WC is 15.77 Mag^{-3} and that of the final composite is 12.3 Mgm^{-3}. Assuming that all of the pores are filled with copper, and given $r_c = 8.94$ Mgm^{-3}, calculate

   (i) the volume fraction of copper in the composite,

   (ii) the volume fraction of pores in WC compact before infiltration and

(iii) the original density of WC compact.

[*Ans.* (a) 0.507; (b) 0.507; (c) 7.775]

9. An epoxy matrix is reinforced with 40% by volume E-glass fibres to produce a 20 mm diameter composite to carry a load of 25 kN. Calculate the stress acting on the fibre elements. The modulus of epoxy is 3 Gpa and that of glass fibre is 72.4 Gpa.

[*Ans.* 187.3 MPa]

10. In the design problem of Example 11, if one uses high strength carbon instead of the high modulus carbon, what will be the changes as compared to the pure epoxy member?

[*Ans.* Diameter = 7.3 mm; Weight = 0.179kgf]

11. If Kevlar fibres are used instead of carbon fibres in Example 11, show that the volume fraction of fibre needed would be 0.8, and the diameter of the member would be 13.1 mm, and the weight 0.57 kgf.